P9-CJT-036

Microfluidic Reactors for Polymer Particles

Microfluidic Reactors for Polymer Particles

EUGENIA KUMACHEVA
University of Toronto, Canada
and
PIOTR GARSTECKI
Institute of Physical Chemistry,
Polish Academy of Sciences, Poland

A John Wiley and Sons, Ltd, Publication

This edition first published 2011

Registered office
John Wiley & Sons Ltd, The Atrium, Southern Gate, Chichester, West Sussex, PO19 8SQ, United Kingdom

For details of our global editorial offices, for customer services and for information about how to apply for permission to reuse the copyright material in this book please see our website at www.wiley.com.

Library of Congress Cataloging-in-Publication Data

Kumacheva, Eugenia.
Microfluidic reactors for polymer particles / Eugenia Kumacheva and Piotr Garstecki.
p. cm.
Includes bibliographical references and index.
ISBN 978-0-470-05773-5 (cloth) — ISBN 978-0-470-97923-5 (ePDF) — ISBN 978-0-470-97922-8 (obook) — ISBN 978-1-119-99028-4 (ePub)
1. Microreactors. 2. Microfluidic devices. 3. Emulsion polymerization. I. Garstecki, Piotr. II. Title.
TP159.M53K86 2011
668.9—dc22

2010042340

A catalogue record for this book is available from the British Library.

Print ISBN: 9780470057735
ePDF ISBN: 9780470979235
oBook ISBN: 9780470979228
ePub ISBN: 9781119990284

Set in 10.5pt/12.5pt Times by Thomson Digital, Noida, India.
Printed and bound in Singapore by Markono Print Media Pte Ltd

Contents

Preface

The manipulation of fluids in channels with dimensions in the range of from tens to hundreds of micrometers – microfluidics – has recently emerged as a new area of science and technology. Microfluidics has applications spanning the analytical chemistry, organic and inorganic synthesis, cell biology, optics, and information technology fields. Many of these applications have been demonstrated over the last two decades. During the past six or seven years, microfluidic synthesis has shown very promising applications in the continuous production of high value materials, including inorganic nanoparticles, polymers, organic compounds for positron emission tomography, and polymer particles. Microfluidic synthesis of micrometer-size polymer beads with precisely controlled dimensions and a variety of compositions, shapes and morphologies, has rapidly attracted great interest from scientists and technologists with very different backgrounds and occupations, ranging from polymer colloids to cell biology and drug delivery.

The motivation behind writing this book was that it would: (i) serve as a comprehensive introduction to this rapidly developing field, (ii) guide scientists and engineers working in the area of microfabrication and microfluidics toward new applications of microfluidic devices, in particular, microreactors for the synthesis of polymer particles, and (iii) serve as a source of information for the those wishing to join the field. This book is intended for a broad audience, including undergraduate and graduate students, postdoctoral research associates, and researchers and engineers in industry.

We met in 2002 at Harvard University in the laboratory of Professor George Whitesides. The exploratory spirit of this remarkable research group inspired us to work, firstly, toward the fundamental understanding of two-phase microflows and, later, toward applications of microfluidic technologies. The Whitesides' laboratory strongly supports interdisciplinary collaboration, an attitude that was especially valuable to us, as we have very different backgrounds. We believe that we were able to combine and use our expertise in chemistry, physics and engineering, in order to develop microfluidic techniques for the formation of emulsions and of polymer particles.

The structure of this book is very straightforward. Our complementary backgrounds in polymer and materials science (E.K.) and in fluid mechanics and microfluidics (P.G.) helped us shape the book into a compehrensive review. From the applications of polymer particles and the current methods used for their synthesis (Chapters 1 and 2, respectively) we introduce the basics of microfluidics pertinent to the subject of the book (Chapter 3). The fundamental aspects of the physics of flow of immiscible liquids are covered in Chapter 4. From a detailed review of the current state-of-the-art for the methods of formation of droplets (Chapter 5) and a review of high-throughput microfluidic droplet and bubble generators (Chapter 6), we move to the synthesis of various types of polymer colloids (Chapters 7–10). The organization of the material in Chapters 7–10 is somewhat arbitrary. We believe that two separate chapters about the microfluidic synthesis of rigid polymer particles and gel microbeads (Chapters 7 and 8, respectively) provide important guidelines for the synthesis of these types of polymer particles. On the other hand, in Chapters 9 and 10 we did not make a distinction between the types of polymer colloids; instead, we discuss separately the microfluidic production of particles with capsular morphologies (Chapter 9) and particles with different shapes and structures (Chapter 10). The book is concluded with a brief summary of the future directions of research in the microfluidic production of polymer colloids.

We wish to thank Professor George Whitesides (Harvard University) for giving us the inspiration to work in this exciting field. While working at Harvard, we enjoyed the stimulating collaboration with the members of the research group and we wish to particularly thank Dr Michinao Hashimoto, Dr Michael Fuerstman, Dr Irina Gitlin, Dr Douglas Weibel (currently Professor at the University of Wisconsin-Madison), and Professor Shoji Takeuchi (currently at the University of Tokyo). We greatly appreciate scientific discussions with Professor Howard Stone (then at Harvard University now at Princeton University) and his assistance in the interpretation of experimental results. He contributed to our work with his unique knowledge and understanding of fluid hydrodynamics, and was an invaluable collaborator.

After completing research at Harvard University, the authors continued their work in their own laboratories. E.K. thanks her research group at the University of Toronto for the hard work in the area of continuous microfluidic synthesis of polymer particles. Rapid progress achieved in this fast developing field would not be possible without long hours spent in the laboratory, intensive discussions, fruitful collaborations and friendly competition within the research group. In the E.K. group, this work was pioneered by Dr Shengqing Xu (currently, a Senior Scientist at Dow Corning) and Dr Zhihong Nie, who has recently become an Assistant Professor at the University of Maryland. Patrick Lewis and Dr Minseok Seo have made the first steps toward the synthesis of copolymer particles and foam-templated materials, respectively. Dr Hong Zhang and Ethan Tumarkin paved the way for the microfluidic preparation of physically crosslinked microgels. Dr Wei Li and Jai Il Park worked on the microfluidic synthesis of particles with an interpenetrating network structure and bubble-templated particles, respectively. A large number of graduate students, postdocs, and undergraduate students participated in various research projects, including the synthesis of porous particles, particles for modeling the behavior of cells, the encapsulation of cells, microfabrication of microfluidic reactors, and characterization of monomer mixtures. The contributions of Jesse Greener, Stanislav Dubinsky, Neta Raz, Lindsey Fiddes, Lucy Siyon Chung, Chantal Paquet, Dinesh Jagadeesan, Raheem Peerani, Edmond Young, Danut Voicu, Monika Kleczek, Alexander Kumachev, and Micelle Mok are greatly appreciated. Fruitful collaborations with Professors Gilbert Walker, Peter Zandstra, Axel Guenther, Craig Simons, and Aaron Wheeler (all at the University of Toronto) helped to address the various aspects of microfluidic synthesis of polymer particles, which included fundamental fluid hydrodynamics and the applications of polymer particles. P.G. thanks the members of his research team, who continue to work on a wide range of fundamental and applied aspects of multiphase microflows. E.K. and P.G. are grateful to Anna Lee and Dr Neil Coombs for preparing the cover art for this book.

Finally, we are greatly indebted to our families who were always there for us. In particular, we thank our spouses, Boris Kumachev and Justyna Garstecka. They gave us their continuous support when we spent long hours at work and whilst we were writing this book. Without their patience and understanding the book would have never been written.

1

Applications of Polymer Particles

Polymer particulate materials have important applications in fundamental research, in industry, biology, and medicine, and in the environmental sciences. The range of applications of polymer particles is so broad that it could be the subject of a separate book. Here we focus on the most important areas of science, technology, and medicine that currently use polymer colloids. The emphasis of the present section is on the application of polymer particles with a narrow size distribution.

The ability to synthesize monodisperse spherical polymer particles aids in the development of theoretical models for the mechanisms controlling the chemical and physical properties of particulate materials. Currently, only silica and polymer submicrometer-size microbeads can be synthesized with a truly narrow size distribution. A detailed investigation of the interactions of polymer colloids in aqueous and nonaqueous media contributed significantly to the current understanding of crystal nucleation and growth and to the mechanisms of phase transitions in colloid systems. In addition, studies of the rheology of concentrated dispersions of polymer particles and their sedimentation and self-assembly, chromatography, drug delivery and medical diagnostics, encapsulation of biologically active species (e.g., proteins or cells), and the utilization of electronic, optical, magnetic, electrokinetic properties, and film-forming properties of polymer microbeads are far from an exhaustive list of the applications of polymer colloid particles.

Polymer particles can be used as stable suspensions in polar or non-polar liquids or as solid materials, such as close packed arrays or films. Both systems find a broad range of applications due to the presence of functional groups on the particle surface. Surface

Microfluidic Reactors for Polymer Particles. Eugenia Kumacheva and Piotr Garstecki.

functional groups can be introduced during particle synthesis with an initiator, a monomer, or with a chain-transfer agent. Examples of surface functional groups include $–SO_3^-$, –CHO, –OH, or COOH groups. Post-synthesis reactions of surface groups may occur inadvertently or with the objective of chemical modification of the surface of the polymer colloids. For instance, in poly(vinyl acetate)-based particles, hydrolysis of surface acetate ester groups transforms them into hydroxyl and acetic acid groups. Derivatization of surface functional groups is driven by the targeted application of polymer particles: heterogeneous catalysis, such as acid-catalysed hydrolyses, decarboxylation, chemical analysis, or biomedical applications. Examples of derivatization of surface functional groups include reactions of benzyl chloride groups with ammonia to produce surface amino groups, or redox reactions of surface aromatic groups to reduce them to amine groups.

Colloidal polymer catalysts have been used for organic chemical transformations, for example, hydrolysis, substitution, oxidation, hydrogenization and C–C bond formation (Ford and Miller, 2002; Arshady *et al.*, 2002). Catalytic groups on the surface of particles include quaternary ammonium, pyridine, carboxylic and sulfonic acids, and transition metal chelates. Polymer particles typically used for heterogeneous catalysis are either relatively large microspheres (with a diameter of 40 μm and larger), or submicrometer-size latex particles. Polymer catalytic supports can be rigid (e.g., polystyrene microbeads) or soft (e.g., polysaccharide or polyacrylamide microgels).

Polymer particles find important applications in medical diagnostics, cell separation, and drug delivery (Arshady *et al.*, 2002). Most of the biomedical applications of polymer particles exploit the binding of biological molecules, for example, proteins, DNA fragments, or antibodies, to the polymer surface when it carries specific functionalities such as aldehyde, amine, thiol, epoxy, or acid groups. For example, surface epoxy groups readily react with thiol or amine groups in biomolecules. Alternatively, carbodiimide coupling includes activation of surface carboxyl groups, which is followed by a rearrangement reaction and the formation of a covalent bond between the particle surface and a biological molecule. Chemical attachment of biomolecules ensures high specificity and irreversible adsorption, and it can be used for the sensing or separation of biological molecules.

Polymer particles have been used for immunospecific, that is, immunomagnetic or immunofluorescent, cell separation. In the first method, magnetic latex particles coated with a layer of a hydrophilic gel carry surface-attached antibodies, and thus they can attach only to the cells carrying a complementary antigen. Following the attachment, particle-coated cells are removed from the system by applying a magnetic field. In the second method, immunospecific latex particles are labeled with a fluorescent dye. Subsequent to the attachment of the particles to the cells, the cells are removed using flow cytometry.

Cell separation based on immunospecificity can be used in the therapy of cancer of the nerve cells, lymphoma, and leukemia. For example, in clinical treatment of a cancer of the nerve cells, a large sample of bone marrow is mixed with polymer magnetic particles that are immunospecific for the cancer cell surface. The separation of cancer cells is achieved by applying a strong magnetic field, then a suspension of healthy cells is introduced to the patient's bones. This step is followed by chemotherapy and radiation.

Clinical diagnostics applications of polymer particles also rely on interactions of polymer particles carrying antibodies and proteins that are characteristic of certain diseases. The detection may be in the form of coagulation (or "agglutination"), which is monitored visually or by turbidimetry, fluorescence, or colorimetrics. Other diagnostic applications are based on interactions of particles with DNA molecules or their fragments to obtain information about inherited diseases, viruses, and bacteria.

Radiolabeled polymer particles find applications in nuclear medicine, in imaging, therapy, and laboratory studies (Arshady *et al.*, 2002). Particle size in these applications varies from several nanometers to tens of microns. Examples of radiolabeled particles are ^{99m}Tc-labeled albumin microspheres, ^{166}Ho-polylactide microspheres or ^{131}I-polystyrene nanospheres. The advantage of radiolabeled particles versus molecularly soluble radiolabeled species is that they maintain the radiolabel in a specific body location for diagnostic or therapeutic purposes. This can prevent or substantially reduce the spreading of radioactivity to other body parts. Another important feature of radiolabeled particles is the size selectivity of sequestering of these particles by various organs. For example, 10–90 μm-diameter radiolabeled albumin particles will be trapped in the lungs, thereby providing the ability to image this organ.

Nanometer-size and submicrometer-size polymer colloids – lyposomes, polymer microgels, polymersomes, capsules, and solid polymer particles – can also be used for drug delivery, and, in particular, targeted, site-specific drug delivery (Oupicky, 2008). A drug can be loaded in the particle interior, be a part of the particle, or it can be attached to the particle surface. The site-specific delivery of the drug is achieved by attaching biological molecules to particle surfaces, as discussed above. The problem of clearing the particles by the reticulorendothelial system is partly solved by coating them with poly(ethylene oxide).

The encapsulation of cells in polymer particles with dimensions in a range of from tens to several hundred micrometers was proposed in the early 1960s as a method of reducing the effects of immune rejection, thereby forming the basis of "cell therapy" for the treatment of various diseases. The encapsulation of cells and other biological molecules such as peptides and proteins in microgels has led to the development of systems for the study and treatment of hormone or protein deficiencies (Ross *et al.*, 2000), hepatic failure (Liu and Chang, 2006), cancer (Cirone, Bourgeois, and Chang, 2003), and diabetes (Lim and Sun, 1980). Typically, polymer microgels encapsulating cells are prepared from biological polymers, such as alginate or agarose.

Polymer particles find applications in size-exclusion chromatography (SEC), or gel-permeation chromatography (GPC), a chromatographic method in which polymer molecules in solution are separated and analyzed based on their hydrodynamic volume. This method is generally used for the purification and analysis of synthetic and biological polymers. Polymer chemists typically use either silica, or crosslinked polystyrene particles.

Porous polymer particles are also used as ion-exchange resins (Okay, 2000). The beads have a highly developed network of pores on the surface and can easily trap and release ions, with a simultaneous release of surface-bound ions. Ion-exchange resins are widely used in separation, purification, and decontamination technologies. Water softening and purification are the most common examples of such processes. Typically, ion-exchange resins are based on crosslinked polystyrene beads. Crosslinking prolongs the time needed to accomplish ion exchange; however, crosslinked polymer particles are mechanically stronger and are more stable. The required functional groups can be introduced during or after polymerization, as discussed above. Typical groups include strongly acidic, weakly acidic, strongly basic, and weakly basic functionalities. Particle dimensions and pore size distribution strongly impact the performance of ion-exchange resins: smaller, highly porous particles have a larger surface area available for ion exchange.

Solid particle-derived materials can be formed by randomly organized particles, or by particles assembled in highly periodic, crystalline arrays. A typical example of the first system includes latex films and coatings. The need for environmentally friendly organic coatings has sparked a lot of interest in water-borne paints based on polymer latex particles. The formation of films from latex particles in the coating industry is one of the most important practical applications of polymer colloids. A broad range of specific applications includes the production of printing inks, adhesives, varnishes, and water-borne paints. Coatings are formed by applying a liquid dispersion of the polymer particles onto the substrate. Other than polymer particle additives, paints typically contain pigments, particle stabilizers, and photo- and corrosion protection agents (Ottewill and Rowell, 1997 and references therein).

The ability of monodisperse polymer colloids to assemble into regular arrays in concentrated dispersions has been known for a long time (Pieranski, 1983; van Megan, 1984; Gast and Russel, 1998). The formation of liquid colloid crystals was driven by the balance of electrostatic forces acting between the particles. Therefore the presence of surface-charged groups in polymer colloids and the ionic strength of the dispersion medium had a great influence on the ability of polymer colloids to crystallize. Recently, Geerts and Eiser (2010) reported spontaneous crystallization of micrometer-diameter polystyrene particles coated with long double-stranded DNA molecules. The DNA was weakly attracted to the oppositely charged substrate and, as a result, these two-dimensional colloidal crystals floated several microns above the surface.

Liquid colloid crystals are iridescent owing to constructive interference of light beams reflected from regularly spaced rows of particles at the Bragg angle. For a particular system the spectral position of the diffracted light depends on the crystal lattice constant, which, in turn, is determined by the properties of the continuous phase. Thus colloid crystalline arrays have found applications as diffraction gratings and chemical and biochemical sensors (Ottewill and Rowell, 1997). For example, the change in the concentration of analytes in the dispersion medium was monitored by the change in the position of diffraction peak (Weissman *et al.*, 1996; Holtz *et al.*, 1998). Other applications of fluid colloid crystal arrays include photothermal nanosecond light-switching devices, which change their properties in response to electric fields, and sensors to mechanical deformation (Ottewill and Rowell, 1997 and references therein).

The ability to preserve the regularity of liquid colloid crystals in the solid state has led to the development of a new class of organic and inorganic materials with properties that originated from their periodic structure, and from chemical composition (Xia *et al.*, 2000; Paquet and Kumacheva, 2008). Two-dimensional hexagonal lattices of polymer colloidal spheres have been used as ordered arrays of optical microlenses in image processing (Hayashi *et al.*, 1991), as masks for fabricating periodic micro- or nanostructures (Roxlo *et al.*, 1987; Hulteen and Duyue, 1995; Hulteen *et al.*, 1999), and as relief structures to cast elastomeric stamps (Xia *et al.*, 1996). Three-dimensional colloid crystals of polymer particles have also been utilized as templates for the fabrication of ordered macroporous materials (Park, Qin and Xia, 1998, Yan *et al.*, 1999; Velev *et al.*, 1999), as filters, switches, and photonic band gap (PBG) materials.

To summarize, an expanding range of applications of polymer colloids imposes requirements on the synthesis and assembly of particles with multiple functionalities and precise control of particle size, shapes, and morphologies. Development of new methods or modification of conventional methods of particle production are required to address these requirements in the most effective and cost-efficient way.

References

Arshady, R., Corain, B., Zecca, M., Jayakrishnan, A., and Horak, D. (2002) Amphiphilic Functional Microgels. In: *Functional Polymer Colloids & Microparticles*, Vol. 4 (eds. R. Arshady and A. Guyot), The MML Series, Citus Books, London, pp. 203–251.

Cirone, P., Bourgeois, J.M., and Chang, P.L. (2003) *Hum. Gene Ther.*, **14**, 1065–1077.

Ford, W.T., and Miller, P.D. (2002) Functional Polymer Colloids as Catalysts. In: *Functional Polymer Colloids and Microparticles*, Vol. 4 (eds. R. Arshady and A. Guyot), The MML Series, Citus Books, London, pp. 171–202.

Gast, A.P., and Russel, W.B. (1998) *Phys. Today*, 24.

Geerts, N., and Eiser, E. (2010) *Soft Mat.*, **6**, 664–669.

Hayashi, S., Kumamoto, Y., Suzuki, T., and Hirai, T. (1991) *J. Colloid Interface Sci.*, **144**, 538.

Holtz, J.H., Holtz, J.W., Munro, C.H., and Asher, S.A. (1998) *Anal. Chem.*, **70**, 780–790.

Hulteen, J.C., and Duyue, R.P.V. (1995) *J. Vac. Sci. Technol.*, **A13**, 1553–1558.

Hulteen, J.C., Treichel, D.A., Smith, M.T., Duval, M.L., Jensen, T.R., and Duyne, R.P.V. (1999) *J. Phys. Chem., B*, **103**, 3854–3863.

Lim, F., and Sun, A.M. (1980) *Science*, **210**, 908–910.

Liu, Z.C., and Chang, T.M.S. (2006) *Liver Transplant.*, **12**, 566–572.

Okay, O. (2000) *Prog. Polym. Sci.*, **25**, 711–779.

Ottewill, R.H., and Rowell, R.L. (1997) Order–Disorder Phenomena. In: *Polymer Colloids: A Comprehensive Introduction*, Academic Press, San Diego, pp. 250–275.

Oupicky, D. (2008) *Adv. Drug Delivery Rev.*, **60**, 957.

Paquet, C., and Kumacheva, E. (2008) *Mater. Today* **11**, 48–56.

Park, S.H., Qin, D., and Xia, Y.X. (1998) *Adv. Mater.*, **10**, 1028–1032.

Pieranski P. (1983) *Contemp. Phys.*, **24**, 25.

Ross, C.J.D., Bastedo, L., Maier, S.A., Sands, M.S., and Chang, P.L. (2000) *Hum. Gene Ther.*, **11**, 2117–2127.

Roxlo, C.B., Deckman, H.W., Gland, J., Cameron, S.D., and Chianelli, R.R. (1987) *Science*, **235**, 1629–1631.

van Megan, W., and Snook, I. (1984) *Adv. Colloid Interface Sci.*, **21**, 119.

Velev, O.D., Tessier, P.M., Lenhoff, A.M., and Kaler, E.W. (1999) *Nature*, **401**, 548.
Weissman, J.M., Sunkara, H.B., Tse, A.S., and Asher, S.A. (1996) *Science*, **274**, 959–960.
Xia, Y., Tien, J., Qin, D., and Whitesides, G.M. (1996) *Langmuir*, **12**, 4033.
Xia, Y., Gates, B., Yin, Y., and Lu, Y. (2000) *Adv. Mater.*, **12**, 693–713.
Yan, H., Blanford, C.F., Holland, B.T., Parent, M., Smyrl, W.H., and Stein, A. (1999) *Adv. Mater.*, **11**, 1003–1006.

2

Methods for the Generation of Polymer Particles

CHAPTER OVERVIEW

2.1 Conventional Methods Used for Producing Polymer Particles

This chapter describes the methods that are currently used to generate polymer particles, with a particular focus on microbeads in the micrometer-size range. Particle-forming processes occur in two-phase systems, in which the starting reagent(s) and/or the resulting polymer(s) are in the form of a fine dispersion in an immiscible fluid. Classification of processes used for producing polymer particles relies on: (i) the initial state of the system (e.g, single-phase state versus a multi-phase state); (ii) the mechanism of particle formation, including chemical and physical methods; and (iii) the size of the resulting polymer particles. We will focus on the first and the second features of the classification, with emphasis on the polymerization methods.

Based on the initial state of the system, the polymerization methods can be divided into two groups. In the first group of methods, particle synthesis begins in a one-phase solution that contains molecules of monomers or reactive pre-polymers, initiators, and stabilizers. In the course of the polymerization, the system becomes heterogeneous, due to the nucleation of seed particles. Growth of primary particles occurs through the attachment of molecules or their clusters to the seeds. A classical example of this

Microfluidic Reactors for Polymer Particles. Eugenia Kumacheva and Piotr Garstecki.

process is *dispersion polymerization*, in which the monomer and the initiator are soluble in the polymerization medium, but the medium is a poor solvent for the resulting macroradicals and macromolecules (Barret, 1975; Ober, Lok, and Hair, 1985; Arshady, 1992). Phase separation leads to the formation of primary particles that are swollen by the polymerization medium and/or the monomer. Polymerization proceeds largely within the particles, leading to the formation of spherical particles in a size range of from 0.1 to 10 μm. Dispersion polymerization is typically used for the synthesis of vinyl monomers, such as styrene, and acrylic monomers in hydrocarbons or in alcohol–ether or alcohol–water mixtures.

Precipitation polymerization resembles dispersion polymerization in the initial state of the reaction mixture: it is also a homogeneous solution of the monomer, initiator, and stabilizer in the polymerization medium. However, in contrast with dispersion polymerization, primary particles are not in the swollen state. A sharp distinction between dispersion and precipitation polymerizations may not always exist, however the quality of the medium as a solvent serves as a useful guide: in precipitation polymerization both initiation and polymerization processes occur largely in the homogeneous medium. Continuous nucleation and coagulation of the resulting nuclei (primary particles) lead to the formation of larger polymer particles. Examples of precipitation polymerization include the synthesis of tetrafluoroethylene in water (Suwa *et al.*, 1979) or the preparation of poly(*N*-isopropylamide) microgels by polymerizing *N*-isopropylamide monomer dissolved in water in the presence of a chemical crosslinker (Pelton, 2000).

In classical *emulsion polymerization* a monomer that is scarcely soluble in the polymerization medium is emulsified in it using a surfactant (Gilbert, 1996), thus the synthesis begins in a heterogeneous system. The initial diameter of the monomer droplets is in a range of from 1 to 20 μm or larger. Excess surfactant creates micelles. The initiator is soluble in the medium, and not in the monomer. A small amount of monomer diffuses through the medium to the micelles where it reacts with the monomer, and the monomer-swollen micelles become the main loci of polymerization. In the next stage, more monomer molecules from the droplets diffuse to the growing particles and react with initiator molecules. Eventually, the monomer droplets disappear and all remaining monomer is localized in the particles where polymerization continues, until all monomer is polymerized. Typically, the final product is a dispersion of polymer particles in water, known as latex particles. The size of the latex particles thus produced is usually in the range 50–500 nm. Larger particles are generated via a multi-step polymerization process (see later in this chapter).

Microemulsion polymerization begins in a thermodynamically stable emulsion of nanometer-size monomer droplets. Microemulsions are spontaneously formed in the presence of high concentrations of surfactants, which reduce the value of interfacial tension at the monomer–continuous phase to close-to-zero values. Polymerization yields small (in the order of 5–50 nm in size) polymer particles that coexist with empty micelles formed by surfactant molecules (Chang *et al.*, 1998); Candau, Pabon, and Anquetil, (1999).

Miniemulsion and suspension polymerizations occur in monomer droplets containing a monomer-soluble initiator (Hopff, Lussi, and Hammer, (1965; Yuan, Kalfas, and

Ray, 1991; Landfester, 2006). Nucleation occurs in the monomer or droplets of a monomer solution, so that each droplet behaves as an individual small reactor. After polymerization, the original droplets are converted directly into polymer particles of approximately the same dimensions. The differences between the suspension and miniemulsion polymerizations include the size and stability of the droplets. Suspension polymerization takes place in large (from 1 μm to 1 mm diameter) droplets, whereas miniemulsion polymerization occurs in significantly smaller, typically, submicrometer-size droplets. In suspension polymerization, droplets are generally stabilized against coalescence using polymeric stabilizers such as polyvinylpyrrolidone and poly[(vinyl alcohol)-*co*-(vinyl acetate)] or using biopolymers, for example, natural gums or cellulose ethers. In miniemulsion polymerization, the instability of the system is dominated by Laplace pressure in the droplets, which governs monomer diffusion from the droplets, and the growth of larger droplets at the expense of the smaller ones in the process of Ostwald ripening (Ostwald, 1990; Higuchi, 1962). Stabilization is achieved by adding to the monomer phase a substance with a very poor solubility in the continuous phase, thereby exploiting stabilization based on building a counteracting osmotic pressure in the system. For oil-in-water miniemulsion polymerization, the addition of hydrophobic solvents, such as hexadecane, or strongly non-polar monomers results in greatly increased stability of the droplets (Lowe, 2000).

Alternatively, polymer particles can be formed by physical means. For example, heat-induced phase separation of the solution of linear poly(*N*-isopropylamide) leads to the loss in solubility of this polymer and the association of polymer molecules into particles (microgels) (Deng, Xiao, and Pelton, 1996; Chan, Pelton, and Zhang, 1999). In another approach, emulsification of polymer solutions and subsequent removal of the solvent from the droplets by extraction or evaporation also produces polymer particles (Gañán-Calvo *et al.*, 2006). In this method, the size of the particles is precisely controlled by tuning the concentration of the initial polymer solution; however, using a high polymer content (and hence, generating large, micrometer-diameter polymer beads) is problematic because of the high viscosity of such solutions and the difficulties in their emulsification.

Among the synthetic methods, three types of polymerization can be used to produce polymer particles with dimensions exceeding 10 μm, namely: the multi-stage interfacial polymerization, the swelling (Ugelstad) method, and the suspension polymerization described above. (Dispersion polymerization is generally used for the synthesis of particles with dimensions of up to approximately 10 μm.)

The *swelling method* can be combined with the particle polymerizations described above (Ugelstad *et al.*, 1980, 1985, 1999; Jorgedal, 1985). In this method, relatively small monodispersed seed particles prepared by, for example, emulsion polymerization, undergo swelling with a monomer or a monomer mixture. This step is followed by polymerization of the monomer(s) taken up by the seeds. To grow large polymer particles with dimensions of up to tens of micrometers, the procedures of swelling and polymerization may be repeated several times. The method yields large polymer particles with a narrow size distribution. The multi-stage Ugelstad method can produce monodipersed polymer particles with dimensions of up to 100 μm; however this method is time consuming and material specific. It is challenging to generate particles loaded with low molecular weight molecules or nanoparticles, and microbeads with complex morphologies and non-spherical shapes.

Multi-stage interfacial polymerization also exploits polymer seed particles synthesized by, for example, emulsion polymerization (O'Callaghan, Paine, and Rudin, 1995). Interfacial polymerization on the surface of seeds results in increasing microbead diameter. Typically, the particles are grown to a certain size, after which they serve as seeds in a subsequent stage, and the dimensions of polymer beads gradually increase with the number of polymerization stages. This process may be complicated by the secondary nucleation process in which a monomer polymerizes in the continuous phase rather than on the surface of seeds, thereby leading to the nucleation of secondary small particles. Typically, this complication, as well as possible aggregation of the large particles, is overcome by diluting the dispersion of the particles in each subsequent stage and by using a mixed initiator approach, that is, a combination of the oil-soluble and water-soluble initiators (O'Callaghan, Paine, and Rudin, 1995a; O'Callaghan, Paine, and Rudin, 1995b; Kalinina and Kumacheva, 1999). The process is time-consuming, and generally, it is not used for the synthesis of polymer microbeads with dimensions exceeding several micrometers.

Suspension polymerization is a straightforward, simple, and cost-effective method, however, generally, it produces polymer particles with a broad distribution of sizes. This drawback originates from the limited ability to emulsify monomers in monodispersed droplets and to suppress subsequent coalescence of these droplets prior to their polymerization. As a result, fractionation is used to narrow the distribution of the sizes of the resulting particles. This step is time-consuming and it results in the loss of material. The polydispersity of polymer microbeads produced by suspension polymerization can be significantly improved (i) by generating monomer droplets with a narrow size distribution and (ii) by minimizing coalescence between these droplets before they solidify or gel.

We leave a detailed discussion of the emulsification to Chapter 4. Here we note that conventional emulsification methods, such as sonication, stirring, or ultra-turrax, do not yield droplets with a narrow size distribution; however, several promising methods exist that produce relatively monodisperse "precursor" droplets for suspension or miniemulsion polymerization.

Firstly, membrane emulsification has proved effective in preparing droplets of monomers with dimensions in the micrometer-size range and polydispersity close to 10% (Yuyama *et al.*, 2000). Emulsification of monomers in the immiscible continuous phase occurs by pressing a liquid monomer or a crude pre-emulsion through membranes with a particular, well-defined size of pores. The size of the droplets is determined by the size of the pores in the membrane, the pressure, the viscosity of the continuous and droplet phases, and the value of the interfacial tension between the liquids. The subsequent polymerization step is similar to conventional suspension polymerization. Examples of particles produced by using membrane emulsification include polystyrene microbeads (Omi *et al.*, (1994); Omi *et al.*, 1995) or biodegradable polylactide microbeads (Ma, Nagai, and Omi, 1999).

Emulsification in the Bibette process is achieved by shearing a polydisperse emulsion at a well-defined low shear rate in a narrow gap between the two vertical, coaxial cylinders in a Couette apparatus (Mason and Bibette, 1997; Mabille *et al.*, 2000). At the optimized viscosity of the initial emulsion, controlled dissipation of mechanical energy yields droplets with a relatively narrow distribution of sizes. The final dimensions of the droplets are determined by the applied shear rate, the viscosity of the original emulsion, and the interfacial tension between the droplet and the continuous phases. The distribution of droplet sizes narrows with a reducing width of the gap in the Couette apparatus.

Emulsification can be also achieved by forcing fluids into the bulk continuous medium through a nozzle (Berkland, 2001; Loscertales, 2002) or a vibrating orifice (Partch, 1985; Esen and Schweinger, 1996). For example, photopolymerization of aerosol droplets of acrylate monomers, which were generated using a vibrating-orifice droplet generator, was utilized to generate spherical microbeads (Esen and Schweinger, 1996).

Generally, following emulsification, micrometer-size droplets are transferred into the reactor and polymerized or gelled in a batch process. Polymerization takes from tens of seconds to several hours, depending on the type of monomer. During this process, droplets collide and coalesce, especially, if their concentration in the emulsion is high. As a result, polydispersity of the resultant polymer particles is typically substantially higher than that of the original precursor droplets, even if the latter were produced under optimized conditions. Continuous polymerization of droplets under unbound conditions narrows particle polydispersity; however, in comparison with a batch process, it does not completely suppress coalescence of droplets.

One of the main advantages of microfluidic synthesis of polymer particles, as will be discussed in Chapters 7–10, is the ability to generate droplets with a very narrow size distribution *and* to avoid coalescence between them, by solidifying or gelling droplets as they flow through microchannels and are separated by the well-defined gap of the continuous phase. Frequently, microfluidic synthesis is limited to the emulsification step, so that subsequent to emulsification, precursor droplets are polymerized or gelled in a batch process, where their coalescence is unavoidable.

2.2 Microfluidic Generation of Polymer Particles

Microfluidic synthesis and assembly of polymer particles is described in details in Chapters 7–10. Here we explain the most important general features of the microfluidic production of polymer microbeads. Particles can be generated via a continuous single phase (Dendukuri *et al.*, 2006) or a multiphase microfluidic synthesis (Xu *et al.*, 2005; Dendukuri et al., 2005). Currently, the latter synthesis dominates the field; therefore here we focus on the mutiphase microfluidic reactors. Typically, a microfluidic reactor for the multiphase synthesis of polymer particles contains two or three parts (corresponding to the stages of the synthesis), as shown in Figure 2.1: a mixing compartment, a droplet generator, and a compartment for polymerization, gelation or solvent withdrawal. A mixing compartment is required when reagents have a high reactivity, in order to avoid

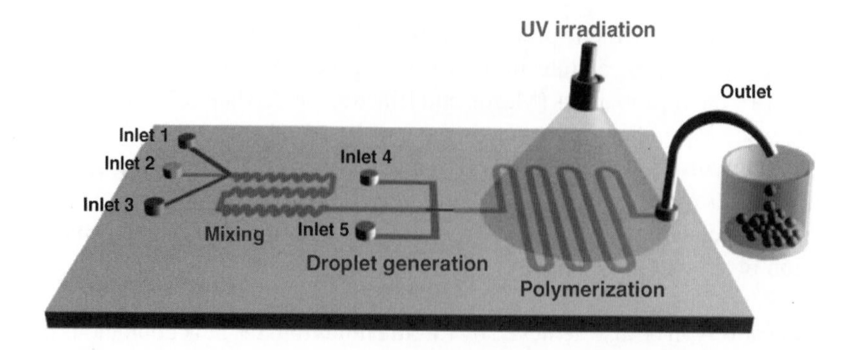

Figure 2.1 *Schematic of the microfludic reactor for multiphase synthesis of polymer particles. The monomers are supplied to the mixing compartment through* inlets *1–3. A continuous phase is supplied to* inlets *4 and 5. Droplets of the monomer mixture move to the polymerization zone where they are exposed to UV-irradiation to initiate photopolymerization.* Adapted with permission from Li, W., et al., Multi-Step Microfluidic Polymerization Reactions Conducted in Droplets: The Internal Trigger Approach, *J. Am Chem. Soc.*, 130, 9935–9941. Copyright (2008) with permission from American Chemical Society.

chemical changes in the tubing supplying the liquid mixture to the reactor. In the droplet generator, two immiscible liquids – the droplet phase and the continuous medium – are introduced into the separate microchannels or capillaries. The stream of the liquid that is to be dispersed periodically breaks up into droplets, which are emulsified in the continuous phase. The droplets move to the extension channel in which they are subjected to irradiation, heating, addition of chemical agents, or solvent extraction.

Microfluidic emulsification can be carried out in droplet generators with varying geometries, such as T-junctions, co-axial capillaries, flow-focusing geometry, and terrace-like microchannels. A detailed discussion of existing microfluidic droplet generators is given in Chapter 5. Regardless of the type of droplet generator, for a particular combination of continuous and droplet phase liquids, the size of the droplets depends on the dimensions of the microchannels or capillaries and the flow rates of the liquids. A highly periodic manner in which the stream of the droplet phase (a monomer, an oligomer, or a polymer solution) breaks up, determines the narrow size distribution of the "precursor" droplets and a well-defined distance between them when they are moving in the downstream channel.

The transformation of precursor droplets into polymer particles is achieved by either chemical or physical mechanisms, such as polymerization, crosslinking, thermally induced gelation, self-assembly, evaporation of the solvent from the droplets, or phase separation. Currently, free-radical and condensation polymerization are the two methods that are most frequently used to generate rigid polymer particles with a uniform or a capsular structure (see Chapters 7–10).

As in conventional suspension polymerization, in microfluidic synthesis every droplet performs as a small isolated reactor with a volume in the range of from pico- to nanoliter. Therefore, in addition to the synthesis of polymer particles, droplets generated by microfluidic emulsification can be used for solution-based synthesis. Such an application of droplets can be useful for polymer synthesis, which is difficult to perform in a single-phase microfluidic format due to the gradual increase in the viscosity of the liquid with the increasing molecular mass of the polymer.

Continuous microfluidic synthesis and assembly of polymer particles offer several beneficial and, in some ways, unique features, which are listed below and described in detail in Chapters 7–10:

- The ability to generate polymer particles with polydispersity below 5% (and under particular conditions, below 1%).
- The capability to generate polymer particles with non-conventional, non-spherical shapes. One of the methods utilizes solidification or gelation of droplets confined by microchannels with dimensions smaller than the unperturbed droplet diameter. Another method uses multiphase droplets with non-symmetric morphologies, from which, following polymerization, one of the phases is removed (Nie *et al.*, 2005) A powerful, although less frequently used method exploits one-phase polymerization using projection lithography, in which the shape of the particles is determined by the features of the mask.
- The capability to produce polymer particles with complex architectures, which originates from the ability to control the morphology of precursor droplets by hydrodynamic means. Examples of such particles include core-shell particles (capsules), Janus particles, or photonic balls.
- Control of internal structures of polymer particles realized by the uniform supply of energy e.g., UV radiation to precursor droplets.
- For multiple reactions or physical processes leading to the transformation of droplets into particles, microfluidics allows control of the location and timing of a particular chemical or physical process by using a "time-to-distance" transformation. This feature allows control of particle composition and internal structure.
- Shear imposed on precursor droplets can play an important role in controlling their shape and architecture, thereby influencing the shape and structure of the resulting polymer particles.
- Owing to the narrow distribution of sizes of particles generated by the microfluidic methods, they can be used for the controlled encapsulation of cells. The average number of cells per particle is accurately described by the Poisson distribution.

Along with the advantages, successful production of polymer particles has several requirements. These requirements include: (i) efficient microfluidic generation of precursor droplets from liquids with well-defined macroscopic properties such as interfacial tension and viscosity (this requirement being important in multiphase polymer synthesis); (ii) fast transformation of precursor droplets into polymer particles; and (iii) the ability to scale-up the microfluidic production of the particles by using multiple parallel microfluidic reactors.

In the following chapters we describe the characteristic features of the continuous microfluidic production of polymer particles: the basics of microfluidic emulsification, the production of rigid and soft polymer particles with controlled shapes and morphologies, and the generation of polymer particles with various compositions.

References

Arshady, R.R. (1992) *Coll. Polym. Sci.*, **270**, 717–732.

Barret, K.E.J. (ed.) (1975) *Dispersion Polymerization in Organic Media*, John Wiley & Sons, Ltd, Chichester, p. 322.

Berkland, C., Kim, K., and Pack, D.W. (2001) *J Controll. Release* **73**, 59–74.

Candau, F., Pabon, M., and Anquetil, J.-Y. (1999) *Colloid Surf A: Physicochem. Engng. Asp.*, **153**, 47–59.

Chang, H.-C., Lin, Y.-Y., Chern, C.-S., and Lin, S.-Y. (1998) *Langmuir*, **14**, 6632–6638.

Chan, K., Pelton, R., and Zhang, J. (1999) *Langmuir*, **15**, 4018–4020.

Deng, Y., Xiao, H., and Pelton, R. (1996) *J. Colloid Interface Sci.*, **179**, 188–193.

Dendukuri, D., Tsoi, K., Hatton, T.A., and Doyle, P.S. (2005) *Langmuir*, **21**, 2113–2116.

Dendukuri, D., Pregibon, D.C., Collins, J., Hatton, T.A., and Doyle, P.S. (2006) *Nat. Mater.*, **5**, 365–369.

Esen, C., and Schweinger, G.J. (1996) *J. Colloid Interface Sci.*, **179**, 276–280.

Gañán-Calvo, A.M., Martín-Banderas, L., González-Prieto, R., Rodríguez-Gil, A., Berdún-Álvarez, T., Cebolla, A., Chávez, S., and Flores-Mosquera, M. (2006) *Intern. J. Pharm.*, **324**, 19–26.

Gilbert, R.G. (1996) *Emulsion Polymerization: a Mechanistic Approach*, Academic Press, London, 362 pp.

Higuchi, W.I., and Misra, J. (1962) *J. Pharm. Sci.*, **51**, 459–466.

Hopff, H., Lussi, H., and Hammer, E. (1965) *Makromol. Chem.*, **82**, 175–184.

Jorgedal, A., Hansen, F.K., and Nustad, K. (1985) *J. Polym. Sci., Polym Symp.*, **72**, 225–240.

Kalinina, O., and Kumacheva, E. (1999) *Macromolecules*, **32**, 4122–4129.

Landfester, K. (2006) *Ann. Rev. Mater. Res.*, **36**, 231–279.

Li, W., Pham, H.H., Nie, Z., MacDonald, B., Guenther, A., and Kumacheva, E. (2008) *J. Am Chem. Soc.*, **130**, 9935–9941.

Loscertales, I.G., Barrero, A., Guerrero, I., Cortijo, R., Marquez, M., Gañán-Calvo, A.M. (2002) *Science* **295**, 1695–1698.

Lowe, K.C. (2000) *Art. Cells, Blood Subs. Immob Biotechnol.*, **28**, 25–38.

Ma, G.H., Nagai, M., and Omi, S. (1999) *Colloid Surf A: Physicochem. Eng. Asp.*, **153**, 383–94.

Mabille, C., Schmitt, V., Gorria, P., Caldeon, F.L., Faye, V., Deminiere, B., and Bibette, J. (2000) *Langmuir* **16**, 422–429.

Mason, T.G., and Bibette, J. (1997) *Langmuir*, **13**, 4600–4613.

Nie, Z., Xu, S., Seo, M., Lewis, P.C., and Kumacheva, E. (2005) *J. Am. Chem. Soc.*, **127**, 8058–8063.

Ober, C.K., Lok, K.P., and Hair, M.L. (1985) *J. Polym. Sci. A*, **23**, 103–108.

O'Callaghan, K.J., Paine, A.J., and Rudin, A. (1995a) *J. Appl. Polym. Sci.*, **58**, 2047–2055.

O'Callaghan, K.J., Paine, A.J., and Rudin, A. (1995b) *J. Polym. Sci., Part A: Polym. Chem.*, **33**, 1849–1857.

Omi, S., Katami, K., Yamamoto, A., and Iso, M. (1994) *J. Appl. Polym. Sci.*, **54**, 1–11.

Omi, S., Katami, K., Taguchi, T., Kaneko, K., and Iso, M. (1995) *Macromol. Symp.*, **92**, 309–420.

Ostwald, W.Z. (1900) *Phys. Chem.*, **34**, 495–503.

Partch, R.E., Nakamura, K., Wolfe, K.J., and Matijevic, E. (1985) *J. Colloid Interface Sci.*, **105**, 560–569.

Pelton, R. (2000) *Adv. Colloid Interface Sci.*, **85**, 1–33.

Suwa, T., Watanabe, T., Seguchi, T., and Okamoto, J. (1979) *J. Polym. Sci. Polym. Part A: Polym. Chem.*, **17**, 111–127.

Uglestad, J., Kuggerud, K.H., and Fitch, R.M. (1980) Swelling of aqueous dispersions of polymer-oligomer particles. Preparation of polymer particles of predetermined particle size

including large monodisperse particles. In: *Polymer Colloids II* (ed. R.M. Fitch), Plenum Press, New York, pp. 83–93.
Ugelstad, J., Mfutakamba, H.R., Mork, P.C., Ellingsen, T., Berge, A., Schmid, R., Holm, L., Jorgedal, A., Hansen, F.K., and Nustad, K. (1985) *J. Polym. Sci. Polym. Symp.*, **72**, 225–240.
Uglestad, J., Muftakamba, H.R., Mork, P.C., Ellingson, T., Berge, A., Shmid, R., Holm, L., Velev, O.D., Tessier, P.M., Lenhoff, A.M., and Kaler, E.W. (1999) *Nature*, **401**, 548.
Xu, S., Nie, Z., Seo, M., Lewis, P.C., Kumacheva, E., Garstecki, P., Weibel, D., Gitlin, I., Whitesides, G.M., and Stone, H.A. (2005) *Angew. Chem. Int. Ed.*, **44**, 724–728.
Yuan, H.G., Kalfas, G., and Ray, W.H. (1991) *Polymer Rev.*, **C31**, 215–299.
Yuyama, H., Watanabe, T., Ma, G.-H., Nagai, M., and Omi, S. (2000) *Colloid Surf. A: Physicochem. Eng. Asp.*, **168**, 159–174.

3

Introduction to Microfluidics

CHAPTER OVERVIEW

In Chapter 2 we gave an overview of the applications of polymeric microparticles and their methods of formation. These include the classical and industrially implemented methods and the microfluidic techniques that are the focus of our book. Microfluidics itself is a relatively young branch of chemical engineering, spanning only the last 20 years. The use of microfluidic systems for the formation of droplets is even younger, as it has only been of interest for the last few years. The rapid development of the understanding of multiphase microflows and of the wide range of applications of these systems to the generation of droplets and polymeric particles forms the basis of commercial implementations. However, as it stands, this set of techniques does not yet constitute a technology. There are important scientific and engineering problems that still need to be tackled. We will give an overview the current state-of-the-art academic achievements, and demonstrate their strengths and potential, along with the remaining challenges.

3.1 Microfluidics

Microfluidics is a branch of chemical engineering that studies the design, fabrication, and operation of systems of microscopic channels that conduct fluids. If compared with

Microfluidic Reactors for Polymer Particles. Eugenia Kumacheva and Piotr Garstecki.

standard fluidic systems, the microfluidic channels are typically small, having widths (or diameters) ranging from single micrometers to tens or hundreds of microns. The pumping of liquids and gases through microducts typically occurs at small speeds. In this context "small" means that the viscous forces dominate over the inertial ones. As a result, the flow of liquids at the microscale can most often be described by equations of flow based on a simple proportionality between the speed of the flow and the magnitude of the force that drives the flow (Squires and Quake, 2005). The resulting flow is laminar, i.e. the sub-volumes of liquid flow side-by-side, following the field of the gradient of pressure, and the streamlines never cross each other (Figure 3.1). This characteristic of the flow provides for extensive control: the speed of flow obeys the simple Hagen–Poiseuille equation, which predicts the speed of flow as a quotient of the pressure drop through the particular capillary and its hydraulic resistance. Importantly, the resistance to flow is a function of the dimensions of the channel and the viscosity of the fluid. This allows networks to be designed that distribute the flow of liquids in accordance with the desired pattern. This property, when combined with typically large values of the Peclet number (Stone, Stroock, and Ajdari, 2004), reflecting the fact that diffusional transport is typically slow in comparison with the flow, makes it possible to control the profiles of the concentration (Jeon *et al.*, 2000) of chemicals and also the profiles of the temperature (Lucchetta *et al.*, 2005) in the channels, all with minute consumption of the fluids.

A crucial contribution to the explosion of the research activity in the field of microfluidics was the development of accessible procedures for microfabrication

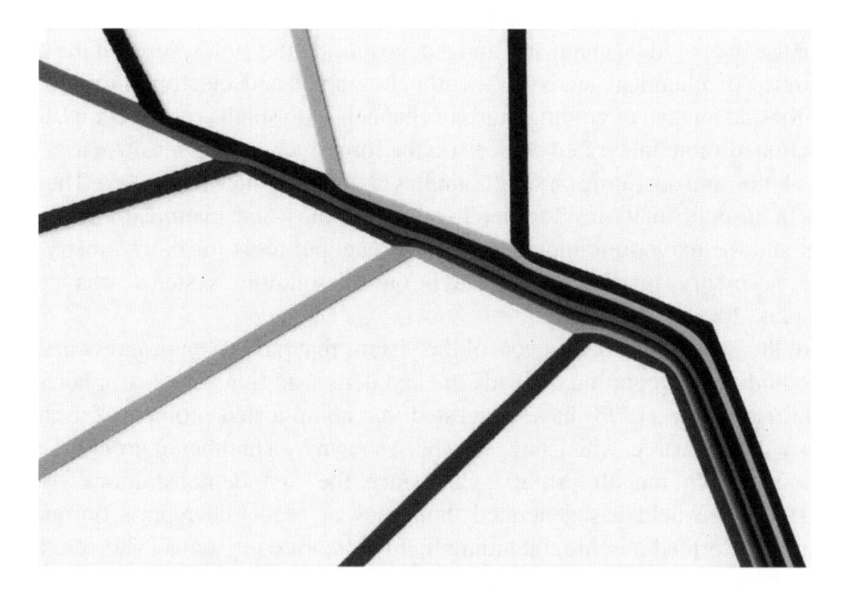

Figure 3.1 *A micrograph by Felice C. Frankel of laminar streams of aqueous dyes in microchannels. Photograph by Felice C. Frankel, reproduced with permission.*

(Whitesides, 2006). A judicious choice of the microfabrication techniques – either lithography, milling or etching – allows for the facile preparation of both planar and truly three-dimensional systems of microchannels (Becker and Locascio, 2002; Becker and Gartner, 2008; Desai, Hansford, and Ferrari, 2000; Tseng, 2004; Voldman, Gray, and Schmidt, 1999; Weibel, DiLuzio, and Whitesides, 2007). In the planar systems, all the ducts share a common plane and all have the same height, while the dimensions in the plane can vary across the system. Spincoating allows for the tuning of this height across the lengthscales: from single nanometers, through micrometers to fractions of a millimeter. It is also possible to form systems in which the heights of the channels change along the line of flow of the fluids (the so-called 2.5-dimensional systems) (Tseng, 2004), or to prepare systems that are truly three-dimensional, such as the axi-symmetric systems (Ganan-Calvo, 1998). The technique that has probably played the most important role in expanding academic interest in microfluidics is fast prototyping via lithography and replication of the masters in polydimethylosiloxane. This technique – often referred to as soft-lithography–makes it possible to go from the idea to the fabricated chip within a day, with facile reproduction of the existing masters for multiple experiments (Duffy *et al.*, 1998). The extensively developed techniques of fabrication of micro-electro-mechanical systems (MEMS) provide a vast and readily available set of tools for integration of actuators, electrodes, and waveguides with microfluidic channels (Verpoorte and De Rooij, 2003). Such integration opened the vistas to systems based on electrostatic forcing of flow, such as via electro-osmosis or electrophoresis, and for the simultaneous readout of the results of on-chip separations and reactions (Dittrich, Tachikawa, and Manz, 2006).

The character of flow, the small volumes of liquids used in experiments, and the facile access to microfabrication, all prompted a vision of the development of microfluidic chips for use in analytical chemistry and diagnostics. In the 1990s, some of the existing technologies of chemical analysis – chromatography and electrophoresis – which already took advantage of guiding fluids in channels with small cross-sections, inspired construction of more integrated devices (in the form of chips) for sensitive assays with high resolution and operating on small samples of fluids (Whitesides, 2006). The intense interest in in-field analytics for defense against bio- and chemical-terrorism and warfare, and the exploding interest in high-throughput tools for biochemistry, meant that the necessary funding for research on microfluidic systems was provided (Whitesides, 2006).

One of the most important aspects of the visions that has driven progress in the area of microfluidics is integration. Already the first demonstrations of electrophoresis on a chip (Harrison *et al.*, 1993) have suggested that complicated protocols for chemical analyses will be feasible, which have now been shown by a number of groups (Erickson and Li, 2004). In the almost 20 years since the first demonstrations (Harrison *et al.*, 1993), the field has generated thousands of academic reports on analytical techniques performed on-chip, including highly integrated systems (Thorsen, Ismagilov, and Zheng, 2002) and commercial applications. The area of microfluidics has gone through a phase of rapid expansion and is now maturing (Figure 3.2). The exponential explosion of interest has slowly saturated the field, which is now transiting into the phase

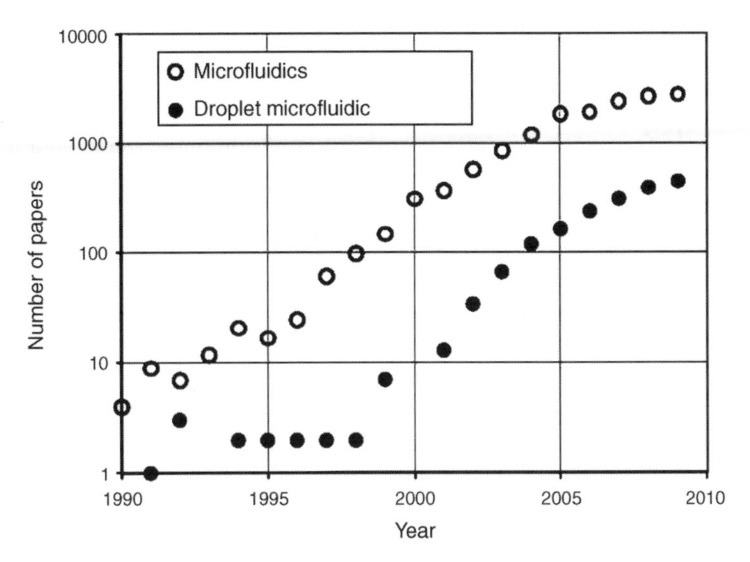

Figure 3.2 *Number of scientific articles related to microfluidics (citing the word "microfluid" ○) and of those pertaining to droplet microfluidics (citing both of the words "microfluid" and "drop" ●) plotted as a function of the year of publication (extracted from the ISI Web of Knowledge database). One can clearly see the exponential growth of the number of publications up to approximately 2005, and then the slow saturation of the increase in the number of articles*

of research more oriented towards appliations. This is possible because the fundamental concepts and understanding, although still being areas of active investigation, have already been laid.

3.2 Droplet Microfluidics

At the beginning of this century Thorsen *et al.* (2001) demonstrated the use of a simple microfluidic device for the formation of monodisperse droplets. This report opened up a completely new wave of interest in microfluidic systems that operate the flow of immiscible liquids. The use of micron-scale capillaries for construction of devices that generate monodisperse droplets (Ganan-Calvo, 1998) were known earlier. However, the report by Thorsen, Maerkl, and Quake (2001) was still critical to the establishment of the new field of on-chip droplet microfluidics. The fact that it was possible to control the flow of immiscible liquids with a similar fidelity as that possible for the flow of simple liquids (Kenis, Ismagilov, and Whitesides, 1999) was not trivial. As we will discuss in detail in the next chapter, the interfacial forces introduce complicated interactions, with their magnitudes depending on the curvature and surface area of the interface. Despite these complications, multiphase flows do subject themselves to extensive control. Chapter 5 will focus on the aspect of this control that is central to the microfluidic synthesis of polymeric particles, that is, on the processes of formation of droplets, bubbles, and more complicated objects, such as multiple droplets and capsules.

The ability to form monodisperse droplets in microfluidic systems, and to control virtually all the parameters of the process of formation, that is, the volume and distribution of volumes of the droplets, and the frequency of their formation, forms the basis for a number of potential applications. These applications can be divided into two general classes. The first, which is explored in this book, is comprised of preparatory techniques for the formulation of new materials for the pharmaceutical, cosmetic, and food industries. The second, which lies largely outside the scope of this book, but one that has implications for the synthesis of polymeric particles, encompasses systems for analytical and synthetic chemistry performed within individual microdroplets. Originally, the interest in the use of microdroplets as reaction chambers stemmed from the opportunity to minimize the volume of liquid samples and, at the same time, obtain multiple measurements from identical or continuously varied (Tice, Ismagilov, and Zheng, 2004) compositions. The feature that, in practice, the droplets can be formed at frequencies of hundreds or thousands droplets per second provides for reliable statistics.

Shortly after the first reports (Zheng, Roach, and Ismagilov, 2003) it became clear that the use of microdroplets as reaction chambers offers additional attractive features (Song, Chen, and Ismagilov, 2006). These features include the avoidance of dispersion of time of residence of compounds contained in droplets, which is inherent to pressure driven flow-through reaction chambers. In droplet microfluidics the reaction mixtures are enclosed, and stirred constantly (Song *et al.*, 2003a), within the droplet. In addition, droplet systems provide for rapid mixing (Song *et al.*, 2003a), which is otherwise one of the problems in single-fluid microfluidics associated with the laminar nature of flow. Finally, because the droplets translate through channels at a well-determined speed, there is a simple correspondence between the time and position of the droplet on the chip, offering a convenient temporal resolution of the analytical measurements and the ability to control the kinetic conditions of reactions (Song and Ismagilov, 2003b) via, for example, titrations at defined spots on the chip. The use of microdroplets as reactors has been reviewed extensively (Gunther and Jensen, 2006; Song, Chen, and Ismagilov, 2006; Teh *et al.*, 2008). The microfluidic systems that achieve formation of droplets of different chemical compositions, their synchronization, merging, and processing of chemical reactions can be used in the development of new processes for formulation of materials, that is, synthesis of monodisperse nanoparticles (Shestopalov, Tice, and Ismagilov, 2004) or aqueous gels (Um *et al.*, 2008) and we will refer to a number of these techniques in the remaining parts of the book.

In the next chapter we will review the fundamental concepts of fluid mechanics that are required to understand the processes of formation of droplets discussed in Chapters 5 and 6. Chapters 7–10 discuss in detail reports that are available on the application of microfluidic systems to the synthesis of polymeric and hydrogel particles and of microcapsules.

References

Becker, H. and Locascio, L.E. (2002) *Talanta* **56**, 267–287.

Becker, H. and Gartner, C. (2008) *Anal. Bioanal. Chem.*, **390**, 89–111.

Desai, T.A., Hansford, D.J., and Ferrari, M. (2000) *Biomol. Eng.*, **17**, 23–36 (2000).

Dittrich, P.S., Tachikawa, K., and Manz, A. (2006) *Anal. Chem.*, **78**, 3887–3907.

Duffy, D.C., McDonald, J.C., Schueller, O.J.A., and Whitesides, G.M. (1998) *Anal. Chem.*, **70**, 4974–4984.

Erickson, D. and Li, D.Q. (2004) *Anal. Chim. Acta*, **507**, 11–26.

Ganan-Calvo, A.M. (1998) *Phys. Rev. Lett.*, **80**, 285–288.

Gunther, A. and Jensen, K.F. (2006) *Lab Chip*, **6**, 1487–1503.

Harrison, D.J., Fluri, K., Seiler, K., Fan, Z., Effenhauser, C.S., and Manz, A. (1993) *Science*, **261**, 895–897.

Jeon, N.L., Dertinger, S.K.W., Chiu, D.T., Choi, I.S., Stroock, A.D., and Whitesides, G.M. (2000) *Langmuir*, **16**, 8311–8316.

Kenis, P.J.A., Ismagilov, R.F., and Whitesides, G.M. (1999) *Science*, **285**, 83–85.

Lee, C.C., Sui, G.D., Elizarov, A., Shu, C.Y.J., Shin, Y.S., Dooley, A.N., Huang, J., Daridon, A., Wyatt, P., Stout, D., Kolb, H.C., Witte, O.N., Satyamurthy, N., Heath, J.R., Phelps, M.E., Quake, S.R., and Tseng, H.R. (2005) *Science*, **310**, 1793–1796.

Lucchetta, E.M., Lee, J.H., Fu, L.A., Patel, N.H., and Ismagilov, R.F. (2005) *Nature*, **434**, 1134–1138.

Shestopalov, I., Tice, J.D., and Ismagilov, R.F. (2004) *Lab Chip*, **4**, 316–321.

Song, H., Bringer, M.R., Tice, J.D., Gerdts, C.J., and Ismagilov, R.F. (2003a) *Appl. Phys. Lett.*, **83**, 4664–4666.

Song, H. and Ismagilov, R.F. (2003b) *J. Am. Chem. Soc.*, **125**, 14613–14619.

Song, H., Chen, D.L., and Ismagilov, R.F. (2006) *Angew. Chem. Int. Ed.*, **45**, 7336–7356.

Squires, T.M. and Quake, S.R. (2005) *Rev. Mod. Phys.*, **77**, 977–1026.

Stone, H.A., Stroock, A.D., and Ajdari, A. (2004) *Annu. Rev. Fluid Mech.*, **36**, 381–411.

Teh, S.Y., Lin, R., Hung, L.H., and Lee, A.P. (2008) *Lab Chip*, **8**, 198–220.

Thorsen, T., Roberts, R.W., Arnold, F.H., and Quake, S.R. (2001) *Phys. Rev. Lett.*, **86**, 4163–4166.

Thorsen, T., Maerkl, S.J., and Quake, S.R. (2002) *Science*, **298**, 580–584.

Tice, J.D., Ismagilov, R.F., and Zheng, B. (2004) Forming droplets in microfluidic channels to indexing concentrations in droplet-based assays. Paper presented at 228th American Chemical Society National Meeting, Philadelphia, PA, USA, (106-CHED) August 22–26, 2004.

Tseng, A.A. (2004) *J. Micromechan. Microeng.*, **14**, R15–R34.

Um, E., Lee, D.S., Pyo, H.B., and Park, J.K. (2008) *Microfluid. Nanofluid.*, **5**, 541–549.

Verpoorte, E. and De Rooij, N.F. (2003) *Proc. IEEE*, **91**, 930–953.

Voldman, J., Gray, M.L., and Schmidt, M.A. (1999) *Annu. Rev. Biomed. Eng.*, **1**, 401–425.

Weibel, D.B., DiLuzio, W.R., and Whitesides, G.M. (2007) *Nat. Rev. Microbiol.*, **5**, 209–218.

Whitesides, G.M. (2006) *Nature*, **442**, 368–373.

Zheng, B., Roach, L.S., and Ismagilov, R.F. (2003) *J. Am. Chem. Soc.*, **125**, 11170–11171.

4

Physics of Microfluidic Emulsification

CHAPTER OVERVIEW

Microfluidic methods for the generation of polymer particles all, or almost all, rely on the integration of a two-step process, which includes: (i) on-chip formation of simple or more complicated core-shell droplets, and (ii) polymerization, gelation or, more generally, solidification of these droplets, cores, or shells into particles or capsules. Solidification can be done *in situ*, on the same microfluidic chip that is used for the formation of droplets, or *off* the chip. Importantly, except for a few examples of polymerization of sub-volumes of the continuous simple fluid [single-phase microfluidic synthesis (Dendukuri *et al.*, 2007)] all of these processes are based on multiphase flows with a free interface between the immiscible fluids. Thus, the dynamics of formation of simple and structured droplets lies at the very core of the microfluidic technologies for formulations of polymeric particles and will be introduced in the next chapter. Here, we wish to present the rudimentary concepts of

Microfluidic Reactors for Polymer Particles. Eugenia Kumacheva and Piotr Garstecki.

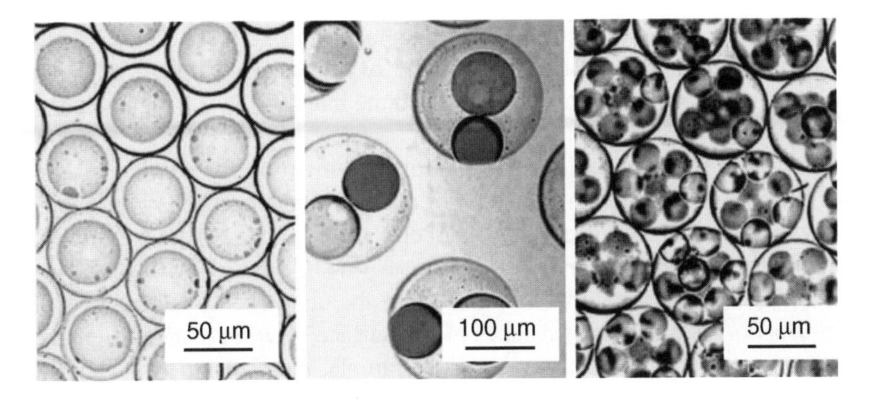

Figure 4.1 *The well-defined shapes of inner and outer droplets critically depend on interfacial dynamics.* Reprinted with permission from Soft Matter, *Controlled formulation of monodisperse double emulsions in a multiple-phase microfluidic system by T. Nisisako, S. Okushima and T. Torii,* ***1****, 1, 23–27 Copyright (2005) Royal Society of Chemistry, (http://dx.doi.org/10.1039/B501972A)*

the physics of flow and the formation of droplets, and provide the basis for understanding the hydrodynamics of microfluidic emulsification.

Bubbles, droplets, and multiple droplets are composed of fluids and are surrounded by fluids, and yet have the well defined shapes of spheres or, if squeezed between the walls of the microchannels, of ellipsoids or disks. Figure 4.1 shows a few examples of structured droplets with well-defined spherical shapes. The origin of the stability and definition of the shapes of these segments of fluids lies in the interfacial tension. The sole existence of the interface that separates two immiscible fluids – be it two immiscible liquids, such as oil and water, or a liquid and a gas, such as water and air – introduces an additional cost to the free energy of the system. As is described below, this energetic contribution is proportional to the surface area of the interface separating the immiscible phases. As all systems, on their way to either local or global equilibrium, tend to exchange their energy into entropy, the surface area of the interface is normally minimized and the condition for minimization of the interfacial area defines the shapes of the droplets.

4.1 Energy of the Interfaces Between Immiscible Fluids

Liquids present a condensed state due to the attractive electromagnetic interactions between the molecules. Each molecule in the bulk of the liquid experiences an attractive potential of interaction with the surrounding molecules. The observation that these molecules condense into a liquid, as opposed to them staying in the vapor phase, is equivalent to the statement that their free energy is lower in the condensed state, that is, when they are in close contact with other molecules of their own type, than when they are separated from each other. The molecules at the interface have approximately half the number of contacts with molecules of the same type, and are also exposed to contacts with molecules of the other liquid (or more generally fluid). If these other contacts were favored in terms of energy of interaction then the two fluids would mix. In other words, if the fluids do not mix, the molecules exposed at the interface have a higher energy than those contained in the bulk.

The attraction between molecules commonly stems from the van der Waals interactions that provide a potential in the order of the thermal energy (kT) per molecule. These, when scaled to the surface density of energy, give an order of magnitude of tens of $\mathrm{mJ\,m^{-2}}$, as observed for most interfaces (de Gennes, Brochard-Wyart, and Quere, 2002). It is important to note that often the component of the free energy of the system associated with the existence of an interface between two immiscible phases has a significant entropic contribution. For example, molecules of water in the bulk phase can freely probe all possible spatial configurations of the hydrogen bonds, yielding large rotational and configurational entropy. The same molecules, when exposed to an alkane chain, or a non-polar surface, need to form a hydration shell, which strongly restricts the number of possible configurations of the network of hydrogen bonds and as a result increases the entropic contribution to the free energy.

It is important to bear in mind that the value of the interfacial energy depends on temperature and generally decreases with increasing temperature. For example, the surface tension of water (interfacial tension between water and air) decreases approximately linearly with the temperature from $76\,\mathrm{mN\,m^{-1}}$ at $0\,°\mathrm{C}$ to zero at $370\,°\mathrm{C}$. The dependence of the magnitude of interfacial tension on temperature is the basis of a set of interesting phenomena – Marangoni effects – which may lead to, for example, motion of droplets in a gradient of temperature, or an instability of thin layers of liquids, leading to the appearance of convective cells.

4.2 Surfactants

The value of the interfacial tension can be lowered by addition of surface-active agents (surfactants). Surfactants are molecules that covalently bind moieties with two different chemistries, for example, a hydrophilic polar group that is readily wetted by water with a hydrophobic, alkyl chain. The fact that these two complementary parts are bound covalently effectively removes the energy of contact between the different groups from considerations of the fluid mechanics. That is to say, the energies of covalent bonds are much larger than thermal energies or the energies associated with the pressure applied to the fluids and are thus inaccessible in normal processing of immiscible fluids. Most surfactants form a molecular solution in the liquid in which they are dissolved only up to a critical concentration (termed the critical micelle concentration, or CMC), above which they start to self-assemble into supramolecular aggregates (micelles). Below the CMC, the value of the interfacial tension depends on the concentration of the surfactant (and decreases with an increasing concentration of the surface-active agent). Above the CMC the value of surface tension stabilizes at a constant value. Surfactants are also used to stabilize foams and emulsions against coalescence. When two interfaces approach each other to within nanoscopic separations, thermal fluctuations can lead to formation of a capillary bridge between the two bubbles (or two droplets), followed by fusion of the two, initially separated, volumes of fluid. Surfactants protect the interfaces either through steric interactions between the alkyl chains in oil or, for example, pegylated chains in water, or through electrostatic repulsion between charged ionic segments. This, either steric or electrostatic, repulsion at small separations of the two interfaces effectively reduces the probability of formation of a capillary bridge and thus fusion of the droplets.

It is important that the surfactant has to be chosen correctly for a given pair of immiscible fluids. There is a wide selection of effective surfactants for water–hydrocarbon and water–air interfaces. However, it is more difficult to find an effective surfactant for a water–silicon oil interface or an interface with a fluorocarbon fluid.

When discussing the processes of formation of droplets it is often essential to consider the kinetics of absorption of the surfactant at the interface. The process of formation of droplets involves creation of an interface, and in order for this interface to be protected by surfactants, the molecules of the surface active agent need to diffuse and absorb onto the interface. When the flow and process of formation of a fresh interface is fast in comparison with the typical time for surfactants to diffuse to the interface, the value of the interfacial tension can be time-dependant and can change along the axis of a breaking jet.

4.3 Interfacial Tension

Formally, interfacial tension is defined as the coefficient of proportionality between the interfacial energy and the surface area of the interface. Although this term strictly applies to all interfaces between immiscible fluids (either liquid–liquid, or liquid–gas), the interfacial energies of the latter are commonly referred to by the term "surface tension".

Conveniently, most, if not all, of the problems involving flow of immiscible fluids that are introduced in this book can be treated within continuous mechanics. This preposition is based, among other considerations with regard to the physics of flow, on the assumption that the simple proportionality between the interfacial energy and the surface area of the interface holds down to very small scales. Interestingly, as shown by Moseler and Landman (2000) this is true down to the scale of single nanometers, that is, down to length-scales of the same order as the size of the molecules sitting at the interface (Figure 4.2).

4.4 Laplace Pressure

So far we have been discussing the interfacial effects in the language of energies. In fluid dynamics and for the description of the processes and mechanisms of formation of droplets in microfluidic systems, it is more convenient to use the language of forces and pressures. Similarly, the simple relationship between, for example, gravitational potential energy and gravitational force:

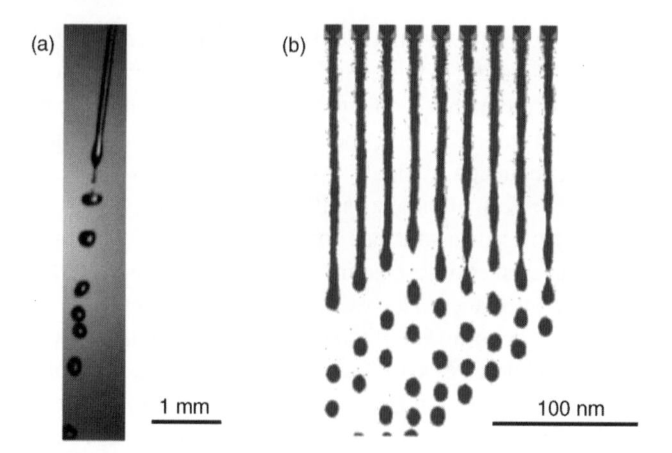

Figure 4.2 *A macroscopic falling jet of water undergoing a Rayleigh–Plateau instability (a), and a series of snapshots from a molecular dynamics simulation of a nanoscopic jet of benzene (b), reprinted with permission from Moseler, M., and Landman, U., Formation, Stability, and Breakup of Nanojets, Science, 289, 1165–1169 (2000) AAAS*

$$U = -\int F \mathrm{d}x \tag{4.1}$$

or conversely

$$F = -\mathrm{d}U/\mathrm{d}x \tag{4.2}$$

yields the force as a momentary coefficient of proportionality between the increase (decrease) of the energy (δU) of the system and a small displacement (δx). For example, (Figure 4.3a), a sphere of radius r has the surface area of $A = 4\pi r^2$. Increasing r by dr changes the interfacial area by $\mathrm{d}A = 8\pi r \mathrm{d}r$, and the interfacial energy by $\mathrm{d}U = \gamma \mathrm{d}A$. An increase in the radius of the sphere increases the interfacial energy, thus the force exerted by the interface is oriented into the interior and is equal to $F(r) = \gamma 8\pi r$.

This can be readily translated into the pressure exerted by the interface, which is equal to $p_L = F/A = 2\gamma/r$. For a bubble of gas in a liquid or for a droplet of one liquid in another, the pressure p_L yields the pressure jump across the interface between the two immiscible phases: the pressure inside the bubble or droplet is higher than the pressure in the surrounding liquid by p_L. The subscript L refers to the name of the famous physicist Pierre-Simon Laplace, who devoted a great deal of his attention to the shapes of soap films and the study of interfacial tension; the pressure difference across the interface is commonly referred to as the Laplace pressure. It is worth noting here that the smaller the bubble or droplet, the higher the pressure inside of it. For example, for a bubble of air in water, with a value for the interfacial tension of $\sim$70 mN m^{-1}, the excess pressures inside bubbles of diameters of 1 mm, 1 μm, and 1 nm are, respectively, three thousandth of an atmosphere, three atmospheres, and three thousands atmospheres! This explains

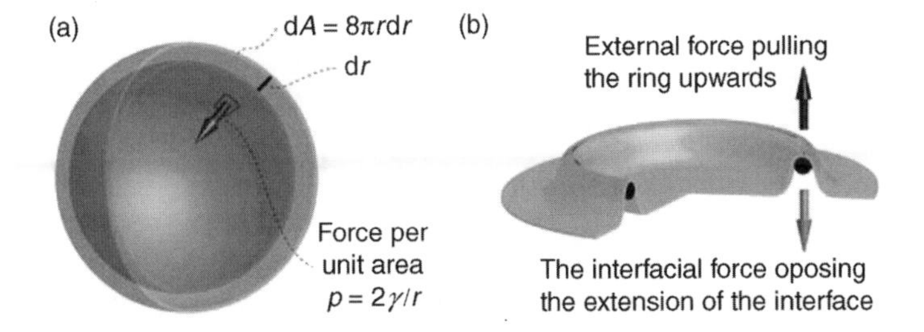

Figure 4.3 *(a) An interface between two immiscible phases acts as an elastic membrane. Extension of the area of the interface results in an increase in the interfacial energy. This is opposed by an interfacial force that can be best represented in terms of the interfacial pressure* p_L. *The force generated by an interface can be measured, for example, via the ring method schematically depicted in inset (b)*

why it is possible to overheat a liquid into a metastable state above the liquid–gas transition temperature: if there are no pockets of gas (for example on the walls of the container) it takes a very long time to nucleate nanoscopic bubbles in the bulk.

The relationship between the interfacial energy and the force needed to extend an interface forms the basis of popular methods of measuring the value of surface (or interfacial) tension. For example, in the ring method (first developed by Pierre Lecomte du Noüy), a metal ring is immersed in the liquid and is slowly drawn upwards, through the interface. As the liquid wets the metal (platinum is usually used because of its low interfacial energy with most liquids) the ring can be withdrawn above the plane of the undisturbed interface with a meniscus extending down from the ring (Figure 4.3b). Pulling the ring up increases the surface area of the liquid–air interface. A measurement of the force needed to withdraw the ring can thus be used to assess the value of the interfacial tension. Similarly, in the Wilhelmy plate method, one measures the force needed to withdraw a plate from a liquid.

We have already derived the excess pressure inside a spherical bubble or a droplet that is due to the action of the interfacial tension. This pressure is proportional to the reciprocal of the radius of the droplet. This quantity is often referred to as the curvature, $\kappa = 1/r$, with r being the radius of curvature. A single radius of curvature defines κ for a line drawn on a plane. As illustrated in Figure 4.4a, at any point of the curve one can draw a tangent circle. The radius of this circle that goes through the point of interest is the radius of curvature, which is a vector pointing into the center of the circle. The curvature of a two-dimensional surface immersed in a three-dimensional space can be expressed in many different ways. For the calculation of the Laplace pressure the most important measure is the *mean curvature*, $\kappa = (^1/_2)(\kappa_1 + \kappa_2)$, where κ_1 and κ_2 are the *principal* curvatures. In order to determine the principal curvatures one considers the curvatures of curves resulting from an intersection of the surface of interest with all possible planes tangent to the vector that is normal to the surface at the point of interest. The largest and the smallest curvatures (which are always given for two planes that are

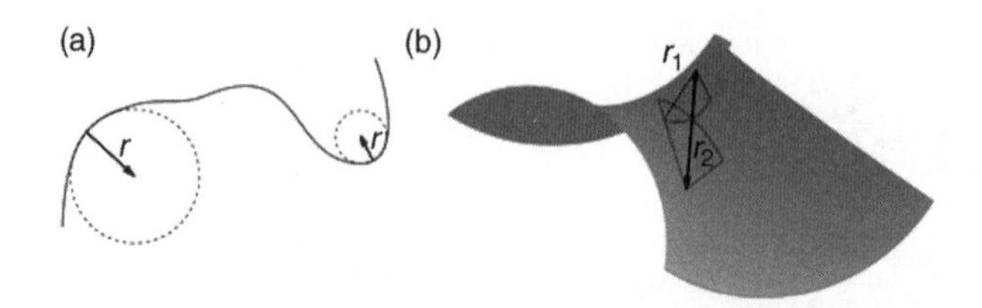

Figure 4.4 (a) The curvature of a line can be defined at any point by fitting a circle that is tangent to the line at the point of interest. Similarly, for a two-dimensional surface immersed in a three-dimensional space, one can calculate the mean curvature by finding the largest and smallest (principal) curvatures of sections of the surface with planes that are parallel to the normal to the surface of interest, at the point of interest. In inset (b) only the curves given by sections of the plane with the surfaces oriented in the directions of principal curvatures are shown, together with the two principal radii of curvature

orthogonal to each other) are principal. Thus, for any surface (as exemplified for a saddle surface in Figure 4.4b) the Laplace pressure at any point is given by $p_L = \gamma(1/r_1 + 1/r_2)$.

If one imagines that the interface between two immiscible fluids is like an elastic membrane, it becomes intuitive as to why the Laplace pressure is proportional to the curvature. The points at which the curvature is the highest are the points at which the membrane is "stretched" the most. Alternatively, the displacement of the interface at the points of highest curvature generates the highest change in the surface area of the interface and, correspondingly, the highest change in the interfacial energy; hence the pressure (which is proportional to the gradient of the interfacial energy) is the highest there. This observation also explains why droplets are spherical. A droplet that is squeezed away from the spherical shape possesses regions characterized by higher and also those characterized by lower curvature. The resulting imbalance of the Laplace pressure at different points on the interface of the droplet drives the liquid towards the perfectly symmetrical sphere.

4.5 Rayleigh–Plateau Instability

Interfacial tension and the Laplace pressure are also responsible for the spontaneous breakup of cylinders of gases and liquids surrounded by an immiscible fluid. A perfect cylinder has equal curvature at any point on its surface. The principal radii of curvature are given by the axial and radial sections of the cylinder, with the axial curvature equal to zero, and the radial curvature defined by the radius of the cylinder. Physical cylindrical morphologies of immiscible fluids, or, more adequately, liquid columns, are never perfect and always exhibit small deformations of the shape due to thermal fluctuations. All these deformations can be decomposed via a series expansion into sinusoidal undulations of different wavelengths. As illustrated in Figure 4.5, at the peaks of these sinusoidal modulations of the diameter of the column, the radial curvature is smaller than at the troughs. As the radial curvature always points into the column, the resulting contribution to the Laplace pressure always drives the fluid from the troughs to the

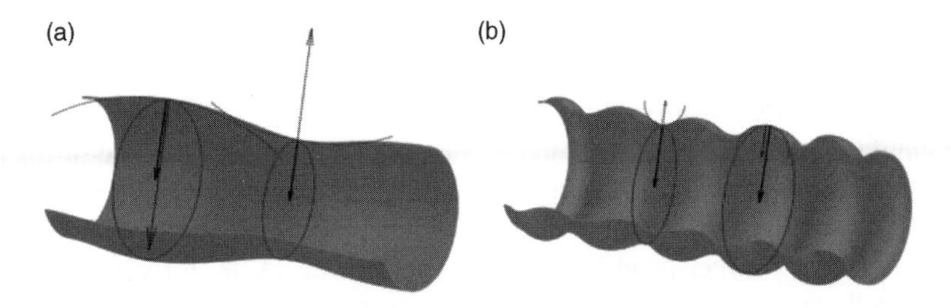

Figure 4.5 Whether perturbations of the diameter of a liquid column amplify or decay depends on their wavelength. Long wavelengths (a) amplify because the radial curvatures (shown with the solid lines) are larger than the axial curvatures, and cause the Laplace pressure to drive the fluid from the troughs to the peaks. Conversely, for short wavelengths (b) the axial curvatures (shown with the dotted lines) dominate and cause the decay of the perturbations

peaks, thus amplifying the perturbation. On the other hand, the axial curvature at the troughs is oriented outside of the cylinder and tends to stabilize the original, unperturbed, shape of the column.

Perturbations characterized by short wavelengths produce axial curvatures that are larger than the radial curvature and thus the Laplace pressure arising from these deformations lead to their decay. In contrast, long wavelengths of the undulations produce small axial curvatures that cannot balance the differences in radial curvatures along the length of the perturbed cylinder (Figure 4.5). As a result, the segments of the perturbed cylinder that have a smaller radius experience higher Laplace pressure than the rest of the slender shape, and the liquid flows from the thinner to the thicker sections.

Thus the perturbation amplifies and ultimately leads to a breakup of the cylinder into droplets. A detailed calculation of the rate of growth of the perturbations as a function of their wavelength, as first derived by Lord Rayleigh, yields the critical wavelength below which the fluctuations decay and the wavelength for which the growth rate is the largest. For a stream of an inviscid liquid (e.g., water) falling through air, the shortest unstable wavelength is equal to the circumference of the jet, and the wavelength providing the fastest growth rate, that is, the wavelength that dominates the breakup and yields the characteristic diameter of the droplets, is equal to $\sim$1.43 times the circumference.

The critical wavelength and the fastest mode of breakup depends on the character of the fluids, for example, an analysis similar to that by Lord Rayleigh but for polymeric liquids was introduced by Tomotika. In addition, the details of the Rayleigh–Plateau instability can be influenced by confinement of the immiscible cylinder with solid walls. For example, a slender droplet surrounded by the continuous fluid and squeezed in between two parallel plates will exhibit an increased stability and the critical wavelength will be longer than for the case of an unconfined cylinder (Son *et al.*, 2003). Also, when a breaking cylinder is confined to a capillary, the viscous effects of transport of the continuous fluid to the collapsing troughs can severely slow down the dynamics of breakup (Hammond, 1983).

4.6 Wetting of a Solid Surface

We have so far only considered interfaces between fluids, be it either liquid–gas or liquid–liquid. Yet, as is especially important in microfluidics, at least one of the fluids needs to contact the solid walls of the microchannels, and thus the subject of proper design of the wetting properties of the two fluids used for emulsification is of utmost importance.

The coefficient of interfacial tension, γ, is a measure of the force with which an interface pulls on a line that terminates it. For example, when three bubbles contact each other, the three soap films that come together at a line make 120° angles between them. This is because all of the three forces exerted by each of the films need to balance exactly. When we replace the bubbles with three distinct phases, one can imagine a droplet of oil floating on an interface between water and air (Figure 4.6a), the three interfacial tensions are no longer equal and the angles need to adapt to values other than 120° in order for the contact line to achieve a mechanical equilibrium. We can write an equation of equilibrium for any of the three directions defined by the interfaces (Figure 6.6a), for example:

$$\gamma_{\text{oil-air}} + \gamma_{\text{oil-water}}\cos\theta + \gamma_{\text{water-air}}\cos\alpha = 0 \tag{4.3}$$

$$\gamma_{\text{oil-air}}\cos\theta + \gamma_{\text{oil-water}} + \gamma_{\text{water-air}}\cos\beta = 0 \tag{4.4}$$

$$\gamma_{\text{oil-air}}\cos\alpha + \gamma_{\text{oil-water}}\cos\beta + \gamma_{\text{water-air}} = 0 \tag{4.5}$$

As the three angles between the interfaces add up to 2π, we need only two such equations to fully determine the geometry of the interfaces at the contact line.

When we replace one of the fluids with a solid surface, the geometry of the problem simplifies (Figure 4.6b) and the set of two equations is reduced to a single one, known as Young's equation, named after the famous English scientist Thomas Young.

$$\gamma_{SC} = \gamma_{SD} + \gamma_{DC}\cos\theta \tag{4.6}$$

In Equation 4.6, discontinuous fluid contained in the droplet (bubble) is represented by the subscript D, the continuous fluid by subscript C, and the subscript S corresponds

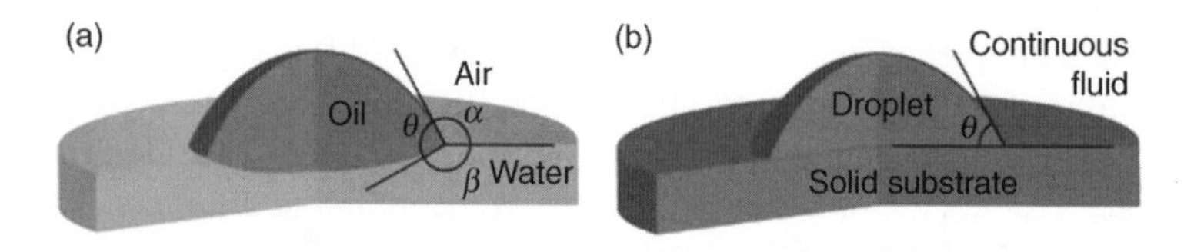

Figure 4.6 *(a) A schematic drawing of a droplet of oil sitting on the water–air interface, and (b) of a droplet of liquid sitting on a solid substrate and surrounded by a second, immiscible fluid*

to the solid. This equation can be solved only if neither γ_{SC} nor γ_{SD} are larger than the sum of the two other interfacial tensions. If this condition is met, the droplet adopts a stable morphology. In contrast, when γ_{SC} is larger than the sum of γ_{SD} and γ_{DC}, the droplet spreads over the solid substrate, a situation that is referred to as complete wetting. At the other end of the spectrum of possible scenarios, when γ_{SD} is larger than γ_{SC} and γ_{DC}, the droplet forms a perfectly spherical shape minimizing the contact with the substrate to a single point. Under such circumstances we say that the fluid contained in the droplet does not wet the substrate at all. All the situations between these two extremes can be qualitatively described with the suffixes *-philicity* and *-phobicity* of the substrate. For a water droplet that makes a contact angle θ smaller than 90°, the substrate is called hydro*philic* (otherwise it is hydro*phobic*). For hydrocarbon liquid one uses the suffixes lyo*philic*/lyo*phobic*. Importantly, when a droplet translates over a surface it may exhibit a hysteresis of the contact angle – the angle at the front (the advancing angle) and the angle at the rear (the receding angle) of the droplet will be different. The interface can also pin to, either topographical or chemical, inhomogeneities of the surface and create rough and irregular shapes.

Owing to the fact that only the characteristics of the outermost molecules influence the wetting properties of a solid, these can readily be altered via modification of the surface chemistry. For example, microcontact printing (Xia and Whitesides, (1998)) can be used to prepare heterogeneously patterned substrates, such as, hydrophilic and hydrophobic patches of the surface. Figure 4.7 shows two examples of such surfaces. As

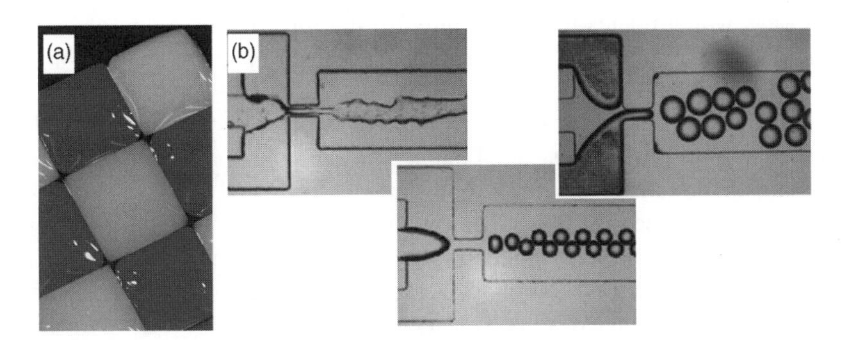

Figure 4.7 *(a) Aqueous solutions of inks deposited on a checkerboard pattern of hydrophilic patches separated by thin hydrophobic stripes (Photograph by Felice C. Frankel, reproduced with permission). In inset (b) we show the irregular stream of water in a microfluidic flow-focusing device that is wetted by both the aqueous phase and the organic phase, formation of droplets of water in oil (the bottom picture in the middle) and formation of oil droplets in water (the most right-hand picture) (Li et al., 2007).* Reprinted with permission from Langmuir, *Screening of the effect of Surface Energy of Microchannels on Microfluidic Emulsification by W. Li et al.,* ***23****, 15, 8010–8014 Copyright (2007) American Chemical Society. See Plate 1*

was demonstrated by the group working with Lipowsky, hydrophilic stripes can be used to guide the flow of water over a patterned surface. The slender droplets that wet such stripes exhibit completely different criteria of stability (Gau *et al.*, 1999) than is the case for liquid columns surrounded by a second immiscible fluid.

In microfluidic systems used for formation of bubbles or droplets it is critical that the fluid-to-be-dispersed *does not* wet the walls of the channels. If both fluids wet the walls, the droplets do not form, and one observes irregular flow patterns (Figure 4.7c). A modification of the surface chemistry of the channels from a hydrophobic to a hydrophilic character can be used to inverse the phases and form droplets of oil in a continuous aqueous phase instead of water droplets in an organic phase (Figure 4.7c).

4.7 Analysis of Flow

So far we have considered only the static shapes of interfaces between immiscible fluids and the relaxation of their shape due solely to the action of the interfacial forces. Microfluidic techniques for formation of droplets take advantage of an interplay between the interfacial dynamics and the dynamics of the flow of the two immiscible phases. For example, a shear flow can distort the shape of a droplet, a droplet blocking a capillary can modify the pressure in that capillary and, conversely, the distribution of pressure and flow can affect the motion of the interface. Conveniently, for all of the experiments described in this book, the standard approximation of incompressible flow is valid and the flow can be described by the Navier–Stokes equation, with a relatively simple inclusion of the viscous terms. The Navier–Stokes equation is, at heart, Newton's Second Law, which relates the acceleration to the force acting on an object. In the following, we will briefly review the notation of acceleration, body, and pressure forces acting on the fluid and the construction of the viscous term.

For fluids, the Lagrangian description of *acceleration* of an object is replaced by the Eulerian description. There, instead of tracing the particular mass point in a fluid, as in the Lagrangian formalism, one considers the momentary speed and acceleration of flow at a given point in space. The transition from the Lagrangian to the Eulerian description simplifies handling of the fields of flow and the notation of forces acting on a fluid, yet it slightly complicates the notation of acceleration, which needs to be represented by a sum of the temporal change of velocity at a given point and a convective derivative that takes into account the acceleration of elementary volumes of fluid as they follow their streamlines (Figure 4.8).

The acceleration of the elemental volumes of the fluid is caused by forces that act on these. The forces can be divided, or categorized, into ones that act on surfaces (pressure and shear stresses) and ones that act on the volumes. The latter, termed *body forces*, can have various origins. One common body force is gravity, which acts on mass. Other examples include electromagnetic forces (e.g., a dielectrophoretic force acting on the body of a polarized material). In microfluidics, owing to the intrinsically small dimensions, body forces are usually dominated by the surface forces and can-be neglected – unless the experiment is intentionally constructed to use them.

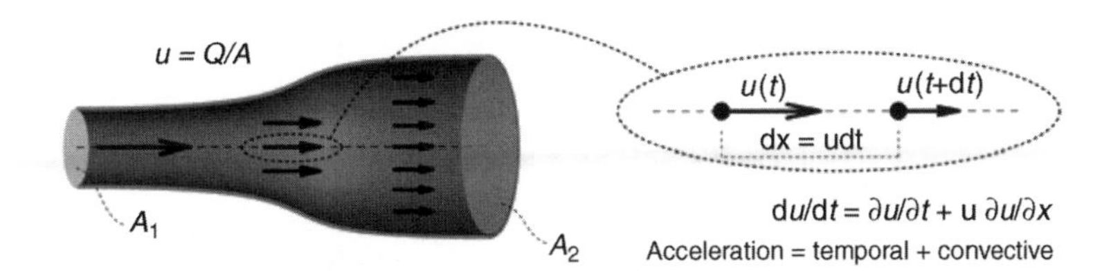

Figure 4.8 *In the Eulerian description of flow the acceleration of particles of fluid needs to take into account both the temporal changes of the speed of flow (u) at a given point (r) in space, and the variation of* u *along the momentary streamline of the elemental volume of fluid passing through* r *at the particular time. The schematic shows a simple example of an expanding pipe in which, even if the rate of flow Q of fluid through the pipe is constant in time and the field of flow is stationary, the fluid decelerates in the expanding section of the tube.*

Most microfluidic flows are caused by the application of a pressure head to the inlet of the device, or, conversely, a negative pressure applied to the outlet, as is sometimes used in devices designed for portability (Garstecki *et al.*, 2006a). It is important to be aware that even if one fixes the volumetric rate of inflow of fluids into the microfluidic chip, this still results in creation of a difference in pressures at the inlet and the outlet. In the case of fixing the pressure, the rate of flow through the device depends on the momentary resistance to flow, conversely, fixing the rate of inflow causes the pressure drop along the microfluidic channel to adapt to its state (e.g., viscosity of the fluids that occupy it). In the Eulerian description of flow, one traces the changes of pressure in space (the pressure field) and quantifies the pressure force acting on the fluid at a given point r as the difference between the pressures acting on the facets of an elemental volume $\mathrm{d}V$. Because the Navier–Stokes equation is expressed in terms of body forces, these small differences in pressures acting on the different facets of the elemental cube are articulated (via the Gauss–Ostrogradsky divergence theorem) in terms of the gradient of pressure: $\mathrm{d}F = -\nabla p \mathrm{d}V$.

Viscosity of fluids is their ability to transfer the momentum in a direction perpendicular to the motion itself. More simply, this is the ability to exhibit friction. The action of viscosity can be easily observed in a simple experiment (Figure 4.9a) of sliding one of two parallel plates separated by some distance (w) with the space between them filled with a liquid. If one of the plates, say the bottom one, is fixed, sliding the top plate at speed (u_p) will require a force tangent to the plate of a magnitude that is proportional to the surface area of the plate, to the speed of translation, and to the viscosity of the liquid. Additionally, this force will be inversely proportional to the separation w: the smaller the separation the harder it will be to slide the top plate (this is why it is possible to effectively "glue" flat substrates, e.g., glass slides, with a tiny drop of liquid compressed between them). Formally, the magnitude of the viscous shear stress (τ) is proportional to the coefficient of viscosity (μ) times the rate of deformation of the liquid: in our simple example, to the rate of shear, which is given by u_p/w. One can note here the fundamental difference between elastic and viscous forces: the former depend on the deformation of

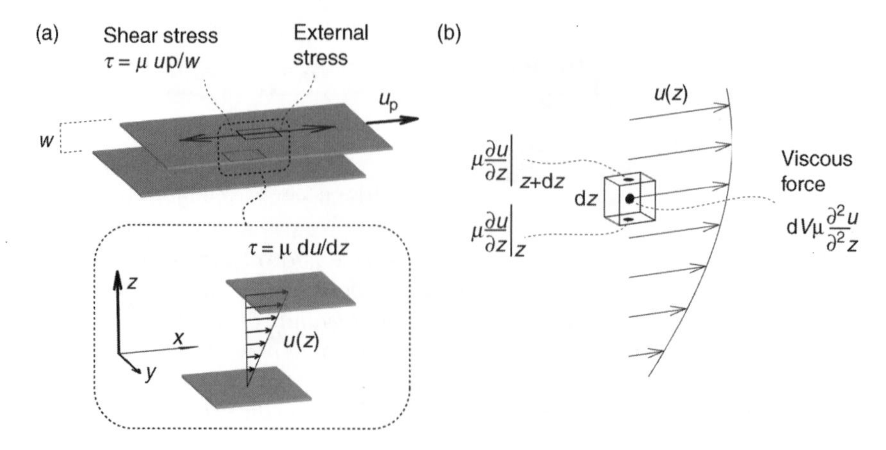

Figure 4.9 *(a) Schematic illustration of the shear stress (τ) opposing tangential force applied to a plate underneath which there is a viscous liquid. In inset (b) we show the shear stress acting on two facets of an elemental volume of liquid in a simple flow field, in which the velocity has only one component* (u) *along the* x*-axis, and depending only on the vertical position* (z)

the object (i.e., within the validity of the Hooke's Law). When the material is stretched, strained or sheared it stores the elastic energy that can be later retrieved from the system when it relaxes to its natural shape. In contrast, the viscous forces depend only on the *rate of deformation* (and in Newtonian fluids they are proportional to this rate). When one removes the external force that deforms the fluid, the motion stops and the fluid does not recover its original shape.

In order to include the viscous forces in the Navier–Stokes equation, again one needs to express the shear stress in terms of a force acting on an elemental volume. Taking into account the difference between the shear stress exerted on the opposing facets of the elemental cube results in terms that include the second spatial derivatives of the speed, as illustrated on a simple example shown in Figure 4.9b. In the general case of a field of velocity (v) that has non-zero components in all three Cartesian directions and varies along all of these, the shear stress is expressed as the sum of the second-order partial derivatives, in short, denoted as $\mu \nabla^2 v$.

Putting all the terms that we have discussed above together, one can construct the Navier–Stokes equation for incompressible flow that combines the forces per unit volume for every point in the flow field, and reads as follows:

$$\rho(\partial \boldsymbol{v}/\partial t + \boldsymbol{v} \cdot \nabla \boldsymbol{v}) = -\nabla p + \mu \nabla^2 \boldsymbol{v} + \boldsymbol{f} \qquad (4.7)$$

where the vectors are typed in bold font, and f stands for the externally applied body force. We do not intend to explain in detail all the algebra for the vectors and tensors here, as this requires a detailed study that would not fit within the scope of this text. We hope that a qualitative understanding of the terms of the Navier–Stokes equation can be acquired on the basis of the simple configurations of flow that were described above.

An analytical description of the flow field in complicated geometries of microfluidic devices and in the presence of an interface is difficult. Numerical methods are increasingly being used to provide a better understanding of the phenomena that are observed experimentally. Despite this, the most important tool in the understanding of the dynamics of microfluidic emulsification is experimentation. The results of experiments are often interpreted with the help of dimensional analysis, which allows to judge the relative importance of the different forces acting on the fluid. It also allows to discriminate between the different potential mechanisms behind the observed behaviors on the basis of the scaling relationships for the observables.

Dimensional analysis is a technique that takes advantage of the observation that all physical quantities are represented by a combination of a few basic units (of length, mass, time, temperature). Further, any physical law or equation that binds a number of physical quantities (e.g., acceleration with force and mass) can be represented through the use of a smaller number of non-dimensional parameters. This procedure is described by the Buckingham's π-theorem [a detailed example of the application of the Buckingham's π-theorem to the problem of formation of droplets in a microfluidic system can be found in De Menech *et al.* (2008)]. Importantly, dimensional analysis not only simplifies the notation of physical equations, but provides insight into what combinations of physical quantities determine the properties of the solutions and decide the dynamics of the systems. These combinations of physical quantities are called dimensional groups or, as is most commonly used in fluid mechanics, non-dimensional numbers.

A good, and a most important in fluid dynamics, example of a non-dimensional number is the Reynolds number that arises from the non-dimensionalization of the Navier–Stokes equation. Reynolds number (*Re*) is defined as:

$$Re = \rho vl/\mu \tag{4.8}$$

where v denotes the characteristic magnitude of speed and l the characteristic dimension in the system. The value of the Reynolds number reflects the relative importance of the inertial and viscous forces. This value is very important for judging the character of flow in pipes and capillaries. At high values of the Reynolds number, the flow becomes turbulent and complex, while at low values of *Re* it is laminar (Figure 4.10).

At low values of *Re*, as is typical for microfluidic devices, the inertial terms in the Navier–Stokes equation can be neglected and it simplifies to the Stokes equation of flow:

$$\nabla p = \mu \nabla^2 \boldsymbol{v} \tag{4.9}$$

Within the regime described by the Stokes approximation, the speed of flow is simply proportional to the forces exerted on the fluid. For example, for a simple pipe flow, the volumetric rate of flow (Q) of a fluid through the pipe (or a microchannel) is given by the Hagen–Poiseuille equation:

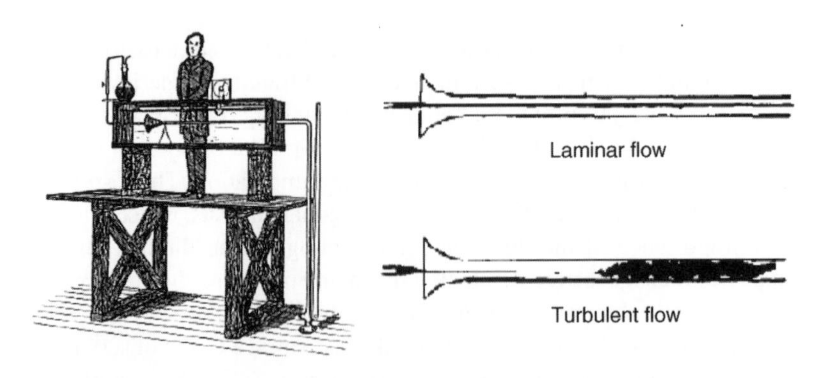

Figure 4.10 Illustration of the experimental setup and the laminar and turbulent flow. Reprinted with permission from (Reynolds, 1883)

$$Q = \Delta p / R \tag{4.10}$$

where R is the hydraulic resistance of the channel and depends on the viscosity of the liquid and the geometry of the duct. For channels of uniform, circular cross-sections, the resistance is equal to:

$$R = 8\,\mu L / \pi r^4 \tag{4.11}$$

where L is the length of the capillary and r is its inner radius.

It is important to observe that the resistance scales with the reciprocal of the fourth power of the radius of the pipe. This is because $1/r^2$ scales the speed of flow ($v = Q/4\pi r^2$) and the same term provides the gauge for the magnitude of the second-order spatial derivatives of the speed of flow (please compare Figures 4.9b and 4.11a). The situation is different in channels with a large aspect ratio for the width to the height, in which the shorter dimension solely determines the speed profile (and thus the magnitude of the viscous term in the Stokes equation). In such channels the resistance will scale as $(wh^3)^{-1}$. The strong dependence of the pressure drop on the shorter distance is an important element of the dynamics of drop formation at low Reynolds numbers (Dollet *et al.*, 2008; Garstecki, Stone, and Whitesides, 2005; Garstecki *et al.*, 2006b). The exact formulae for the hydrodynamic resistance of rectangular and other cross-sections

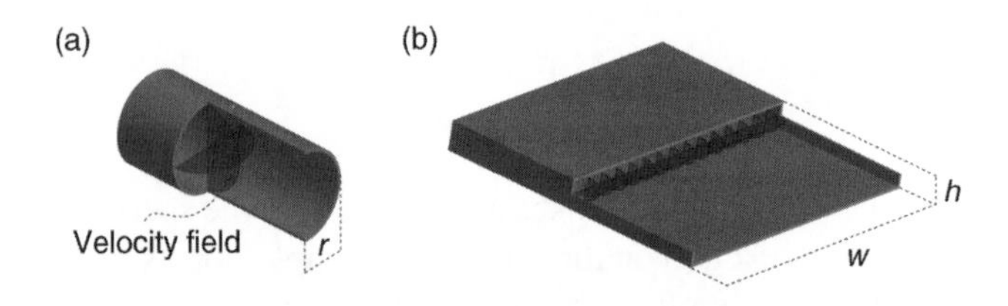

Figure 4.11 Schematic illustrations of the profiles of speed of flow in a channel of a circular (a) and large aspect ratio rectangular (b) cross-sections

of the channels can be complicated, yet amenable to an analytical derivation (Mortensen, Okkels, and Bruus, 2005).

4.8 Flow in Networks of Microchannels

The linear relationship between the pressure difference applied to a channel and the volumetric rate of flow in the Hagen–Poiseuille Law is analogous to the linear relationship between the difference in electrostatic potentials and the intensity of electrical current through a resistor in Ohm's Law. Additionally, the conservation of mass (or flow for an incompressible fluid) of fluid is analogous to the law of conservation of charge and current in electrical systems. These analogies allow the use of Kirchoff's equations of circulation of current through networks of resistance for the calculation of the distribution of volumetric flow of liquid between channels in a microfluidic network (Figure 4.12). It is sufficient to know the resistances of all the channels in the network, and the pressures at the inlet and outlet, to calculate the speed of flow in any part of that system. The phenomenology that arises from dynamics of flow through networks is important in the subject of formation of droplets and bubbles in parallel systems and in the design of systems for high throughput formation of emulsions. We will describe this phenomenology in Chapter 6.

4.9 Dimensional Groups

In processes involving formation of droplets, and thus the motion and evolution of an interface, the most useful are the non-dimensional numbers that compare the interfacial

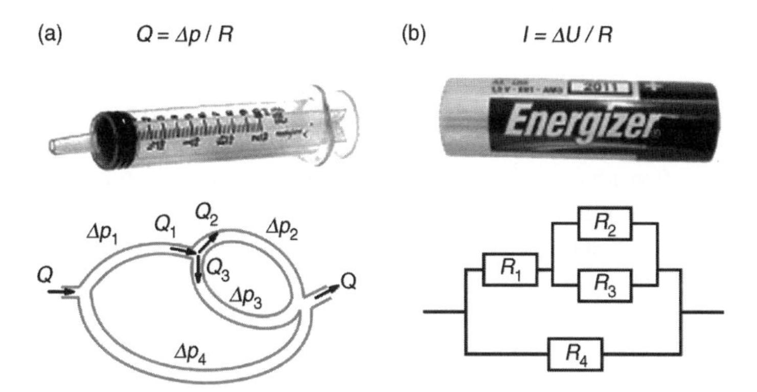

Figure 4.12 The analogy between the distribution of volumetric rate of flow of a simple fluid flowing through a network of microchannels and the Kirchoff equations for the flow of current through a network of conductors. The pressure drop between any two nodes does not depend on the trajectory through the network (i.e., in the example $\Delta p_1 + \Delta p_2 = \Delta p_1 + \Delta p_3 = \Delta p_4$). The conservation of mass, together with the incompressibility of the fluid, guarantee that the sum of inflows and outflows at any node is zero (i.e., $Q_1 - Q_2 - Q_3 = 0$)

forces with other forces acting on the fluids. For example, the *Bond number* (*Bo*), defined as:

$$Bo = \rho a L^2/\gamma \tag{4.12}$$

compares the gravitational force acting on the fluid, approximated by $\rho a L^3$, where ρ is the mass density, a is the gravitational acceleration, and L represents a characteristic length-scale (e.g., the diameter of a jet or a droplet) with the interfacial forces given by $(\gamma/L)L^2 = \gamma L$. The value of the ratio of these two expressions allows for a judgment whether the body force (gravity) dominates the interfacial tension (for $Bo > 1$) or, conversely, the gravitational force is weaker than the interfacial force (for $Bo < 1$).

From the calculation of the Bond number we can estimate what size of droplets will adopt the shape of a section of a sphere (as is true for small droplets and $Bo < 1$) or a flattened shape for large droplets for which the gravitational effects are important ($Bo > 1$). A very large value of *Bo* corresponds to a shape that is determined mostly by the gravitational effects (e.g., a flat puddle). Low values of *Bo* correspond to a spherical shape, while for intermediate values ($Bo \sim 1$) characterization of the shape of the droplet demands a detailed calculation that includes both the interfacial and body forces. In microfluidics the Bond numbers are typically small: a 100 μm droplet of water (density $10^3\,\mathrm{kg\,m^{-3}}$) in oil (of density $800\,\mathrm{kg\,m^{-3}}$) with an interfacial tension of $10\,\mathrm{mN\,m^{-1}}$, can be characterized by a Bond number equal to 2×10^{-3}, a value that guarantees that the droplet is almost perfectly spherical and that the gravitational effects can be neglected in the analysis of the dynamics of the system.

Because the flow in microfluidic devices typically proceeds at low or moderate values of the Reynolds number, the viscous effects are always important. Thus perhaps the most important "interfacial" non-dimensional number is the *capillary number* (*Ca*), which compares the viscous stresses with the interfacial forces:

$$Ca = \mu v/\gamma \tag{4.13}$$

As always in dimensional analysis it is critical to choose the characteristic quantities correctly. For two-phase flows, one usually uses the viscosity μ of the more viscous of the two fluids, the average speed of flow (e.g., for pipe flow it is the superficial speed equal to the volumetric rate of flow divided by the surface area of the lumen of the pipe). Again, as in the case of the Bond number, the definition of the capillary number can be derived from the ratio of the viscous stresses to the capillary stresses. This approach, as exemplified here for a droplet, can lead to a slightly different formula for *Ca*. For example, for a droplet of diameter d in a shear flow in which the speed changes from zero to u over a distance L, the shear rate can be approximated by (u/L), and shear stress by $(\mu u/L)$. The Laplace pressure is given by $(4\gamma/d)$. Combining these two approximations yields a formula for the capillary number:

$$Ca = \mu u d/4\gamma L \tag{4.14}$$

which explicitly incorporates the two different characteristic length scales: one for the shear rate (L) and one for the Laplace pressure, which is scaled by the curvature of the interface ($1/d$).

The capillary number, as calculated above, can be readily used to assess the size of the droplets generated by shearing. Shear emulsification is one of the most common industrial techniques for the preparation of emulsions. In shear flow, large droplets are elongated and undergo a Rayleigh–Plateau instability. Within this process the large droplets are fragmented into smaller ones until their radius is small enough that the curvature (and the associated Laplace pressure) can balance the shear stress. Thus the radius of the droplets can be approximated by setting $Ca = 1$ or, conversely, equating the shear stress ($\mu u/L$) with the interfacial pressure ($4\gamma/d$), which yields $d \sim 4L(\gamma/\mu u)$. In general, a strong dependence of the diameter of the droplets on the value of the capillary number [here: $d \propto (\mu u/\gamma)^{-1}$] signifies the dominating role of shear stress in the emulsification process.

In microfluidic systems, the capillary number can change over wide ranges of values. As we will explain in the next chapter the value of the capillary number determines the mechanism of formation of droplets and bubbles. Although most of the microfluidic emulsification devices operate within the regimes of flow that are dominated by viscous effects, there are systems that use faster flow in which the inertial terms become dominant. For example, in the axi-symmetric devices developed by Ganan-Calvo (1998) the condition for breakup of a bubble or droplet from the jet that issues from the orifice is determined by the balance of the two inertial terms in the material derivative. In these systems the most useful non-dimensional number is the *Weber number* (We), which judges the relative importance of the inertial and interfacial stresses:

$$We = \rho u^2 L/\gamma \tag{4.15}$$

Again, the number can be derived from a direct comparison of the kinetic energy ($\rho u^2 L^3$) with the energy added by the Laplace pressure [$(\gamma/L)L^3 = \gamma L^2$]. Equivalently, the same formula can be derived from a comparison of the forces, expressed as [$\rho L^3(u^2/L)$] for the inertial force, with the term (u^2/L) serving as an approximation of the acceleration, and as (γL) for the interfacial force.

The Weber number can also be used for a quick qualitative assessment of the dynamics of fluids ejected from, for example, a hose – for speeds of flow characterized by large values of We one can expect an elongated jet that breaks up into droplets far away from the nozzle. Conversely, for small values of We, the interfacial tension effects will modulate the flow immediately at the nozzle, leading to dripping of droplets without formation of a pronounced jet.

Finally, in problems in which it is difficult to judge whether the viscous or inertial terms control the interfacial dynamics one can use the *Ohnesorge number* (Oh):

$$Oh = \mu/(\rho\gamma L)^{1/2} \tag{4.16}$$

The Ohnesorge number is often called the "Reynolds number for free surface flows" as it compares the viscous and inertial forces in the motion of an interface between two fluids. Low values of *Oh* suggest that inertial terms are more important than viscous ones and vice versa.

The above description of non-dimensional numbers is meant only as a rudimentary introduction of the concepts that will be most useful in the explanation of the dynamics of microfluidic techniques for the formation of foams and emulsions. For a deeper treatment of dimensional analysis and free-surface flows an interested reader should consult available textbooks (White, 2000) and reviews (Eggers, 1997) on the subject.

References

Abbott, N.L., Folkers, J.P., and Whitesides, G.M. (1992) *Science*, **257**, 1380–1382.

de Gennes, P.G., Brochard-Wyart, F., and Quere, D., (2002) *Capillary and Wetting Phenomena; Drops, Bubbles, Pearls, Waves*, Springer Publishing, New York.

De Menech, M., Garstecki, P., Jousse, F., and Stone, H.A. (2008) *J. Fluid Mech.*, **595**, 141–161.

Dendukuri, D., Gu, S.S., Pregibon, D.C., Hatton, T.A., and Doyle, P.S. (2007) *Lab Chip*, **7**, 818–828.

Dollet, B., van Hoeve, W., Raven, J.P., Marmottant, P., and Versluis, M. (2008) *Phys. Rev. Lett.*, **100**, 034504.

Eggers, J. (1997) *Rev. Mod. Phys.*, **69**, 865–929.

Ganan-Calvo, A.M. (1998) *Phys. Rev. Lett.*, **80**, 285–288.

Garstecki, P., Stone, H.A., and Whitesides, G.M. (2005) *Phys. Rev. Lett.*, **94**, 165401.

Garstecki, P., Fuerstman, M.J., Fischbach, M.A., Sia, S.K., and Whitesides, G.M. (2006a) *Lab Chip*, **6**, 207–212.

Garstecki, P., Fuerstman, M.J., Stone, H.A., and Whitesides, G.M. (2006b) *Lab Chip*, **6**, 437–446.

Gau, H., Herminghaus, S., Lenz, P., and Lipowsky, R. (1999) *Science*, **283**, 46–49.

Hammond, P.S. (1983) *J. Fluid Mech.*, **137**, 363–384.

Li, W., Nie, Z.H., Zhang, H., Paquet, C., Seo, M., Garstecki, P., and Kumacheva, E. (2007) *Langmuir*, **23**, 8010–8014.

Mortensen, N.A., Okkels, F., and Bruus, H. (2005) *Phys. Rev. E*, **71**, 057301.

Moseler, M. and Landman, U. (2000) *Science*, **289**, 1165–1169.

Nisisako, T., Okushima, S., and Torii, T. (2005) *Soft Mat.*, **1**, 23–27.

Reynolds, O. (1883) *Philos. Trans. R. Soc. London*, **174**, 935.

Son, Y., Martys, N.S., Hagedorn, J.G., and Migler, K.B. (2003) *Macromolecules*, **36**, 5825–5833.

White, F.M., (2000) *Fluid Mechanics*, McGraw-Hill, New York.

Xia, Y.N. and Whitesides, G.M. (1998) *Annu. Rev. Mater. Sci.*, **28**, 153–184.

5

Formation of Droplets in Microfluidic Systems

CHAPTER OVERVIEW

Microfluidic Reactors for Polymer Particles. Eugenia Kumacheva and Piotr Garstecki.

5.1 Introduction

At the heart of any emulsification method lies the process of breakup of a large domain of fluid into smaller portions, either droplets or bubbles, depending on the fluid that is being dispersed. This can be achieved via the Rayleigh–Plateau instability, as we described in Chapter 4. It is sufficient to force the fluid-to-be-dispersed into an elongated and unstable shape that spontaneously breaks into droplets or bubbles. Forcing of liquids into unstable shapes does not necessarily need any specified geometrical constraints. For example, in a waterfall the gravitational energy forces water into elongated sheets and jets that break into droplets. The size of these drops is not directly linked to the dimensions of the waterfall. Similarly, when one uses shear to elongate large droplets of vinegar in olive oil to make vinaigrette dressing, the diameters of the droplets is not a function of the size of the cup in which they are stirred, but rather of the vigor of stirring. These and other such processes are effective in the sense that they produce foams, aerosols, and emulsions in a high-throughput manner and are often used in industry (e.g., in the high-throughput, continuous flow shear emulsificators). The effectiveness, however, comes at the expense of a limited control over the distribution of the diameters of the droplets, and the foams and emulsions produced via these classical vistas are typically polydisperse.

The control over the process of formation of drops or bubbles can be improved by the introduction of geometrical constraints that either only set the characteristic size of the jets or streams of fluids, or they also regulate the flow field of the fluids, and in this way control the mechanism of breakup. For example, in a technique termed membrane emulsification (Figure 5.1) the liquid that is to be dispersed into droplets is pressed through a membrane comprising small pores. The diameter of the pores, together with the tangential flow of the continuous phase along the surface of the membrane, determine the distribution of diameters of the droplets.

5.1.1 Geometrical Confinement

This chapter overviews the microfluidic methods for formation of droplets, bubbles, and the more complicated architectures of capsules and multiple emulsions. The main difference between conventional, bulk, emulsification, and microfluidic methods of formation of drops lies in the presence of the walls of the microchannels. This feature

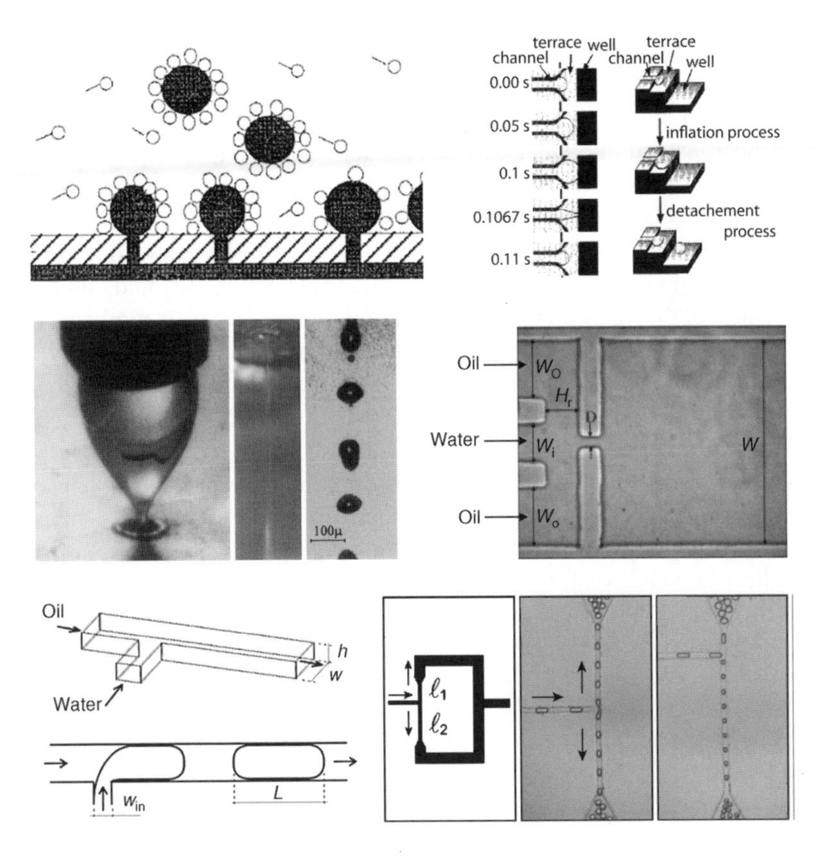

Figure 5.1 *From left to right, top to bottom: a schematic representation of membrane emulsification (Schroeder, Behrend, and Schubert, 1998), the micro-terrace system (Sugiura et al., 2002), capillary flow-focusing (Ganan-Calvo, 1998), planar flow-focusing (Anna, Bontoux, and Stone, 2003), T-junction (Garstecki et al., 2006), and diverging junction (Link et al., 2004). Reprinted with permission from Journal of Colloid and Interface Science, Effect of Dynamic Interfacial Tension on the Emulsification Process using Microporous, Ceramic Membranes by V. Schröder, O. Behrend and H. Schubert,* ***202****, 2, 334–340 Copyright (1998) Elsevier Ltd. Reprinted with permission from Langmuir, Prediction of Droplet Diameter for Microchannel Emulsification by S. Suguira et al.,* ***18****, 10, 3854–3859, Copyright (2002) American Chemical Society. Reprinted with permission from Phys. Rev. Lett., Generation of Steady Liquid Microthreads and Micron-Sized Monodisperse Sprays in Gas Streams, Ganan-Calvo.,* ***80****, 285–288 Copyright (1998) American Physical Society. Reprinted with permission from Applied Physics Letters, Formation of dispersions using flow focusing in microchannels by Anna, S.L., Bontoux, N. and Stone, H.A.,* ***82****, 3, 364–366 Copyright (2003) American Institute of Physics. Reprinted with permission from Lab on a Chip, Formation of droplets and bubbles in a microfluidic T-junction—scaling and mechanism of break-up by Garstecki, P. et al.,* ***6****, 437–446 Copyright (2006) Royal Society of Chemistry. Reprinted with permission from Phys. Rev. Lett., Geometrically Mediated Breakup of Drops in Microfluidic Devices, Link et al.,* ***92****, 5, 054503 Copyright (2004) American Physical Society*

has pronounced consequences on the parameters that define the dynamics of flow. Geometrical confinement defines the length scale set by the cross-sections of the channels and introduces the interactions between the fluids and the walls, making the wetting properties and interfacial stresses fundamentally important.

Because body forces, such as gravity or inertia, are proportional to the volume of the fluid and surface forces are proportional to the surface area, small scales naturally favor the latter. For example, keeping all other parameters constant, lowering the characteristic length scale of the flow field lowers the value of the Reynolds number (*Re*). Typically, most of the microfluidic systems operate at low or at most moderate values of *Re* and the viscous effects dominate over inertial ones. The flow is usually laminar, which allows for precise design of the flow field and execution of even complicated protocols of flow, for example for controlled formation of not only single droplets, but also multiple drops with a number of immiscible shells.

At the same time, small length scales determine that interfacial tension is of paramount importance. Confinement introduces three different effects that are all important, and which will be discussed in this chapter. Firstly, the small scales of microfluidic channels usually translate into typically small values of the capillary number (*Ca*), which compares the viscous stresses with the Laplace pressure. Most microfluidic devices for formation of droplets operate in a regime in which the capillary number is either very small ($Ca \sim 10^{-5}$–10^{-2}) or, at most, moderate (up to $Ca \sim 1$). If interfacial stresses dominate, the droplets or tips of the phase that is to be broken into droplets can obstruct most of the cross-section of the channel. When this happens the flow of the continuous phase around the droplet or tip is confined to thin films between the interface and the wall, or, in the channels with rectangular cross-sections, also to the corners of the channel. As we will discuss in detail, blocking of the cross-section of the channel plays a fundamental role in the processes of breakup at small values of the capillary number and leads to a process of formation of droplets that is unique to microfluidics.

Secondly, the geometry of the junction at which the immiscible fluids co-flow can be designed in a way that stabilizes the shape of the interface against the capillary instabilities. In this situation, breakup does not happen as a result of a Rayleigh–Plateau type of instability, as is the case in unbounded flows, but as a result of a pressure jump across the interface and "squeezing" of the interface by the continuous fluid.

Finally, the relative importance of interfacial tension makes the aspects of wetting critical. For stable and controllable operation of the microfluidic systems that generate emulsions it is a must that the continuous fluid wets the walls of the channels, while the fluid that is to be dispersed into droplets does not. If any wetting of the walls by the droplet phase occurs, the flow becomes erratic and impossible to control. We will discuss these effects and vistas of controlling the type of flow via appropriate tuning of the surface chemistry of the walls of microchannels.

A small scale of flow also enhances the role of diffusion, which may be important in heat transfer, extraction, gelation, and polymerization. This is often important in solidification of droplets into particles and will be discussed in other chapters.

5.1.2 The Cost of Confinement

Emulsification itself is always an endo-energetic process. Breaking a fluid into bubbles or droplets creates new interfacial area and this is associated with an increase in the interfacial energy. Even when the process undergoes in an unbounded (or weakly constricted) flow, as for example in shear emulsificators, or in a jet ejected from a fire hose, and the breakup of larger domains of fluid into small bubbles or droplets happens spontaneously via a Rayleigh–Plateau type of instability, work has to be first put into transformation of the domains of the fluid into unstable shapes. This is always done against the interfacial forces and often requires additional work put into the motion of the fluid (that is either further dissipated via viscous action, or stored in inertia). Emulsification in microfluidic systems carries the additional cost of dissipation of energy associated with pumping of fluids through channels of small cross-sections. Such pumping is subject to large resistance to flow. As we will discuss in this chapter, this cost comes with the revenue of the ability to tune the distribution of diameters of the droplets with ultra-high precision, not attainable in traditional emulsification techniques.

5.2 Microfluidic Generators of Droplets and Bubbles

Probably the first micro-engineered system for the formation of aerosols and droplets was the capillary flow-focusing device introduced in 1998 by Ganan-Calvo (1998) (Figure 5.1). This device has been shown to produce monodisperse aerosols of droplets (Ganan-Calvo, 1998) and suspensions of microbubbles in liquid (Ganan-Calvo and Gordillo, 2001). The condition of breakup in these systems is based on the balance of two contra-acting inertial terms that arise during formation of a droplet or a bubble (Ganan-Calvo, 2004). Axi-symmetric systems are also useful in the low capillary number flows, for example, for formation of multiple emulsions (Utada *et al.*, 2005) or capsules (Takeuchi *et al.*, 2005) and we will review these in this chapter. Another interesting technique that we will review in greater detail in the chapter devoted to high-throughput formation of emulsions (Chapter 6) is a micro-terrace method of formation of droplets introduced in 2000 by Sugiura and coworkers (Sugiura *et al.*, 2000; Sugiura, Nakajima, and Seki, 2002) (Figure 5.1). In this system the droplet phase is pushed into a set of small orifices that open onto a microscopic terrace. The edge of this terrace prompts breakup of the streams of liquid into droplets via an interfacial instability.

The most popular microfluidic geometries used for emulsification include a T-junction, introduced by Thorsen *et al.* (2001), and a flow-focusing junction introduced in the planar format by Anna, Bontoux, and Stone (2003). Although there are variations of each of these two geometries, the generic features of their simple architecture are always preserved. The T-junction, shown schematically in Figure 5.1, comprises the main channel that carries the continuous fluid. An additional channel, which usually joins the main one at a right angle, delivers the fluid to be dispersed. At the junction, the two immiscible fluids meet, the fluid-to-be-dispersed enters the main channel, a droplet (or bubble) is broken off, the tip retracts to the junction, and the whole process repeats. The mechanism of breakup and the diameters of the droplets (bubbles) depend on the aspect ratios of the widths and height of the channels, on the rates of flow of the two fluids, and on their material parameters – viscosities and the interfacial tension between them. We will review in detail the introduction of this device, its uses, and the physics of formation of foams and emulsions in the T-junction in the next section. An interesting variation of the T-junction is a system proposed by Link *et al.* (2004), which is useful in splitting already preformed droplets at a diverging junction of channels. Appropriate choice of the hydraulic resistances of the outlet channels from the junction can be used to tune the ratio of the volumes of the two daughter droplets (De Menech, 2006; Jullien *et al.*, 2009; Link *et al.*, 2004). This system found applications in the formation of libraries of chemical reagents in liquid slugs (Adamson *et al.*, 2006).

The other most popular planar geometry, the flow-focusing junction, draws from the axi-symmetric flow-focusing introduced by Ganan-Calvo (1998). Generically, the flow-focusing device comprises a set of three parallel inlet channels that join upstream of a narrow orifice (Figure 5.1). The orifice leads to an outlet channel. The three inlets deliver the fluid-to-be-dispersed (via the central inlet) and the continuous fluid (via the two side channels). At the junction the tip of the fluid-to-be-dispersed periodically enters the orifice, inflates a droplet or bubble, and breaks to release a segment of the dispersed fluid. Similarly, to the T-junction, the mechanism of breakup depends both on the geometry of the junction and the speed and properties of the fluids. We will discuss the different regimes of emulsification in a flow-focusing device in Section 5.4.

In subsequent sections we will review variations of the microfluidic devices used for emulsification, including the more complicated geometries (both planar and axi-symmetric) that can be used to form multiple droplets. Finally, we will discuss other important concepts, including the problem of control of wetting of the channels with the continuous fluid, the use of non-Newtonian fluids, and variations in the dimensions of the microfluidic devices.

5.3 T-Junction

The T-junction deserves its popularity for the simplicity of its architecture and robust performance with most fluids. As we explain below, provided that the continuous fluid preferentially wets the walls of the channels, and that the viscosity of the fluid-to-be-dispersed is not too large, the T-junction can produce monodisperse droplets providing good control over their volume. We begin with a description of the device itself, and a historical overview of the developments in understanding the dynamics of breakup in a T-junction, and in applications of this device to different ranges of properties of the fluids. Then we will discuss the mechanism of formation of droplets and provide a guide for the choice of parameters that lead to the desired characteristics of the emulsions.

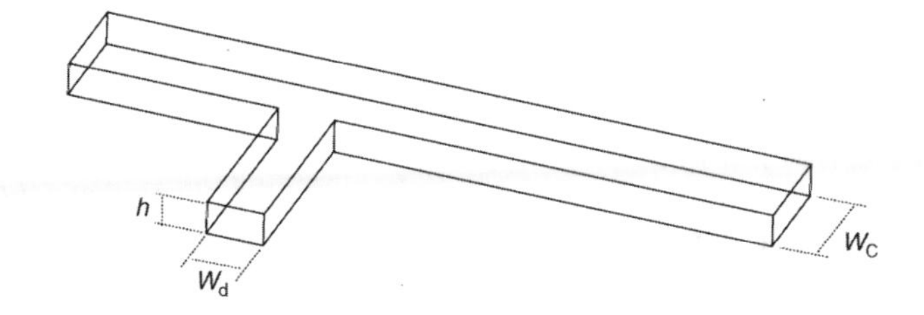

Figure 5.2 Geometry of a planar T-junction of rectangular cross-sections of the channels with all the important dimensions marked on the diagram. Reprinted with pemrission froom Phys. Rev. Letts., Dynamic Pattern Formation in a Vesicle-Generating Microfluidic Device, Thorsen et al., **86**, 18, 4163–4166 Copyright (2001) American Physical Society. Reprinted with permission from Lab on a Chip, Droplet formation in a microchannel network by T. Nisisako, T. Torii and T. Higuchi, **2**, 1, 24–26 Copyright (2002) Royal Society of Chemistry

5.3.1 Parameters that Determine the Dynamics

Within the planar geometry of a T-junction, the geometry of the junction is fully specified (Figure 5.2) by the width (w_c) of the main channel, the ratio w_d/w_c of the width (w_d) of the inlet channel for the dispersed phase to the width of the main channel, and by the ratio (h/w_c) of the height (h) of the channel to the width of the main channel (w_c).

Besides the geometry of the junction, seven parameters completely determine the dynamics of the system. These parameters include the volumetric rates of flow (Q_c, Q_d), viscosities (μ_c, μ_d) and densities (ρ_c, ρ_d) of each of the two immiscible phases, and the interfacial tension (σ) between them. Because of the typically small scale of the channels, the Bond number of the flow is typically vanishingly small and buoyancy can safely be neglected. Thus, both in the case of a large (e.g., bubbles of air in water) and in the case of small contrast of densities (e.g., water in oil) it is sufficient to specify one (the greater of the two) density (which we will denote for simplicity as ρ) in order to judge the relative importance of the inertial and viscous terms.

In the case of the simplest geometry ($w_d/w_c = h/w_c = 1$) the problem is fully specified by seven parameters: w_c, Q_c, Q_d, μ_c, μ_d, ρ, σ. Thus, according to Buckingham's Pi Theorem, the volume V of the droplets, conveniently rescaled by $v = V/w_c^3$ can be expressed as a function of four dimensionless parameters. The Reynolds number, that can be calculated, for example, for the flow of the continuous phase, and which is typically faster than Q_d, $Re = Q_c\rho/w_c\mu$ is the first of these four. Knowing that Re usually assumes small values, that is, viscous effects are typically more important than the inertial ones, the second dimensionless group should capture the relative importance of the capillary forces to the shear stress, and this group is well represented by the capillary number (Ca). The Ca can be calculated in various ways and, as we will discuss in detail, a well-chosen definition can capture the dependence of V on the geometry of the junction. For simplicity, in most of the reports calculation of Ca is based solely on the speed of flow and viscosity of the continuous phase: $Ca = \mu Q_c/w_c^2\sigma$. Re and Ca are supplemented by the contrast of viscosities of the two phases $\lambda = \mu_d/\mu_c$, and the ratio of the rates of flow $q = Q_d/Q_c$. For more complicated geometries, the two aforementioned geometrical aspect ratios (w_d/w_c, h/w_c) need to be included in the description.

5.3.2 First Reports

The T-junction geometry was first introduced in 2001 by Thorsen *et al.* (2001). They utilized soft lithography (Duffy *et al.*, 1998) to prepare the simple system of channels in PDMS (polydimethylsiloxane). The heart of the device is the junction at which the main channel that carries the continuous fluid is joined by a channel which delivers the fluid to be dispersed. Figure 5.3 shows the original scheme and a micrograph of the system reported by Thorsen *et al.* (2001).

Owing to the lithographic procedure and reflow of the resist, the channels had a non-rectangular, dome-like cross-section and were truly micro-scale: at the junction, both the main channel and the channel that delivered the fluid-to-be-dispersed had a width of 35 μm and a height of 6.5 μm. As can be seen in Figure 5.3 the junction has a complicated geometry, the junction itself is wide and has the largest height. As a result, via the capillary forces, the droplet phase always stays in the junction, and the droplets are broken off from a tip that extends downstream from this bulge. Thorsen *et al.* (2001) reported formation of both monodisperse and bi-disperse droplets.

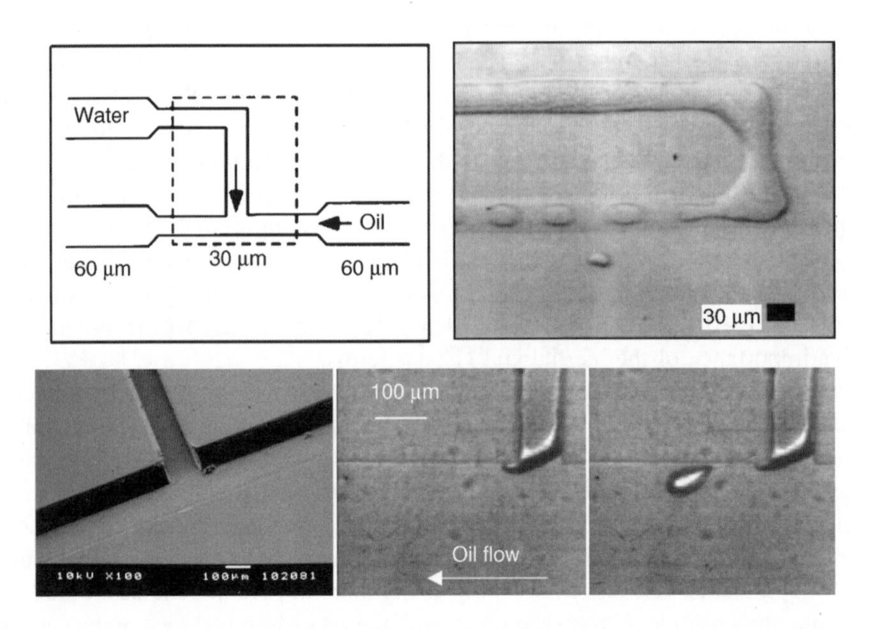

Figure 5.3 *Top panel: a schematic of the T-junction and a micrograph of the system forming aqueous droplets in a continuous phase of oil from the first report on the use of the T-junction (Thorsen et al., 2001). Bottom panel: T-junction system fabricated in polymethylmethacrylate by Nisisako, Torii, and Higuchi (2002). The channels had rectangular cross-sections ($w_c = 500\,\mu$m, $w_d = h = 100\,\mu$m). Reprinted with permission from Phys. Rev. Lett., Dynamic Pattern Formation in a Vesicle-Generating Microfluidic Device, Thorsen et al.,* ***86****, 18, 4163–4166 Copyright (2001) American Physical Society. Reprinted with permission from Lab on a Chip, Droplet formation in a microchannel network by T. Nisisako, T. Torii and T. Higuchi,* ***2****, 1, 24–26 Copyright (2002) Royal Society of Chemistry*

Thorsen *et al.* used pressurized containers to deliver the fluids (water as the discontinuous phase and various alkanes with the addition of surfactant as the continuous phase). When the pressure applied to water was much less than the pressure applied to the oil, the tip of the aqueous phase rested at the junction. At higher pressures it periodically entered the main channel to form a droplet. Increase of the pressure applied to water (and the resulting increase in the rate of flow of the aqueous phase) resulted in an increase in the volumes of the droplets. They described formation of droplets in their system with the classical condition of breakup in shear that uses the balance of the interfacial forces σ/r (where r is the radius of the droplet) and the shear forces. It was pointed out that the droplet does not obstruct completely the cross-section of the channel, and that the flow through the corners between the interface and the walls of the channel contribute to the effect of shearing-off the tip of the continuous phase. Nisisako, Torii, and Higuchi (2002) followed in 2002 with a report on the formation of aqueous droplets in sunflower oil in a planar T-junction system of rectangular cross-sections of the channels (Figure 5.3) fabricated in polymethylmethacrylate. As the width of the main channel was much larger than the width of the inlet for the dispersed phase ($w_d/w_c = 1/5$) the droplets were not obstructing the main channel during their formation and were broken off by the shear stresses exerted by the flow of the continuous phase.

Very quickly it was realized that formation of droplets in microfluidic systems offers an attractive opportunity for performing chemistry within the microdroplets. Already by 2003 the group of Professor Ismagilov had published their first articles (Song *et al.*, 2003a; Song and Ismagilov, 2003b; Song, Tice, and Ismagilov, 2003c; Tice *et al.*, 2003; Zheng, Roach, and Ismagilov, 2003) on the subject. These reports showed that it is possible to enclose into a droplet a set of reagents, mix them quickly, and process them in a flow-through reactor without the introduction of any dispersion in the time of residence of the reagents in the device.

The characteristic of rapid mixing within the droplets is also of importance in the use of microfluidic technologies for formation of polymer and gel particles, as it is often the case that reagents need to be mixed within the droplet in order to start the polymerization or gelation process.

Tice *et al.* (2003), even in their first article, noticed that the volume of the droplets depended predominantly on the ratio of the volumetric rates of flow of the two immiscible phases and not on the total speed of flow within the range of capillary numbers between 10^{-3} and 10^{-2} (Figure 5.4). Because the shear stresses are proportional to the value of the rate of flow this observation suggested that the mechanism of breakup in the T-junction is different from the classical competition between the Laplace pressure and the shear stress, and that it cannot be well explained within the context of the capillary number alone.

Soon after, the same group reported experiments on the formation of droplets out of a compound stream that contained fluids of different viscosities. These Authors reported that the contrast of viscosities in the droplet enhances the process of mixing, but also that the viscosity of the fluid-to-be-dispersed affects the dynamics of formation of the droplets. Tice, Lyon, and Ismagilov (2004) demonstrated that it is possible to form monodisperse plugs for different values of the contrast of viscosities of the fluids that

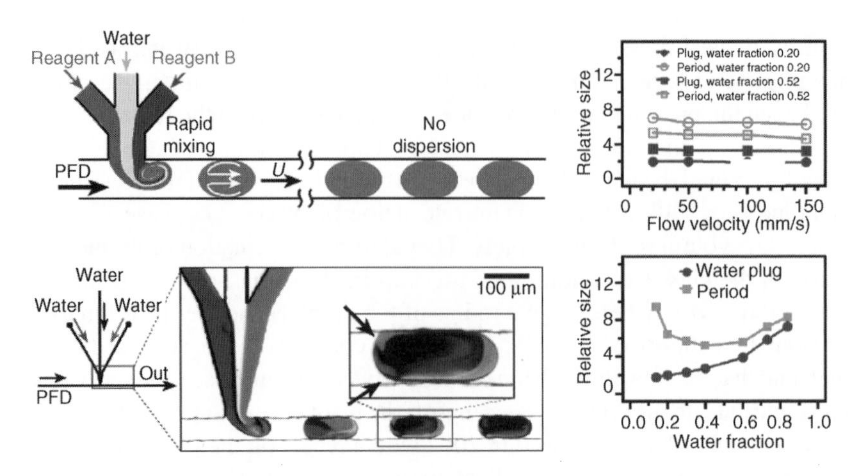

*Figure 5.4 Left panel: a scheme and micrograph of a T-junction system that combines the flow of three different aqueous streaks and breaks the combined stream into droplets. The solutions mix effectively within the droplets, allowing for processing of the chemical reaction within droplets without any dispersion of time of residence and with superior control over the kinetics of reaction. Right panel: the length of the droplets non-dimensionalized by the width of the channel, as a function of the speed of flow, and of the ratio of the rates of flow (Tice et al., 2003). Reprinted with permission from Langmuir, Formation of Droplets and Mixing in Multiphase Microfluidics at Low Values of the Reynolds and the Capillary Numbers, J.D. Tice et al., **19**, 22, 9127–9133 Copyright (2003) American Chemical Society. See Plate 2*

constituted the droplets, even for viscosities significantly greater than that of water. The major qualitative change is that for higher viscosities of the droplet phase (or rather when the contrast λ is high) the system crosses over to a jetting or laminar regime at lower speeds of flow, or alternatively, at smaller capillary numbers.

5.3.3 Mechanism of Operation of the T-Junction System

Generally, the T-junction can operate in three different modes: dripping, jetting, and laminar co-flow (Figure 5.5), which appear sequentially as the rate of flow of the fluids through the device is increased. At the smallest rates of flow the system operates in the dripping regime: the tip of the discontinuous phase enters the main channel and propagates downstream occupying most, or a substantial fraction, of the cross-section of the channel, and is later broken at the junction. The droplet so released flows downstream, while the tip of the discontinuous phase recoils to the junction and the process repeats. Above a critical rate of flow the point of breakup gradually moves downstream of the junction – the stream of the fluid-to-be-dispersed always extends into the main channel and the droplets are broken off several widths of the channel away from the junction. At even higher rates of flow the point of breakup is completely washed away, the two immiscible streams co-flow in a laminar, side-by-side, fashion, and no droplets are generated.

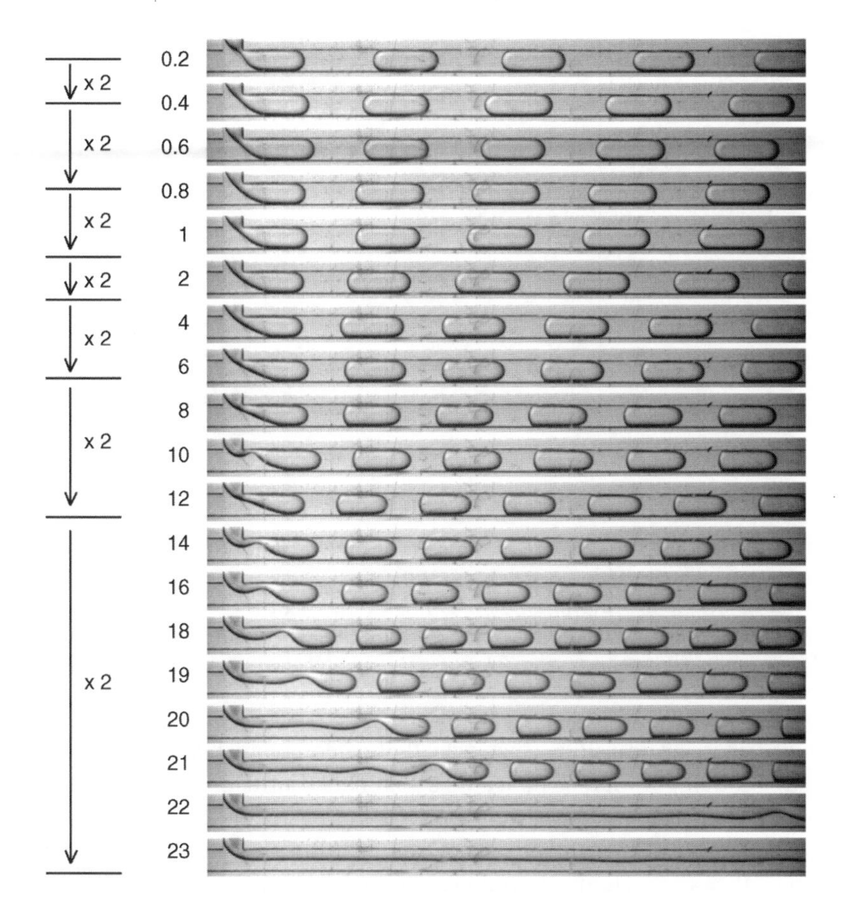

Figure 5.5 *Micrographs illustrating the transition from squeezing to jetting in a microfluidic T-junction. A stream of water is introduced into the flow of a continuous phase of solution of surfactant (Span 80) in hexadecane. The cross-section of the channels is set to $400 \times 400\,\mu m$. The numbers to the left of the micrographs indicate the rate of flow of the each of the liquids ($Q_d = Q_c$). The horizontal lines mark approximately each twofold increase of the total rate of flow through the system. The value of the capillary number is proportional to the rate of flow and ranges from 2×10^{-4} to 2×10^{-2}. The ratio of the rates of flow is held constant and the increase in the total rate of flow has little effect on the volume of the droplets up to $Ca \sim 10^{-2}$. For higher values of the capillary number the system quickly crosses over to the jetting regime and laminar co-flow of the two immiscible phases. Images captured by Judyta Węgrzyn at the Institute of Physical Chemistry, PAS*

Guillot and Colin (2005) analyzed in detail the transitions between the three regimes in a system similar to the T-junction. In their system the two immiscible fluids flew from opposite channels to meet at a wide junction that leads to a perpendicular outlet channel (Figure 5.6). This geometry favored co-flow of the two phases in the junction and breakup either at the entrance into the outlet channel, or within the outlet channel, downstream of its entrance. They used combinations of liquids of different viscosities

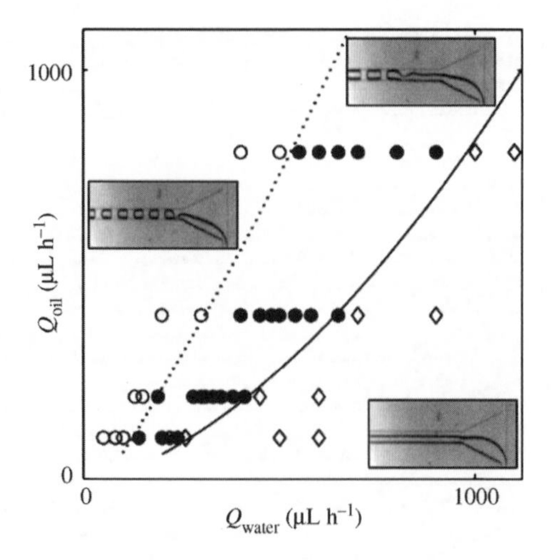

Figure 5.6 *Diagram of the mode of operation of a system similar to a T-junction, showing the regions of the values of Q_c (Q_{oil} on the graph) and Q_d (Q_{water} on the graph) within which the system operates in the dripping, jetting, and laminar regimes (Guillot and Colin, 2005). Reprinted with permission from Physical Review E, Stability of parallel flows in a microchannel after a T junction by Guillot and Colin,* ***72****, 066301 Copyright (2005) American Physical Society*

and showed that the transitions between the different breakup regimes cannot be attributed to a unique value of the capillary number. Furthermore, they were able to show that the viscous shear forces were too weak to cause significant deformation of the shape of the stream of the aqueous phase, and that the pinching mechanism was not governed by the competition between the interfacial and viscous forces, but rather by conservation of mass of the two fluids. In addition, these Authors used the observation that the tip of the stream of the aqueous phase blocks the cross-section of the channel to construct a model that correctly predicted the transition between the jetting regime and the laminar co-flow of the two phases. These observations further supported the hypothesis that the breakup in the confinement of microchannels cannot be described by a simple competition between the interfacial and viscous forces.

In 2006 Garstecki *et al.* (2006) reported a detailed study of the formation of droplets and bubbles in a standard T-junction comprising channels of rectangular cross-section. This study examined a wide range of viscosities of the continuous phase and of the rates of flow.

Qualitatively, in the case of formation of aqueous droplets in oil, the system typically produced monodisperse droplets and their volume could be effectively controlled with the rates of flow of the two immiscible phases (Figure 5.7). Increase of the rate of flow of the aqueous phase resulted in an increase in the volume of the droplets, and – conversely – an increase in the rate of flow of the continuous phase resulted in a decrease in the volumes of the droplets.

The process of breakup proceeds through the following stages. Firstly, the tip of the discontinuous fluid penetrates into the main channel and inflates a droplet downstream

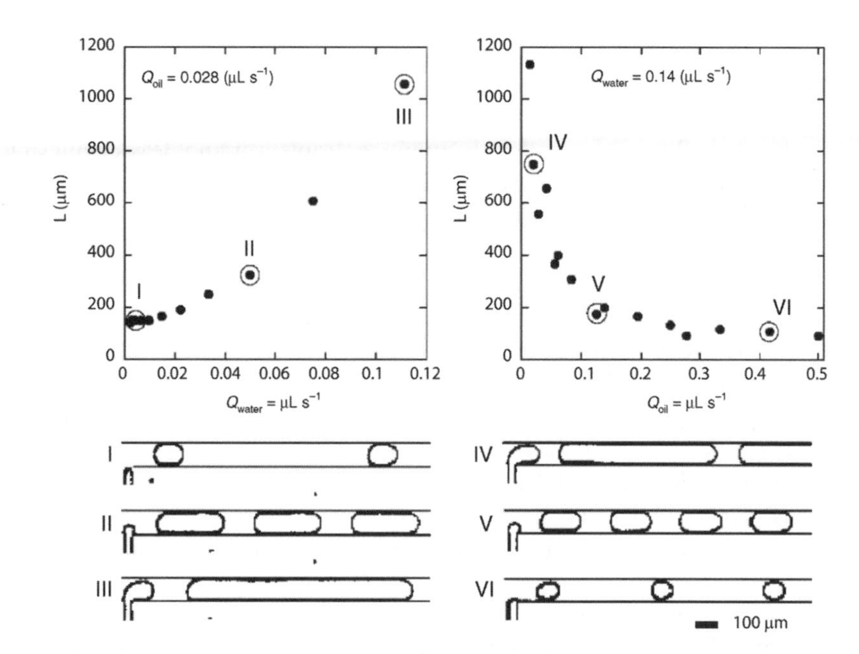

Figure 5.7 *The effect of the increase of the rate of flow of the droplet phase Q_d (Q_{water} on the graphs), and of the rate of flow of the continuous phase Q_c (Q_{oil} on the graphs) on the length of the droplets produced in a planar T-junction of rectangular cross-sections of the channels (Garstecki et al., 2006). Reprinted with permission from Lab on a Chip, Formation of droplets and bubbles in a microfluidic T-junction—scaling and mechanism of break-up by Garstecki, P. et al.,* ***6****, 437–446 Copyright (2006) Royal Society of Chemistry*

of the junction. The interface at the front of the droplet assumes a curvature associated with the width and height of the main channel. The shape at the rear part of the droplet is more complicated: while the radial curvature at the beginning of the process is equal to half the height of the channel, the axial curvature is much smaller. The rear of the droplet gradually moves downstream, which closes the neck that connects the input channel for the droplet phase with the growing droplet. The axial curvature decreases while the radial curvature increases, and finally the neck breaks and the whole process repeats.

There are three types of forces that can act on the growing tip of the discontinuous fluid and which can play a role in the breakup process. These forces include: (i) the interfacial tension force (F_σ), (ii) 'the force arising from the shear stress exerted by the continuous fluid flowing past the growing droplet (F_μ); and (ii) the force arising from the pressure drop along the growing droplet (F_p). The net interfacial force acting on the growing droplet is related to the difference in the Laplace pressure at the front and at the rear end of the droplet. At the beginning of the process of breakup once the droplet blocks most of the cross-section of the main channel the neck that connects the droplet to the inlet of the fluid-to-be-dispersed is wide and the curvature at the front is larger than at the rear of the droplet. Thus the interfacial forces stabilize the tip and act against thinning of the neck and against breakup. Only at the end of the process when the radial curvature of the neck becomes large, does the Laplace curvature drive the fluid in the growing droplet forward and accelerate the collapse of the neck (van Steijn, Kleijn, and Kreutzer, 2009).

Garstecki *et al.* (2006) argued that the growing droplet (or bubble) occupies most of the cross-section of the main channel and constricts the flow of the continuous fluid past the tip of the droplet phase to thin films between the interface and the walls of the channels and to gutters (Fuerstman *et al.*, 2007; Wong and Morris, 1995a; Wong, Radke, and Morris, 1995b) in the corners of the rectangular cross-section of the channel (Figure 5.8). Limitation of the available cross-section for the flow of the continuous fluid increases the viscous resistance to such flow and it was argued that only a small fraction of the continuous fluid passes the tip. Also the shear stress exerted by this flow generates a smaller force on the tip than the resulting pressure drop along the tip. As a result, the continuous fluid, pumped at a constant rate into the junction, squeezes the neck that connects the stream of the fluid-to-be-dispersed with the growing droplet rather than passing the tip by. As a consequence, the condition for breakup is not determined by the balance of the restoring capillary force and distorting shear stress: a balance that could be reflected in the characteristic linear dimension of the droplet being inversely proportional to the capillary number. Rather, the volume of the droplet should only be a function of the rates of flow of the two phases. This hypothesis was confirmed via measurements of the length of the droplets produced in the T-junction to be determined predominantly by the ratio of the

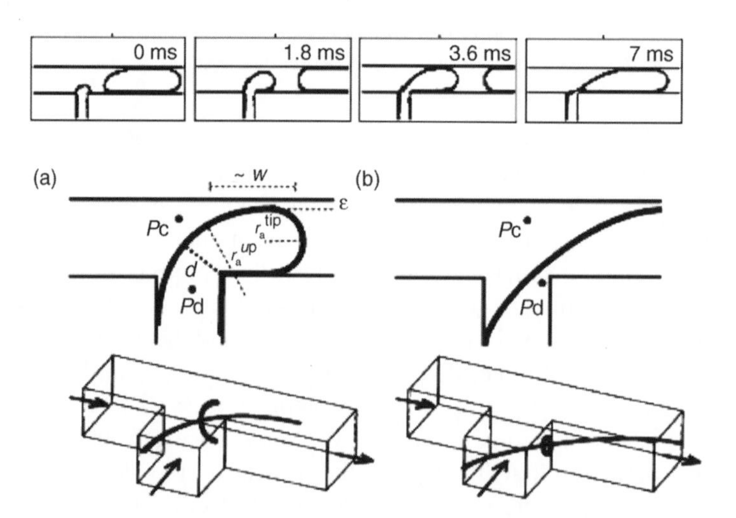

Figure 5.8 *Top panel: micrographs illustrating the process of formation of a droplet in a T-junction. Bottom panel: schematic representations of the shape of the interface of the tip of the droplet phase at the intermediate phase of formation of a droplet (a) and at the final phase (b). (Garstecki et al., 2006). Reprinted with permission from Lab on a Chip, Formation of droplets and bubbles in a microfluidic T-junction—scaling and mechanism of break-up by Garstecki, P. et al.,* ***6****, 437–446 Copyright (2006) Royal Society of Chemistry*

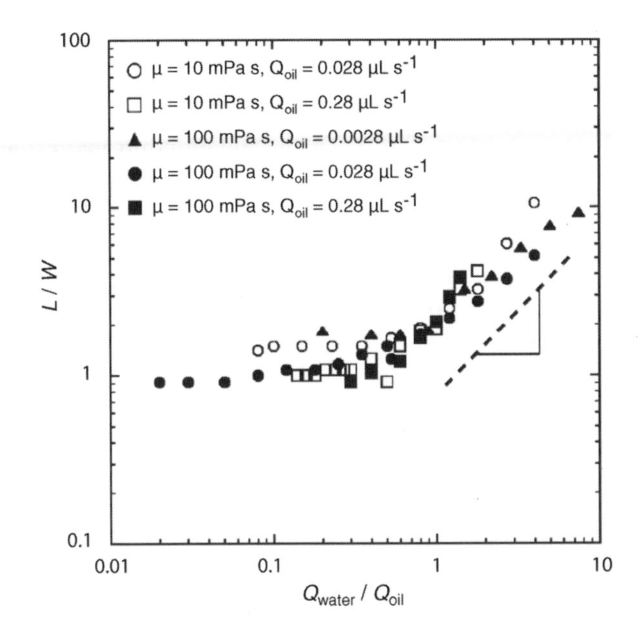

Figure 5.9 *Experimental verification of the blocking–squeezing model of breakup in the T-junction at low values of the capillary number. Even a 100-fold increase in the shear-stress does not alter the scaling of the length (L) of the droplets, which is determined solely by the ratio of the rates of flow (Garstecki et al., 2006). Reprinted with permission from Lab on a Chip, Formation of droplets and bubbles in a microfluidic T-junction—scaling and mechanism of break-up by Garstecki, P. et al.,* **6***, 437–446 Copyright (2006) Royal Society of Chemistry*

rates of flow of the two phases, for different speeds of flow, and different viscosities of the continuous phase. Figure 5.9 shows that even a 100-fold increase of the shear stress exerted on the growing tip does not significantly alter the scaling of the droplet size being a function only of the ratio of rates of flow: $L/w \propto 1 + q$.

The analysis of the forces acting in the breakup process can be translated into a quantitative model for the volume of the droplets. First the tip of the discontinuous phase penetrates and blocks the main channel. This stage is completed when the length of the droplet is similar to the width of the main channel. Once the tip blocks the main channel, the continuous fluid starts to squeeze it, which leads to breakup within a time $t_{squeeze} \sim hw_c/Q_c$. Within this time the droplet elongates by $t_{squeeze}\, Q_d/h\, w_c$. Summing these two contributions leads to the following scaling of the length of the droplet:

$$L/w_c = 1 + \alpha(Q_d/Q_c) \qquad (5.1)$$

As the fraction of the continuous fluid bypasses the droplet and does not contribute to squeezing, and as the speed at which the neck collapses need not be constant, (Garstecki *et al.*, 2006) proposed to treat α as a fitting parameter of order one. This relationship successfully recovers the observed volumes of the droplets, provided that the shear stress exerted by the continuous phase does not significantly distort the shape of the tip, that is, at small values of the capillary number.

In the same report (Garstecki *et al.*, 2006) it was shown that gas introduced into the junction breaks into bubbles according to the same relationship. Interestingly, as the gas was forced into the microfluidic chip at a constant pressure head, the rate of flow of gas depended on the resistance to flow in the outlet channel. The volume of the bubbles was inversely proportional to the viscosity of the continuous phase (as this viscosity linearly increases the resistance to flow in the outlet channel). A similar set of observations was later performed by Xu *et al.* (2006b), who also noticed that the volume of the bubbles scaled in inverse proportion to the product of the rate of flow of the continuous fluid and its viscosity.

In 2007 van Steijn, Kreutzer, and Kleijn (2007) performed detailed experimental analyses of the scaling of the size of droplets and the flow field of the continuous fluid in the vicinity of the tip during breakup in a microfluidic T-junction with square cross-sections of the channels. These measurements confirmed the squeezing model of breakup (Garstecki *et al.*, 2006), including the scaling of the volume of the droplets with the ratio of the rates of flow of the two phases. The observed flow field of the continuous fluid also revealed that up to 25% of the flow of the continuous fluid bypasses the growing droplet. As we noted above, the later analyzes (van Steijn, Kleijn, and Kreutzer, 2009) by the same Authors revealed that the flow in the gutters even reversed during the last stage of the collapse of the neck which connects the stream of the fluid-to-be-dispersed with the growing droplet.

In 2008 De Menech *et al.* (2008) reported numerical simulations of breakup in the T-junction. The results of these investigations confirmed the blocking–squeezing mechanism of breakup at low values of the capillary number (Figure 5.10). In particular, they showed the development of pressure upstream of the growing droplet during the process of breakup. This pressure elevates sharply upon blockage of the main channel by the tip of the discontinuous phase. De Menech *et al.* showed that up to a critical value of the capillary number the volume of the droplets depends only on the ratio of the rates of flow and has no explicit dependence on the value of Ca, and no dependence on the viscosities of the two phases. At faster flow, characterized by values of capillary number larger than the critical value of $Ca \approx 10^{-2}$, the shear stress exerted by the continuous fluid on the growing droplet begins to play an observable role. These Authors called this new regime "dripping" and detailed the transition between squeezing and dripping. In the dripping regime, the tip of the discontinuous phase (or a growing droplet) does not extend into the whole width of the main channel, but, as the speed of flow is increased, this tip is progressively more and more distorted by the flow of the continuous phase. In this regime, both the effects of the pressure drop along a growing droplet and the shear stress are important, and the scaling that takes into account solely the ratio of the rates of flow has to be corrected with a term that includes the value of the capillary number. Much to their surprise, the dependence of the volume of the droplets on the capillary number in the dripping regime continued to be fairly weak. The linear dimension of the droplet scaled as L proportional to

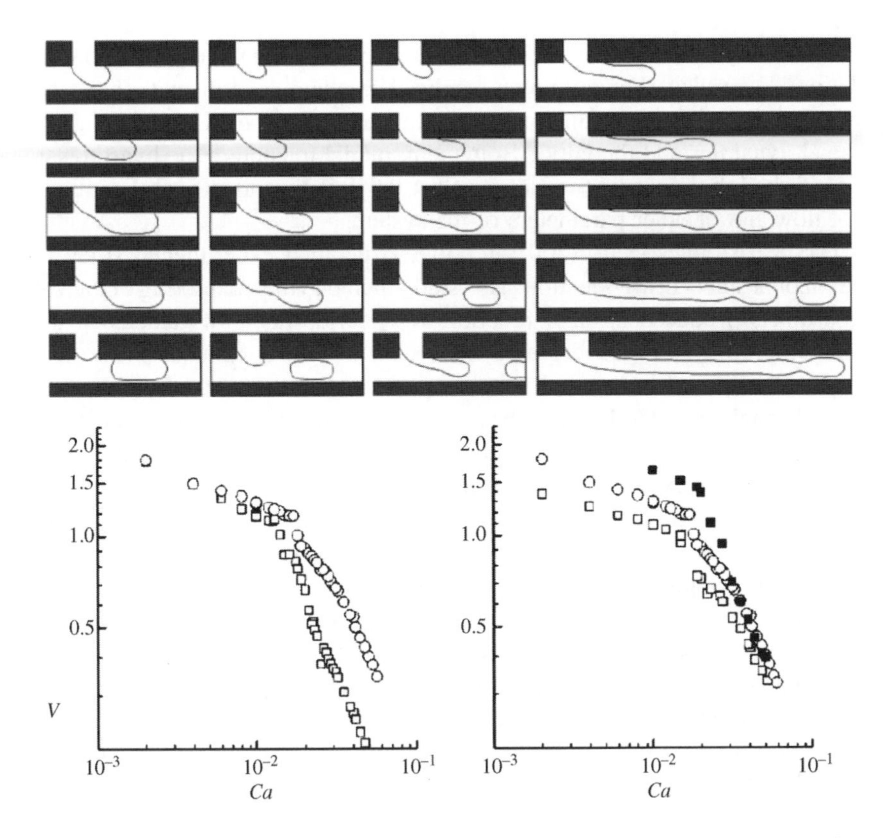

Figure 5.10 *Top panel: snapshots from the numerical simulations (De Menech et al., 2008) illustrating the process of formation of droplets in the squeezing, dripping, and jetting regimes. Bottom panel: the graphs illustrate the dependence of the volume of the droplets (rescaled by the cube of the width of the channel) on the value of the capillary number for various viscosity contrasts and ratios of the rate of flow. One can clearly observe a cross-over from a weak dependence of the volume on the value of Ca for* $Ca < \sim 10^{-2}$*, and a stronger dependence for higher values of Ca. Reprinted with permission from Journal of Fluid Mechanics, Transition from squeezing to dripping in a microfluidic T-shaped junction by M. De Menech, P. Garstecki, F. Jousse and H.A. Stone,* ***595****, 1, 141–161 Copyright (2008) Cambridge University Press*

$Ca^{-0.4}$ (and not as Ca^{-1} as would be the case if shear was the dominant force distorting the droplet).

Similar results were obtained experimentally. For example, van der Graaf *et al.* (2005) observed in a less confined system (the width of the main channel was several times larger than the height of the device and the width of the channel supplying the discontinuous phase) that the diameter of the radius of the droplets scaled as $Ca^{-0.25}$. In 2008, Xu *et al.* (2008) showed that droplets follow the squeezing model at low values of capillary number calculated for the flow of the continuous phase, and that above $Ca \sim 10^{-2}$ the scaling indeed crossed-over to a regime in which the size of the droplets depended on the value of the capillary number with an exponent of -0.2. Other simulations have also shown similar results. For example, Liu and Zhang (2009) reported two-dimensional simulations that included the phase-field model to capture the interfacial dynamics and the lattice Boltzmann model to capture the hydrodynamics. They confirmed the squeezing model for $Ca < 0.018$ independently of the ratio of the rates of flow and of other parameters of the system.

In 2008 Christopher *et al.* (2008) reported a detailed experimental study of the formation of droplets in a T-junction. This study included the monitoring of the volume of the droplets produced in the T-junction over a wide range of values of capillary number ($Ca = 10^{-3}$ to 1), a wide range of viscosities of the continuous phase (between 6 and 350 mPa s), and for a set of different aspect ratios of the T-junction. The width of the main channel was set to 150 μm, the height to 50 μm, while the width of the inlet for the dispersed phase was varied between 65 and 375 μm. The results of this study provided a thorough experimental confirmation of the squeezing model for all combinations of liquids, provided that the value of the capillary number was lower than $Ca \sim 10^{-2}$. For low viscosity contrasts ($\lambda = m_d/m_c = 6$) they observed a well-defined transition to the dripping regime. For higher viscosity contrasts the transition was more gradual. In spite of the complexity of the problem these workers succeeded in construction of an "extended model" that takes into account the details of the geometry and both the squeezing and shearing effects during the process of breakup. This extended model recovered the observed dependence of the volume of the droplets on the rates of flow, viscosities of the fluids, and the aspect ratios of the device (Figure 5.11).

5.3.4 Variations of the Geometry of the T-Junction

A number of reports have presented modifications of the architecture of the T-junction. For example, Xu, Tan, and Wang (2006a) inserted a capillary into the side channel. This allowed them to reduce the width of the inlet for the fluid-to-be-dispersed phase. Hallmark *et al.* (2009) used the same idea to also control the distance between this inlet and the opposite side of the channel. Both devices produce monodisperse droplets. As the width of the main channel is typically much larger than the scale set by the width of the inlet for the droplet phase, the shearing effects usually play a significant role and the scaling of the size of the droplets includes a dependence on the value of the capillary number.

Steegmans, Schroen, and Boom (2009b) reported a study of formation of droplets at moderate to large values of the capillary number (between 10^{-2} and 1) in a Y-junction (Figure 5.12). Within this range of Ca they observed the dripping regime and a power-law scaling of the size of the droplets with the capillary number with an exponent of $-1/2$ (larger than in the case of a conventional T-junction). They also found that the rate of flow of the dispersed phase had minimal effect on the volume of the droplets. This is of interest for the case of up-scaling of this technique for higher throughputs, as it releases the requirement of precise control of the rate of flow for both of the two immiscible phases. Recently Shui, van den Berg, and Eijkel (2009) reported the

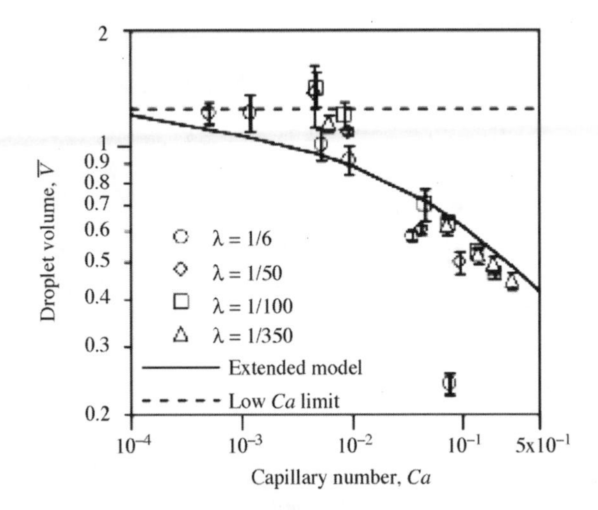

Figure 5.11 *Volume of the droplets (non-dimensionalized by the cube of the width of the channel) as a function of the capillary number for a range of contrasts of viscosities of the two immiscible phases. For $Ca > 10^{-2}$ the squeezing model needs to be corrected by the shearing effects represented via the dependence on the value of Ca (Christopher et al., 2008). Reprinted with permission from Phys. Rev. Lett., Experimental observations of the squeezing-to-dripping transition in T-shaped microfluidic junctions by Christopher et al., **78**, 3, 036317 Copyright (2008) American Physical Society*

formation of droplets in a "head-on" T-junction (Figure 5.13) and observed all three different modes of breakup, in agreement with the earlier observations for the conventional T.

5.3.5 Summary of the Mechanism of Breakup in the T-Junction

In summary, the T-junction systems provide a convenient tool for formation of monodisperse droplets. At low capillary numbers the mechanism of breakup is specific to strongly confined, microfluidic systems: the breakup is governed by the ratio of the rates of flow of the two immiscible phases via the blocking–squeezing model. Within this regime the droplets can be almost arbitrarily large, but not smaller than the typical volume of the main channel, given approximately by $V_{\mathrm{ch}} = hw_{\mathrm{c}}^2$. j Steegmans Schroen, and Boom (2009a) reported in 2009 an interesting statistical analysis of the experimental results of different groups (Garstecki et al., 2006; Nisisako, Torii, and Higuchi, 2002; van der Graaf *et al.*, 2005; van der Graaf *et al.*, 2006; Zhao, Chen, and Yuan, 2006). They showed that all these results can be modeled by a single equation that captures the characteristics of the blocking–squeezing model.

At higher rates of flow the shearing effects come into play and slightly modify the scaling of the volume of the droplets (with a power law dependence on the value of the capillary number, typically with an exponent between -0.2 and -0.4). These effects allow the volume of the droplets to be decreased to below V_{ch}, down to approximately

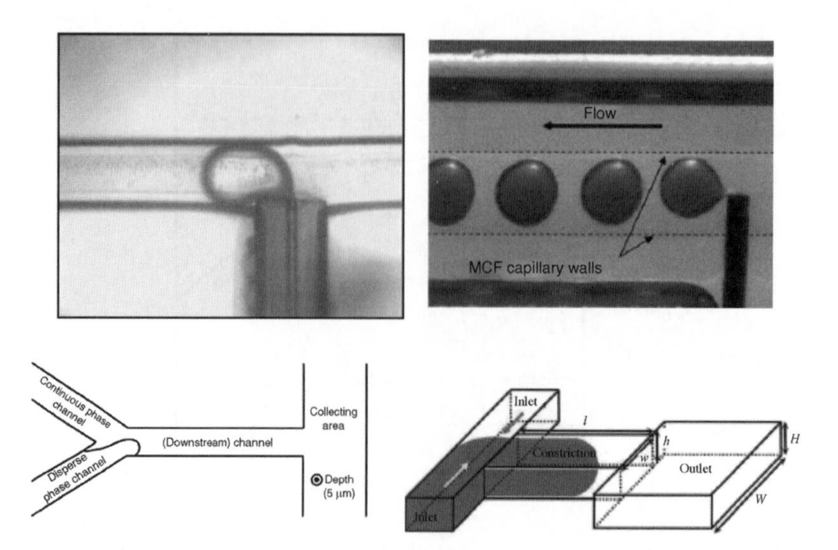

*Figure 5.12 A microcapillary T-junction device (top left) (Xu et al., 2006); a similar device reported by Hallmark et al. (2009) (top right); a Y-junction (Steegmans, Schroen, and Boom, 2009) (bottom left); and "head on junction" (bottom right) reported by Shui, van den Berg, and Eijkel (2009). Reprinted with permission from Langmuir, Controllable Preparation of Monodisperse O/W and W/O Emulsions in the Same Microfluidic Device by J. H. Xu et al., **22**, 19, 7943–7946 Copyright (2006) American Chemical Society. Reprinted with permission from Chemical Engineering Science, The experimental observation and modelling of microdroplet formation within a plastic microcapillarity array by B. Hallmark, C. Parmar, D. Walker et al., **64**, 22, 4758–4764 Copyright (2009) Elsevier Ltd. Reprinted with permission from Langmuir, Characterization of Emulsification at Flat Microchannel Y Junctions by M.L.J. Steegmans, **25**, 6, 3396–3401 Copyright (2009) American Chemical Society. Reprinted with permission from Journal of Applied Physics, Capillary instability, squeezing, and shearing in head-on microfluidic devices by L. Shui, A. van den Berg and J. Eijkel, **106**, 12, 124305 Copyright (2009) American Institute of Physics*

$0.1V_{ch}$. The dripping regime, however, is fairly narrow in the range of rates of flow, or alternatively, in the range of values of the capillary number (see Figures 5.5 and 5.6 and a numerically obtained diagram in Figure 5.13).

Typically, it is difficult to obtain droplets for capillary numbers greater than 0.1 or 1. Increased viscosities of the fluids further decrease the range of rates of flow at which the stream of the fluid-to-be-dispersed breaks into droplets.

5.3.6 Maximum Throughput of a Single Junction

The above considerations allow estimation of the throughput of a single T-junction. Taking $Ca_c = 0.01$ as an upper limit for the speed of flow gives the upper limit on the rate of flow of the continuous phase $Q_c = Ca_c w_c h\sigma/\mu_c$. For the sake of the estimation we can further assume that the ratio of the rates of flow is of the order unity, and that the main channel has a square cross-section, which yields $Q_d \approx Ca_c w_c^2 \sigma/\mu_c$. Because the squeezing regime provides very good reproducibility of the volumes (low standard deviation of the

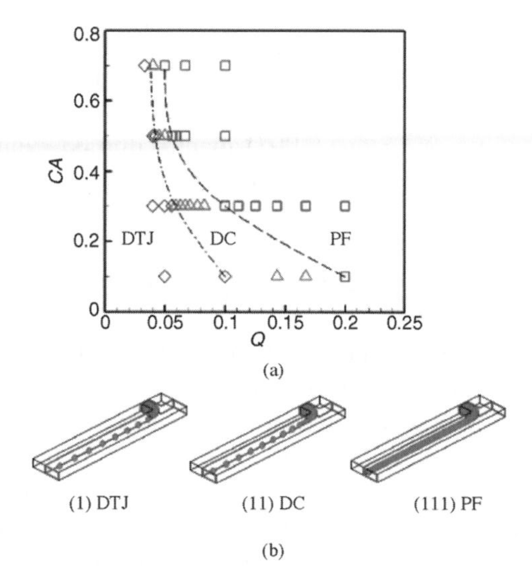

Figure 5.13 A diagram obtained numerically for the mode of breakup [dripping at the junction (DTJ), jetting (DC), and laminar co-flow (PF)] in a microfluidic T-junction. The contrast of viscosities equaled 1/10 (Gupta, Murshed, and Kumar, 2009). Reprinted with permission from Applied Physics Letters, Droplet formation and stability of flows in a microfluidic T-junction by A. Gupta, S. Murshed and R. Kumar, ***94****, 16, 164107 Copyright (2009) American Institute of Physics. See Plate 3*

diameters) and as one usually uses microfluidic systems with the desire to make small droplets, we can further assume that the droplets will be of a volume of the order w_c^3, which leads to the maximum frequency of formation equal to $f_{max} = Ca_c\sigma/\mu_c w_c$. For example, breaking of water in hexadecane in a T-junction of the width of the channel $w_c = 100\,\mu m$, can be performed at a maximum frequency of $\sim 10^3$ Hz and a rate of flow of the aqueous phase of $10^{-9}\,m^3\,s^{-1} = 10^{-3}\,mL\,s^{-1} = 3.6\,mL\,h^{-1}$. Importantly, decreased interfacial tension and increased viscosity will lower the maximum rates of production of droplets.

5.4 Formation of Droplets and Bubbles in Microfluidic Flow-Focusing Devices

5.4.1 First Reports and Observations

The general idea of flow focusing is to focus a stream of one fluid using the flow of another, immiscible fluid. The archetype of this geometry includes a narrow nozzle with both fluids flowing through this small opening. In 1998 Ganan-Calvo (1998) introduced an axi-symmetric flow-focusing (or "capillary flow-focusing") device that comprised a capillary supplying the to-be-dispersed (focused) fluid. The terminus of this capillary was located a small distance upstream of the orifice. From all around the capillary, the outer, continuous (focusing) fluid was flowing, and its flow through the small opening focused the inner fluid into a jet with a diameter smaller than the diameter of the orifice.

This geometry was proven to produce sprays of fairly monodisperse droplets in air and foams of monodisperse bubbles in a liquid (when the inner fluid was a gas, and the outer fluid was a liquid).

In 2003 Anna, Bontoux, and Stone (2003) adapted this axi-symmetric architecture into a planar format of standard microfluidic chips that can be prepared via, for example, a typical lithographic process (Duffy *et al.*, 1998). Anna, Bontoux, and Stone fabricated their system in PDMS (Figure 5.14). The walls of the device were preferentially wetted by an organic phase (a 0.67 wt% of Span 80 in silicone oil of viscosity of 6 mPa s), thus allowing for preparation of water in oil emulsions. In this geometry the two side channels were supplied with the continuous phase of oil, while the inner channel conducted the aqueous phase and was focused by the outer streams into the orifice. This system produced aqueous droplets over a wide range of rates of flow of each of the two phases. Over particular regions of the space spanned by the speed of flow of the fluids, the system produced monodisperse droplets. The diagram of micrographs is reproduced in Figure 5.14. It is evident from this diagram, that the ratio of the rates of flow played the key role in the determination of the volume of the

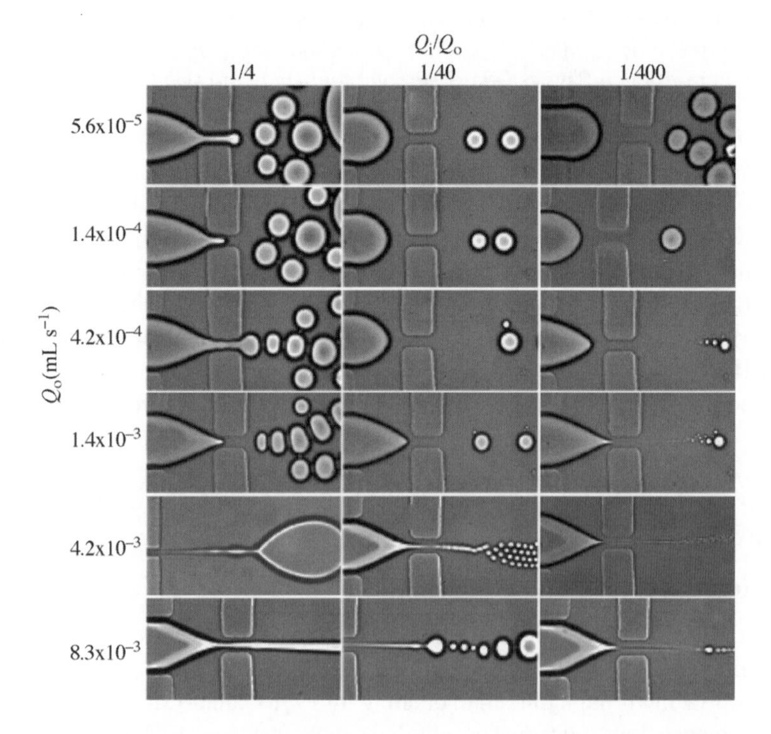

Figure 5.14 *A diagram of micrographs illustrating the operation of the system at different rates of flow of the two fluids (Anna, Bontoux, and Stone, 2003). Reprinted with permission from Applied Physics Letters, Formations of dispersions using a flow focusing in microchannels by S.L. Anna, N. Bonthoux and H.A. Stone,* ***82****, 3, 364–366 Copyright (2003) American Institute of Physics*

droplets produced in the system, while the overall rate of flow was less important in this aspect. In turn, it was the total speed of flow – and correspondingly, the value of the capillary number – that controlled the mode of breakup. Droplets formed only at rates of flow less than a critical value, at which the system crossed over to a jetting regime.

Some time later, in 2004, Garstecki *et al.* (2004) reported the use of a very similar system (Figure 5.15) for formation of bubbles of nitrogen in aqueous continuous phases. This study used three different aqueous fluids – water, and two water–glycerol mixtures of higher viscosity (6 and 11 mPa s). The aqueous phases contained 2% w/w of Tween 20 surfactant to stabilize the bubbles against coalescence. The qualitative observations were that: (i) the system produces monodisperse bubbles over a wide range of the rate of flow (Q) of the liquid and pressure (p) applied to the stream of gas; (ii) that increasing p, decreasing Q, and decreasing the viscosity (μ) of the continuous liquid, all resulted in an increase in the volume (V) of the bubbles; and (iii) that the value of the interfacial tension (changed between 37 and 72 mN m^{-1}) had no impact on V. Measurements of the volume showed that it was related to the above mentioned parameters via a simple relationship $V \propto p/Q\mu$.

It was argued that the lack of dependence of the volume of the bubbles on the value of the interfacial tension excludes the capillary number as the non-dimensional group governing the dynamics of breakup in the orifice. Further, they noted that the speed of collapse of the neck of the gaseous thread was proportional to the rate of flow of the

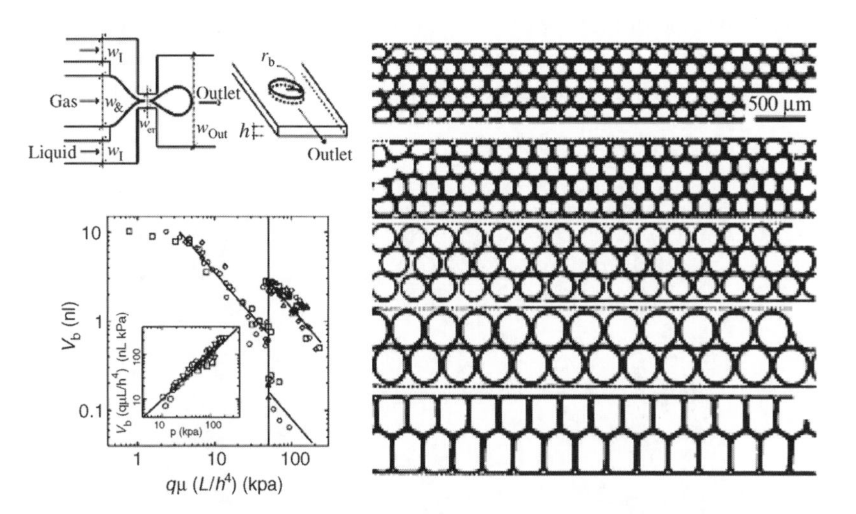

Figure 5.15 *Schematic illustration of the planar flow-focusing device (top left) used to produce bubbles of nitrogen. The volume of the bubbles plotted as a function of the product of the rate of flow of the continuous phase and its viscosity (bottom left). The independent control of volume of individual bubbles and volume fraction of gas in the produced foam allowed for assembly of flowing crystals of bubbles (right panel) (Garstecki et al., 2004). Reprinted with permission from Applied Physics Letters, Formation of monodisperse bubbles in a microfluidic flow-focusing device by P. Garstecki, I. Gitlin, W. DiLuzio et al.,* ***85****, 13, 2649–2651 Copyright (2004) American Institute of Physics*

liquid, yielding the collapse time $t_{collapse} \propto 1/Q$. The rate of growth of the bubble was proportional to the pressure applied to the stream of gas and inversely proportional to the viscosity of the liquid, as it controlled the resistance to flow in the outlet channel. Combining the time of collapse with the rate of growth of the bubbles yields the experimentally observed scaling of the volume of bubbles. Another observation was that the frequency of formation of bubbles was proportional to the product of p and Q. This, together with the scaling of the volume of the bubbles, provided for a simultaneous and independent control of the size of the bubbles and of the volume fraction of gas in the foam. Such independent control allowed for formation of regular assemblies of bubbles in the outlet channel, with the control of the structure of these "flowing crystals" via the adjustment of the volume of individual bubbles (see Figure 5.15). Also in 2004, Cubaud and Ho (2004) reported the use of a flow-focusing device having all channels with a square cross-section ($100 \times 100\ \mu m^2$) to produce bubbles and characterized in detail the regimes of operation of the system and the hydraulic resistance of flow of bubbles, slugs, and annular jets of gas surrounded by liquid. In the same year, Xu and Nakajima (2004) reported the use of a flow-focusing device etched in silicone (Figure 5.16) for formation of droplets of liquid. The channels immediately upstream of the junction had widths of 50 μm and all the channels had a uniform height of 5 μm. The width of the focused jet was determined both by the overall rate of flow of the fluid and the ratio of the rates of flow. For sufficiently large ratios of the rates of flow (Q_d/Q_c) the system produced a jet, which over a range of values of the rates of flow, broke up into highly monodisperse droplets with diameters of single micrometers to small tens of micrometers. At smaller values of the ratio of the rates of flow the jet became unstable and was breaking upstream of the orifice producing polydisperse droplets.

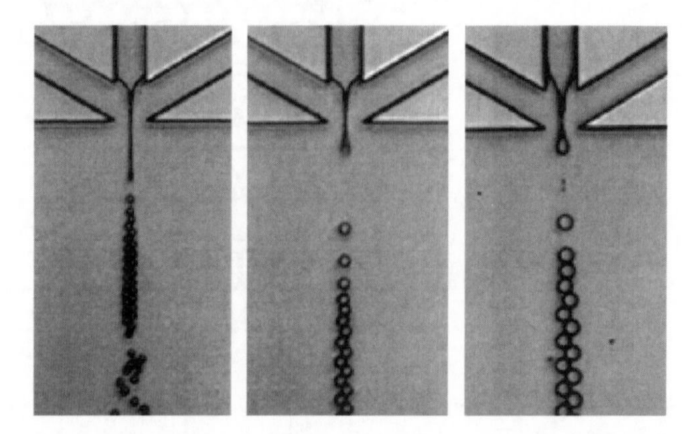

Figure 5.16 *A planar flow-focusing system etched in silicone. The system can be tuned to produce monodisperse droplets of diameters of single micrometers (Xu and Nakajima, 2004). Reprinted with permission from Applied Physics Letters, The generation of highly monodisperse droplets through the breakup of hydrodynamically focused microthread in a microfluidic device by Q. Xu and M. Nakajima, **85**, 17, 3726–3728 Copyright (2004) American Institute of Physics*

These first reports clearly presented the potential of microfluidic flow-focusing devices for preparation of highly monodisperse emulsions and foams. The mechanism of breakup of streams of gaseous and liquid fluids was not clear and demanded further research. We detail these mechanisms below with a division into formation of bubbles and droplets at a low viscosity contrast (viscosity of the dispersed phase being lower than the viscosity of the outer phase) and into breakup of streams at high viscosity contrasts (the viscosity of the inner phase being higher than the viscosity of the continuous liquid), which proves to be a much more complicated process.

5.4.2 Dynamics of Flow-Focusing Systems at Low Contrast of Viscosities

5.4.2.1 Formation of Bubbles

The breakup of a stream of gas in the microfluidic flow focusing device has proven to be an insightful model in understanding of the dynamics of operation of this system at low values of the capillary number and at a low value of the contrast of viscosities. For a gas–liquid system, the value of the contrast of viscosities is very low and the viscosity of gas plays no role in the whole process. The observations of the scaling of the volume of the bubbles formed in a microfluidic flow focusing device (Garstecki *et al.*, 2004) and in particular the lack of dependence of the volume of the bubbles on the value of interfacial tension has inspired more detailed studies of the breakup process. These observations were particularly intriguing, because the experiments (Garstecki *et al.*, 2004) were conducted in a regime of low values of the capillary number. Low values of capillary number, in turn, imply that interfacial forces dominate the viscous stresses. Why, then, does the value of the interfacial tension not influence the scaling of the volume of the bubbles?

In order to investigate this conundrum, in 2005 Garstecki, Stone, and Whitesides (2005) conducted a series of experiments in which they monitored the details of the evolution of the shape of the gas–liquid interface in the process of formation of a single bubble. Figure 5.17 shows the microfluidic geometry and a sketch of the shape of the "neck" that connects the stream of gas with the growing tip during the process of formation of a bubble.

Qualitatively, the whole process comprises the following stages. Firstly, the tip of the stream of gas enters the orifice, penetrates into the outlet channel and starts to inflate a bubble and at the same time it necks within the orifice. One can parameterize the rate of the necking by monitoring the evolution of the width of the neck over time (Figure 5.17). The evolution of the width of the neck over time is generic, that is, the temporal profile of the width of the neck is qualitatively similar for a range of viscosities of the continuous liquid, its rate of flow, pressure applied to gas, and the value of the interfacial tension. Initially, after the tip has penetrated the orifice, the width (w_m) of the neck is simply equal to the width of the orifice. Then it thins at a constant rate, and finally it snaps quickly (Figure 5.17). The generic character of the profile of collapse made it possible to quantitatively compare the rate of thinning of the neck during breakup. Formally, these Authors compared the value of dw/dt measured within the window in which the necking rate was constant. Quite surprisingly, but in line with the scaling (Garstecki *et al.*, 2004) of the volume of the bubbles, this rate of collapse depended only on the value of the rate

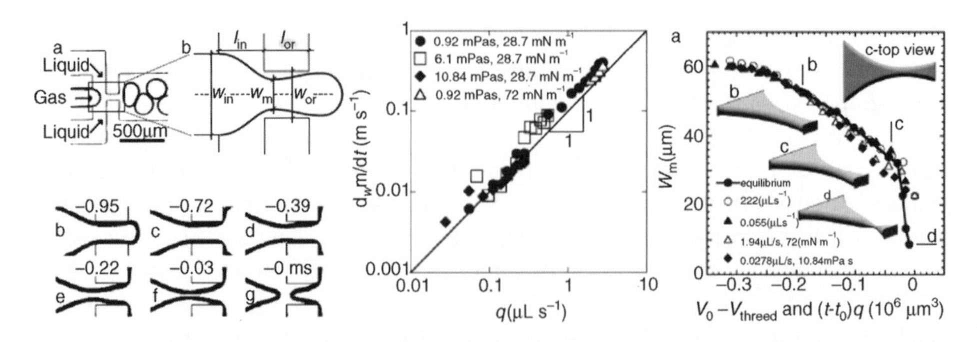

Figure 5.17 *Left panel: schematics that illustrate the definition of the width (w_m) of the neck during breakup and micrographs recorded with a fast camera that show the evolution of the shape of the interface during breakup. Middle panel: shows the speed of collapse measured as the temporal derivative of the width of the neck as a function of the rate of flow of liquid for a range of its viscosities and a range of values of the interfacial tension. Right panel: graph illustrating the overlay of the measured widths of the neck plotted as a function of the volume of liquid pumped into the orifice with those obtained via numerical minimization of the interfacial energy of the neck (Garstecki, Stone, and Whitesides, 2005). Reprinted with permission from Phys. Rev. Lett., Nonlinear dynamics of a flow-focusing bubble generator: An inverted dripping faucet by P. Garstecki et al.,* ***94****, 23, 234502 Copyright (2005) American Physical Society*

of flow of the liquid. That is, neither the change in the viscosity of the continuous fluid, nor the change in pressure applied to the gas nor the value of the interfacial tension produced any change in the rate of collapse of the neck. Additionally dw_m/dt was proportional to the rate of flow of the liquid, and over the range of rates of flow examined in the experiments, dw_m/dt was much smaller than the capillary speed estimated via dimensional analysis (Garstecki, Stone, and Whitesides, 2005). These observations suggested that the collapse of the neck indeed could not be driven by the interfacial pressure and the curvature of the interface. This, in turn, implies that the neck should be stable against a Rayleigh–Plateau type of instability. Indeed, numerical simulations (Garstecki, Stone, and Whitesides, 2005) of a catenoid interface spanned on the terminus of the inlet channel for gas, and on the terminus of the orifice, and subject to the confined geometry of the microfluidic flow-focusing junction have shown that it is stable. This stability means that for any given volume enclosed inside such a catenoidal shape, or equivalently, for any volume left outside of the interface, the shape adopts a stable configuration that minimizes its surface energy. For any such shape (i.e., for any value of the volume enclosed inside, or left outside of the interface) one can measure the width of the neck: w_m.

At low values of the capillary number the interfacial pressure dominates the shear stresses and the tip of the stream of gas also assumes an area-minimizing shape that effectively blocks the cross-section of the orifice. Thus, once the tip penetrates the orifice, any liquid that is pumped into the orifice will squeeze the neck. In the experiment (Garstecki, Stone, and Whitesides, 2005) the liquid was pumped into the orifice at a constant rate and thus the temporal evolution of the width of the neck could be translated into the dependence of the width of the neck on the volume of liquid pumped into the orifice. This allowed for a quantitative comparison of the evolution of the width of the neck recorded in the experiments with the width of the neck for the (equilibrium) shapes that minimize their interfacial energy. The very good agreement obtained in this comparison (see Figure 5.17) suggests that the breakup at low capillary numbers proceeds through a series of equilibrium shapes of the interface. As the speed of collapse is much smaller than the capillary speeds, any perturbations of the shape of the interface are relaxed on time scales much shorter than the interval needed for the whole collapse (equivalently for formation of a single bubble). This separation of time scales explains the observed monodispersity of the bubbles produced in the flow-focusing system at low capillary numbers.

As one can clearly see in Figure 5.17 the quasi-static evolution of the shape of the neck includes two distinct stages. In the first stage, the interface rests on the floor and on the ceiling of the orifice, all along its length. This stage can be called two-dimensional collapse, because the neck thins only in the plane defined by the channels on the chip. In the second stage, the interface detaches from the top and bottom wall of the orifice and forms a truly three-dimensional neck. The radius of this neck thins much faster upon further inflow of the continuous liquid into the orifice. Although also this three-dimensional shape is stable, the picture drawn by the evolution of shapes obtained via minimization of the interfacial area (Garstecki, Stone, and Whitesides, 2005) does not include the dynamics. In 2008 Dollet *et al.* (2008) showed via careful experiments that although indeed the two-dimensional collapse is governed by viscous and pressure terms, the dynamics of the quicker three-dimensional collapse is determined predominantly by the inertia of liquid and gas (Figure 5.18). This analysis formulated important guides for the formation of bubbles in microfluidic flow-focusing devices. Namely, decreasing the aspect ratio of the orifice (defined as the height of the channels divided by the width of the orifice), that is, flatter orifices, extend the temporal span of the 2D-collapse and make up for higher monodispersity of the bubbles. On the other hand, higher aspect ratios (i.e., orifices of cross-sections more similar to square ones) make the contribution of the 3D-collapse more important in determination of the volume of the bubbles. This makes up for higher rates of formation at the cost of slightly higher values of the standard deviation of the volumes of bubbles. This analysis lies in line with earlier experimental observations (Yobas *et al.*, 2006) of the scaling of the volume of aqueous droplets formed in the continuous phase of oil of viscosity of $\sim$20 mPa s, which showed that orifices of rectangular cross-sections produced droplets of volumes inversely proportional to the rate of flow of the continuous phase. At the same time, orifices of circular cross-sections produced droplets whose volumes scaled with the inverse of the third power of the rate of flow of the continuous liquid.

The squeezing mechanism of breakup (Garstecki, Stone, and Whitesides, 2005) has been found to describe the processes of formation of bubbles and droplets, provided that the viscosity of the droplet phase was lower than the viscosity of the continuous phase. For example, Cubaud *et al.* (2005) studied formation of bubbles of air in water and in aqueous solution of surfactant (sodium dodecyl sulfate, 8 mM L^{-1}) in a flow-focusing

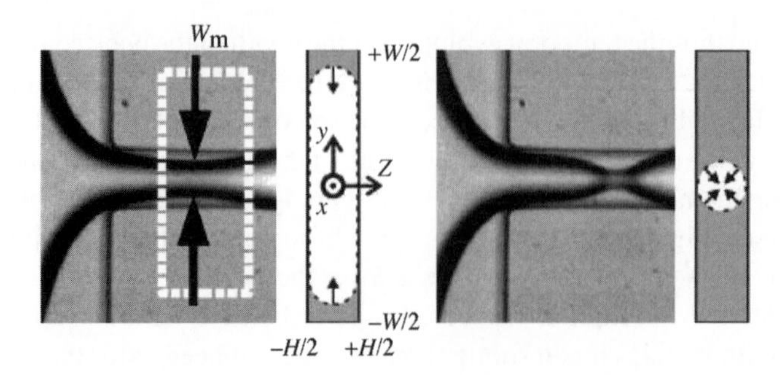

*Figure 5.18 Optical micrographs and schematics of the cross-sectional view of the neck connecting the stream of gas with a growing bubble at the intermediate and late stages of breakup (Dollet et al., 2008). Reprinted with permission from Phys. Rev. Lett., Role of the channel geometry on the bubble pinch-off in flow-focusing devices, **100**, 3, 034504 Copyright (2008) American Physical Society*

device having all channels of uniform, square cross-section ($h \times w = 100 \times 100\,\mu m^2$). The linear size (the length d) of the bubbles, or gaseous slugs, was found to depend only on the fraction of the rates of flow defined as $\alpha = Q_c/(Q_c + Q_d)$. The dependence was very simple and comprised an inversely linear relationship: $d/h = \alpha^{-1}$. The value of the interfacial tension (changed between 73 and 38 mN m^{-1}) did not influence the volume of the bubbles. Interestingly, the relationship reported by this group (Cubaud *et al.*, 2005) is equivalent to the scaling of the volume of droplets derived via the assumption of the squeezing mechanism of breakup (Garstecki *et al.*, 2006): $d/h = 1 + Q_d/Q_c$.

Fu *et al.* (2009) used the same geometry (all channels of square cross-sections $600 \times 600\,\mu m^2$) and extended the study of formation of bubbles to a range of viscosities of the continuous liquid. The results again confirmed the squeezing model (Cubaud *et al.*, 2005; Garstecki *et al.*, 2006) and showed a slight dependence of the time of collapse and of the volume of the bubbles on viscosity of the continuous liquid.

Lorenceau *et al.* (2006) used an axi-symmetric flow-focusing system (Figure 5.19) to form bubbles and found that, in agreement with the blocking–squeezing model, the volume of the bubbles divided by the volume of the constriction scaled linearly with the ratio of the rates of flow of gas and liquid.

Numerical simulations of formation of bubbles in flow-focusing geometries at low capillary numbers generally confirm the squeezing mechanism of breakup. For example, Jensen, Stone, and Bruus (2006) simulated an axi-symmetric system with an orifice of a circular cross-section and found that the volume of the bubbles was proportional to the ratio of the rates of flow of gas and liquid. As the gas was supplied at a constant pressure (not at a fixed rate of flow), the rate of flow of gas – and consequently the volume of the bubbles – was also a function of the resistance of the outlet channel. In particular, the volume of the bubbles was proportional to the fourth power of the diameter of the outlet capillary, and inversely proportional to the viscosity of the liquid. Similarly, Weber and Shandas (2007) analyzed formation of bubbles in a flow-focusing

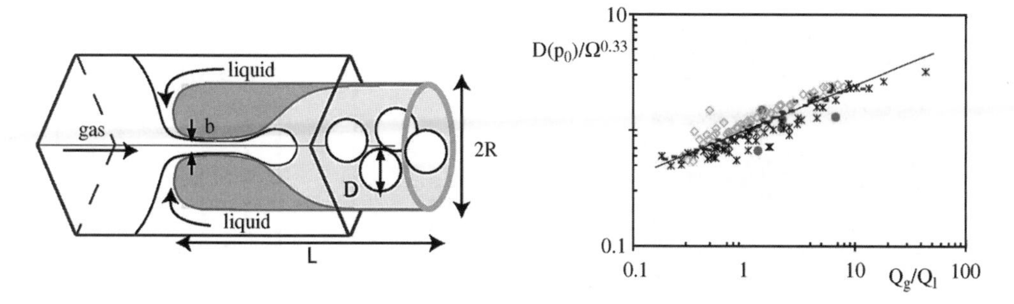

Figure 5.19 *An axi-symmetric system for formation of bubbles and the scaling of the diameter of the bubbles (the normalized cube root of the volume of the orifice) with the ratio of the rates of flow (Lorenceau et al., 2006). Reprinted from Lorenceau, et al., A high rate flow-focusing foam generator. Physics of Fluids **18**/9. Copyright (2006) with permission from AIP. See Plate 4*

device and confirmed the characteristic features of the squeezing model, showing that the volume of the bubbles is directly related to the volume of the orifice. As the mechanism of breakup allows for scaling the process down, this suggests that, in terms of principles, it should be possible to obtain monodisperse nano-bubbles and droplets in nano-scopic fluidic devices.

5.4.2.2 Formation of Droplets

In microfluidic flow-focusing devices formation of droplets is qualitatively different from the formation of bubbles. Liquid in the inner stream supports the shear stress exerted by the continuous phase flowing through the orifice. Thus, the shear stresses that have only marginal influence on the process of breakup of gaseous streams, in this instance, play an important role. The most comprehensive study of the formation of droplets in the planar flow-focusing device was recently performed by Lee, Walker, and Anna (2009). In the experiments these Authors examined a range of geometries of the flow-focusing junction by varying the height of the channels, width of the inlet channels, the orifice and of the outlet channel, and the distance between the end of the inlet channel for the droplet phase to the orifice. Further, the experiments comprised tests of formation of aqueous droplets in organic continuous liquids covering a wide range of viscosities: from 10 to 362 mPa s. This set of experiments spans over a wide range of ratios of viscosity from $\mu_d/\mu_c = 10^{-1}$ to 2.7×10^{-3}. Unfortunately, in the experiments the ratio of the rates of flow was held fixed at a value of $Q_d/Q_c = 0.025$, and only the total rate of flow was varied.

Qualitatively, the experiments revealed four distinct modes of operation of the system: squeezing, threading, dripping, and jetting. In the squeezing regime the tip of the droplet phase effectively blocks the cross-section of the orifice (Figure 5.20, top micrograph). In this regime the droplets are typically larger than the size of the orifice. As the rate of flow of the fluids is increased the shear stresses exerted by the continuous phase deform the tip of the dispersed phase into a more slender, pointed shape with a

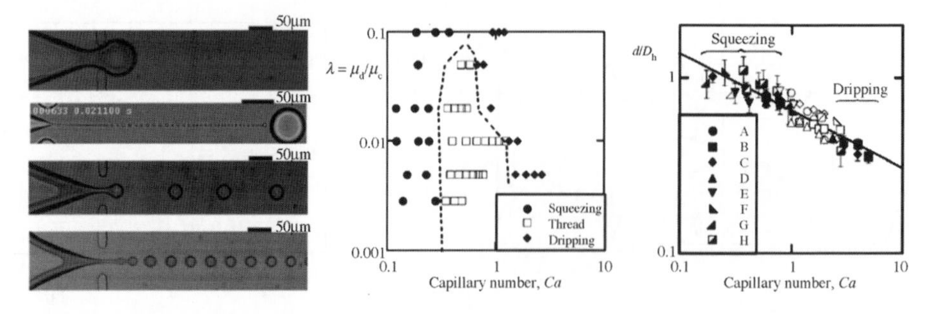

Figure 5.20 *Left panel: from top to bottom – micrographs of the system in the squeezing, threading, dripping, and jetting regimes. Central panel: diagram of the regimes of formation of droplets spanned by the capillary number and the contrast of viscosities of the two fluids. Right panel: scaling of the diameter of the droplets with the capillary number; the solid line follows the relationship* $d/D_h \sim Ca^{-0.33}$ *(Lee, Walker, and Anna, 2009). Reprinted from Lee, W., Walker, L. M. and Anna, S. L. Role of geometry and fluid properties in droplet and thread formation processes in planar flow focusing. Physics of Fluids* ***21****/3. Copyright (2009) with permission from American Institute of Physics*

terminus that is smaller than the width of the orifice. In this – dripping (Figure 5.20, third micrograph) – regime droplets are periodically formed in the orifice and their diameter can be several times smaller than the width of the orifice. Further increase of the rate of flow induces a transition to jetting (Figure 5.20, bottom micrograph), a regime in which the tip of the droplet phase extends downstream of the orifice and small droplets are formed at its terminus.

Additionally, Anna and Mayer (2006) reported that over a finite range of concentrations of surfactant, an additional mode of breakup separates the squeezing and dripping regimes. In this – threading – regime (Figure 5.20, second micrograph) droplets are formed in a manner similar to the one observed in the squeezing/dripping modes, but after each droplet is formed a long and thin thread of the inner liquid is dragged behind the droplet. This thread subsequently breaks up into a group of tiny secondary droplets. As shown in the central panel of Figure 5.20, the squeezing regime extends from the lowest values of the capillary number to $Ca \sim 0.5$, without any significant dependence of the value of the critical capillary number on the contrast of viscosities. Transition to dripping occurs at $Ca \sim 1$, also without much influence of the viscosity of the continuous phase. Notably, these values are larger than in the case of the T-junction, and than in the case of formation of bubbles, where typically the squeezing regime does not extend beyond $Ca \sim 10^{-2}$.

A very important contribution presented in the report by Lee, Walker, and Anna (2009) lies in the construction of an expression for the capillary number that includes the geometrical parameters of the flow-focusing junction. This capillary number proved successful in describing the volume of the droplets produced in the device. In the right panel of Figure 5.20 one can see all the series of data, obtained for a range of aspect ratios of the geometry of the device and for a wide range of viscosities of the continuous fluid group around a line given by $d/D_h \propto Ca^{-0.33}$. This result is very useful from a practical point of view, as it can be treated as a predictive tool that can guide the design of flow-focusing junctions for production of droplets of desired volumes. This form of the scaling of the droplet volume suggests that both the pressure drop along the growing droplet (the effect that dominates in the purely squeezing regime) and the shear stress exerted on the growing droplet by the flow of the continuous fluid, both play a role in the process of formation of droplets.

Because the ratio of the rates of flow of the two phases was held constant, if the droplets formed purely in the squeezing regime, their volume should be constant, independent of the total rate of flow of the two phases. On the other hand, if shear dominated the process of breakup, one would expect a scaling closer to the classical relationship of the diameter of the droplet being inversely proportional to the value of the capillary number ($d/D_h \propto Ca^{-1}$). Interestingly, the observed value of the exponent of (-0.33) is similar to the ones observed in the dripping mode of breakup in T-junctions.

Similar results were reported by Tan, Cristini, and Lee (2006) who used a microfluidic flow-focusing device in which the outlet channel gradually widened downstream of the orifice. In spite of the fact that the device was designed to concentrate the shear stress in the orifice, the scaling of the diameters of aqueous droplets ($\mu_d \sim 1$ mPa s) formed in a continuous phase of oleic acid ($\mu_c \sim 20$ mPa s) were inversely proportional to the rate of flow of the continuous fluid raised to the power of 1/3. Also, Ong *et al.* (2007) reported the same exponent obtained via simulations of formation of aqueous droplets in an oil ($\mu_c \sim 20$ mPa s) in a planar flow focusing device having an orifice with a rectangular cross-section. The same system, however, with a circular orifice produced droplets of diameters proportional to Ca^{-1}. Clearly, the processes of formation of droplets in viscous continuous fluids need further research to provide a thorough understanding of the mechanism of breakup and to provide clear guidelines for experimentalists.

5.4.3 Flow Focusing: Formation of Viscous Droplets

As we will show in this section, the breakup of viscous threads into droplets, in a continuous phase that has a lower viscosity than the droplet phase is more complicated and less understood than the inverse situation described in the last section. It is difficult to draw an exact boundary between the two cases (for the simple dynamics of systems characterized by a low value of the viscosity contrast and the opposite case). When the viscosity of the inner phase is not too high, the breakup of the viscous thread at low values of the capillary number resembles the squeezing model. For example, Funfschilling *et al.* (2009) reported a microPIV study of the velocity field in the continuous aqueous phase ($\mu_c \sim 1$ mPa s) during the formation of droplets comprised of a silicone oil of viscosity of $\mu_d \sim 6$ mPa s. The capillary number characterizing the flow had a small value varying between 10^{-3} and 10^{-2}. The observed evolution of the shape of the interface and the velocity profiles agreed well with the squeezing model of breakup.

To date there are only two comprehensive studies of the effect of the viscosity of the droplet phase on formation of droplets in microfluidic flow-focusing devices. Nie *et al.* (2008) studied formation of organic droplets of viscosities ranging from 10 to 500 mPa s in water (1 mPa s) in a standard flow-focusing geometry (Figure 5.21). They studied the dynamics of the system and the scaling of the volume of the droplets for capillary numbers (defined by the flow of the outer phase) ranging between 5×10^{-4} and 10^{-1}. At low rates of flow the operation of the system resembled the squeezing mechanism of breakup: the tip of the droplet phase blocked the cross-section of the orifice and the droplets were typically larger than the size of the orifice with a narrow distribution of volumes. At moderate rates of flow, the system produced droplets in the dripping regime. In this mode, the diameter of the droplets was comparable to the width of the orifice, and the tip of the droplet phase did not block the orifice completely. The dripping regime also produced monodisperse droplets. Finally, at the highest rates

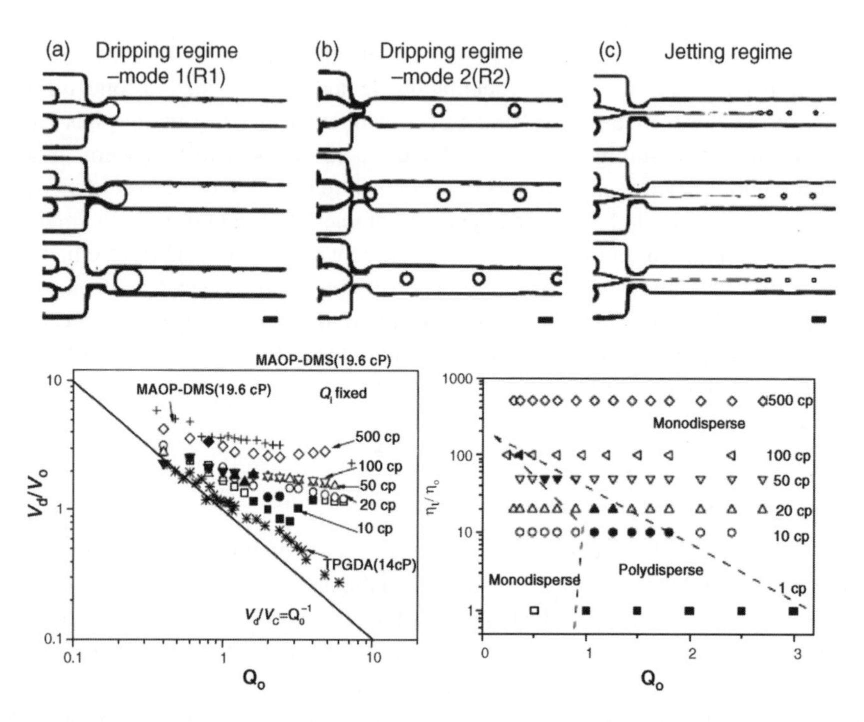

Figure 5.21 *The three modes of breakup observed by Nie et al. (2008) and the scaling of the volume of the droplets with the rate of flow of the continuous liquid (marked as Q_o on the graph). The numbers on the graph indicate the viscosity of the droplet phase. The bottom right diagram shows the ranges of rate of flow of the continuous liquid as a function of the viscosity contrast over which the system produced monodisperse droplets (Nie et al., 2008). Reprinted with permission from Microfluids and Nanofluids, Emulsification in a microfluidic flow-focusing device: effect of the viscosities of the liquids by Z. Nie, **5**, 5, 585–594 Copyright (2008) Springer Science + Business Media*

of flow of the two phases the stream of the inner phase formed a long jet that broke far downstream of the orifice and produced droplets typically much smaller than the width of the orifice. These droplets were monodisperse only over narrow ranges of rates of flow.

For liquids of lowest viscosity (e.g. silicone oil of viscosity 10 mPa s) the scaling of the volume of the droplets showed a relationship of $V \propto Q_c^{-1}$ as expected in the squeezing mode. As the viscosity of the droplet phase increased, the volume of the droplets exhibited progressively smaller sensitivity to the ratio of the rates of flow. For the most viscous dispersed phase (silicone oil of viscosity 500 mPa s) the volume of the droplets was equal to a few times the volume of the orifice, and was almost independent of the rate of flow of the continuous phase. An experiment performed for a fixed ratio of the rates of flow ($Q_c/Q_d = 60$) showed a similar trend for the dependence of the volume of the droplets on the value of the capillary number. For the smallest viscosities, the volumes of the droplets followed approximately a scaling of $V \propto Ca^{-1}$, a result similar to those obtained for example by Lee, Walker, and Anna (2009) ($d \propto Ca^{-0.33}$), suggesting that the pressure drop along the growing droplet (the squeezing effect) plays an important role in the process but that the shear stresses also contribute to the process. For higher viscosities the dependence on *Ca* became gradually weaker, with the volumes of the most viscous silicone oil even increasing slightly with increasing *Ca* (Nie *et al.*, 2008b).

The important, practical, implication of the report by Nie *et al.* (2008) is that all the tested liquids could be used to produce monodisperse droplets. The volume of these droplets could be tuned within a factor of ten from volumes of ~0.2 to ~15 times the volume of the orifice.

Almost simultaneously, Cubaud and Mason (2008) reported experiments on the formation of viscous droplets in a planar flow-focusing system that had all the four channels having the same, square, cross-section of $250 \times 250\,\mu m^2$. The ratio of viscosities was varied over a very wide range [$\mu_d/\mu_c \in (22, 1484)$]. They identified five different regimes of operation of the system. These, roughly in the order of increasing ratio of the rate of flow of the inner phase to the rate of flow of the outer phase included: (i) threading, (ii) jetting, (iii) dripping, (iv) tubing, and (v) viscous displacement.

In Figure 5.2 we reproduced the micrographs of all of these modes together with a phase diagram of the operation of the system, spanned by the capillary numbers calculated for the flow of each of the two phases. In the jetting regime, the width of the jet, and the diameter of the droplets produced in this mode was found to be proportional to the square root of the ratio of the rates of flow (for ratios of viscosity between 22 and 440). In the dripping regime the volume of the droplets was larger than the cube of the width of the channel, and formed slugs squeezed between the walls of the channel (Figure 5.22). This regime is visually reminiscent of the formation of bubbles and droplets in the squeezing regime at low values of capillary numbers. Interestingly, in contrast to other reports, these Authors found that it is the rate of flow (or capillary number) of the inner phase that determines the transition from the dripping to the jetting regime (see Figure 5.22, diagram). Within this range, the scaling of the length of the

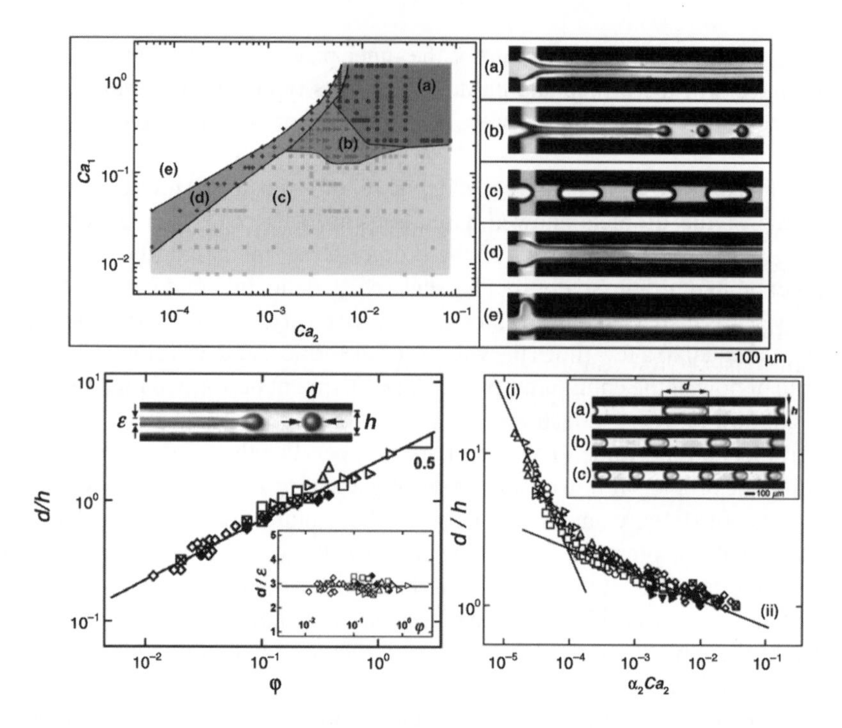

Figure 5.22 *A diagram of the mode of operation of a cross-flow junction (a flow-focusing junction with all the channels having square cross-section). Top left panel: typical capillary number-based flow map with flow patterns: (a) threading (●), (b) jetting (▶), (c) dripping (■), (d) tubing (◆), and (e) viscous displacement (+), fluid pair G38. The micrographs shown in the top right panel illustrate these various regimes. The graphs in the bottom panel illustrate the scaling of the diameter (or length) of the droplets formed in the jetting and dripping regimes as a function of the ratio of rates of flow ($\phi = Q_d/Q_c$) and the factor of the fraction of the rate of flow of the outer phase [$\alpha_2 = Q_c/(Q_c + Q_d)$] and the capillary number calculated for the flow of the continuous phase. Reprinted with permission from Cubaud, T., and Mason, T., Capillary threads and viscous droplets in square microchannels, Physics of Fluids, 20, 053302. Copyright (2008) with permission from American Institute of Physics*

droplets of various viscosities [$\mu_d/\mu_c \in (22, 1484)$] collapsed onto a single master curve when plotted against the product of the fraction α_2 of the rate of flow of the outer phase [$\alpha_2 = Q_c/(Q_d + Q_c)$] and the capillary number calculated for the flow of the outer phase $Ca_2 = Q_c\mu_c/\sigma h^2$. Interestingly, the length of the droplets – and thus the volume of the droplets, which in this regime is approximately linearly related to the length of the droplet – exhibited two distinct regimes of dependence on the value of the product of the liquid fraction and the capillary number. Small droplets ($d/h < 2.5$) had lengths proportional to $(\alpha_2 Ca_2)^{-0.17}$ while the length of large droplets ($d/h > 2.5$) depended much stronger on the value of the product $(\alpha_2 Ca_2)$: $d/h \propto (\alpha_2 Ca_2)^{-1}$. Clearly, as illustrated in the inset of Figure 5.22, formation of viscous droplets in this system is completely distinct from the squeezing, or rate of flow controlled breakup. For a fixed value of the ratio of the rates of flow of the two phases, the volume of the droplets depends critically on the total rate of flow, or, equivalently, on the value of the capillary number.

5.5 Practical Guidelines for the Use of Microfluidic Devices for Formation of Droplets

Microfluidic systems offer convenient tools for formation of ultra-monodisperse foams and emulsions. At low values of the capillary number, the process of breakup is dominated by the blocking and squeezing mechanism and the volume of the bubbles and droplets is predominately determined by the ratio of the rates of flow of the two phases. At capillary numbers above $\sim 10^{-2}$ for the T-junction, and above $\sim 10^{-1}$ for the flow-focusing devices, the shearing effects begin to play a significant role in the process of breakup and the diameter of the droplets typically displays a dependence on the value of the capillary number. The effects of confinement, however, continue to be important and the decrease in the volume of the droplets upon increase of the capillary number is typically characterized by a small exponent: $d \propto Ca^{x}$, with x ranging from -0.25 to -0.5, with most observations centered around $x \approx -1/3$. As was shown in the report by Abate *et al.* (2009) who studied a range of different configurations for the microfluidic drop-generators (from T-junctions, through the variation of the angles at which the channels meet, through a pinned-jet-junction and a set of flow-focusing devices), liquids of similar viscosities follow the squeezing model of breakup in all the types of devices.

5.5.1 Types of Fluids

Volumes of droplets produced in the blocking–squeezing model do not depend on the viscosities of the two phases. This is a consequence of the fact that the squeezing model relies on the separation of time scales for the (fast) equilibration of the flow field and the shape of the interface and for the (slow) process of breakup. This assumption is typically true for fluids of low viscosities (gases, water, and low-viscosity oils). Introduction of a highly viscous phase should slow down the capillary waves and strongly limit the applicability of the blocking–squeezing model. Surprisingly, a high viscosity of the continuous phase does not dramatically modify the dynamics of breakup: the results show only the relatively weak dependence of the volume of the droplets on the value of the capillary number ($d \sim Ca^{-1/3}$). The picture is different when the droplet phase has a high viscosity in comparison with the viscosity of the continuous phase. As the contrast of viscosities increases, the volumes of the droplets show less sensitivity to the rates of flow of each of the two immiscible phases.

Importantly for the applications in production of solid and gel particles, microfluidic systems allow for a high degree of freedom in the choice of the fluids. It is possible to form droplets of metals, both aqueous and organic solutions of monomers, blends of polymers, gels, and even supercritical fluids (Marre *et al.*, 2009). Some of these fluids, especially gels and solutions of polymers, often exhibit non-Newtonian properties. These fluids may exhibit coefficients of viscosity that depend on the rate of shear. Most polymeric liquids are shear thinning, that is, their viscosity become less as the shear rate increases. In addition, these fluids may support normal stresses, or in other words, exhibit elastic behaviors. Such fluids oppose rapid extension, which becomes important in the flow of non-Newtonian fluids through microfluidic contractions and in the processes of breakup of streams of viscoelastic fluids into droplets (Pipe and McKinley, 2009). These additional effects change and complicate the dynamics of the microfluidic droplet generators.

A number of researchers reported observations on formation of droplets of non-Newtonian liquids in microfluidic devices. Husny and Cooper-White (2006) studied formation of droplets of aqueous solution of polyethylene glycol (PEG) of molecular weight of 10^5–10^6 g mol^{-1} in a continuous liquid of silicone oil (of viscosity from 5 to 50 mPa s). They observed clear effects of the non-Newtonian character of the solutions of PEG: prolonged thinning of the necks connecting the stream of the droplet phase with a breaking-off droplet. These necks, or threads, subsequently break up into tiny secondary droplets, in a manner similar to the processes observed in tip-streaming (Anna and Mayer, 2006). As the experiment was conducted in a T-junction that had a very wide channel in comparison with the width of the inlet for the droplet phase, the condition of breakup was determined by the balance of the shear and capillary stresses. These Authors succeeded in constructing an analytical model based on this assumption that recovered the size of the droplets measured experimentally. This report clearly demonstrates that it is possible to form droplets of non-Newtonian fluids and exercise efficient control over the volumes of the primary droplets. The elastic effects are limited to formation of a large number ratio of tiny secondary droplets.

Steinhaus, Shen, and Sureshkumar (2007) used the same set of liquids to study breakup of non-Newtonian streams in a planar flow-focusing device and obtained similar results. The breakup produced primary droplets of well-controlled volumes and a number of secondary droplets created via breakup of the long and thin thread that remained after creation of the primary droplet (Figure 5.23).

In a series of reports Arratia and coworkers. (Arratia, Gollub, and Durian, 2008; Arratia *et al.*, 2009) provided systematic observations of the effect of elasticity of the dispersed phase (aqueous solutions of glycerol and polyacrylamide) on the dynamics of breakup in a continuous organic liquid (mineral oil, shear viscosity ~200 mPa s) in a planar microfluidic flow-focusing junction. They varied the molecular weight of the polymer between 10^3 and 10^7 g mol^{-1}. The concentrations of the polymer were always the same, and so were the shear viscosities of the solutions, including one, reference solution, that did not contain polymer and exhibited a Newtonian viscosity of 230 mPa s. The general observation is that the Newtonian fluid forms monodisperse primary droplets, and in the process of breakup only a short thread behind the droplet is created. This thread typically forms a single secondary droplet. As the molecular weight of the polymer added to the droplet phase increases, the length of the thread left behind the primary droplet also increases. As reported earlier by others (Husny and Cooper-White, 2006; Steinhaus, Shen, and Sureshkumar, 2007) these threads subsequently

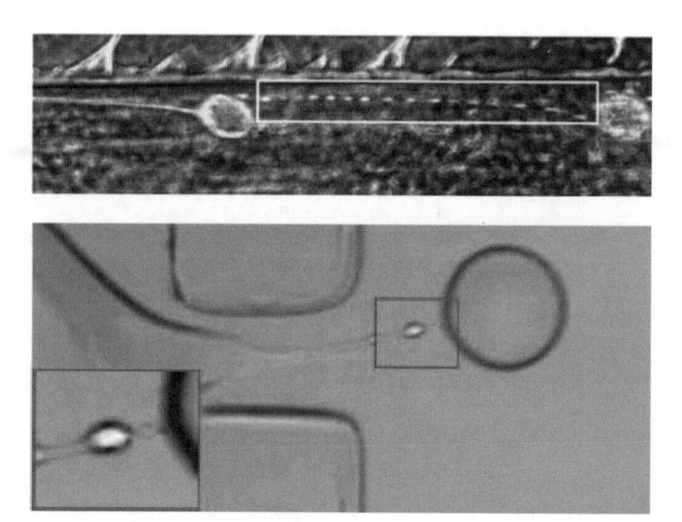

Figure 5.23 *Upper micrograph shows the thread left behind a broken-off droplet in microfluidic T-junction (Husny and Cooper-White, 2006). Reprinted with permission from Journal of Non-Newtonion Fluid Mechanics, The effect of elasticity on drop creation in T-shaped microchannels by J. Husny and J. J. Cooper-White,* ***137****, 1-3, 121–136 Copyright (2006) Elsevier Ltd. The bottom micrograph shows an analogous picture for a flow-focusing system breaking a non-Newtonian liquid into droplets (Steinhaus, 2007). Reprinted with permission from Steinhaus, B., Shen, A., and Sureshkumar, R., Dynamics of viscoelastic fluid filaments in microfluidic devices, Physics of Fluids 19, 073103. Copyright (2007) with permission from American Institute of Physics*

break up into a group of small secondary droplets. As can be seen in the bottom panel of Figure 5.24 the length of the polymeric chains does not substantially alter the values of the rates of flow at which the system crosses between different regimes of operation: from "slugging" at low values of the ratio of the rates of flow of the continuous and droplet phases, through dripping to "tip streaming" – a mode in which a sharply pointed tip of the droplet phase emits small droplets.

5.5.2 Surfactants

At equilibrium, the role of surfactants is only a modification of the value of the coefficient of interfacial tension. Surfactants also slightly modify the viscosity of the liquid in which they are dissolved. Another effect of surfactants is in stabilization of emulsions against coalescence. Surfactants protect the interface and introduce repulsive interactions between adjacent droplets. These interactions can be either steric, arising from the increase in entropy of packing of the tails of the surfactant upon compression of two interfaces against each other, or electrostatic if the surfactants are ionic.

During formation of droplets or bubbles, and during their flow in microchannels, the interfaces are not at equilibrium and additional effects may come into play. For example, during creation of droplets, fresh interface between the immiscible fluid is produced constantly, at a rate that is proportional to the surface area of the droplets and the frequency of their formation. If surfactants are dissolved in the liquids at sufficient

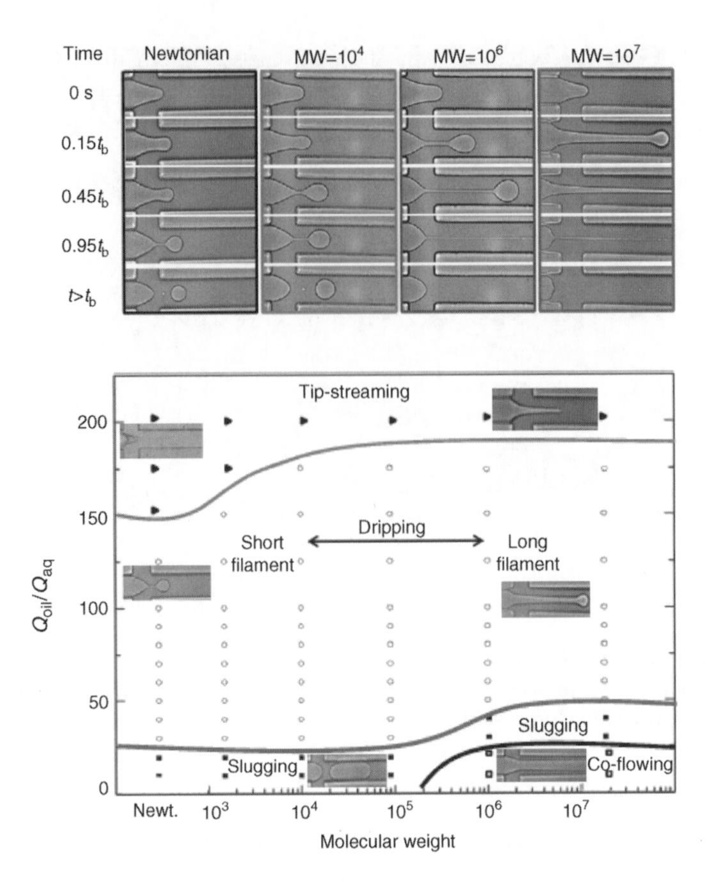

Figure 5.24 *Series of micrographs illustrating the process of formation of non-Newtonian droplets as a function of the molecular weight of the polymer dissolved in the droplet phase. The diagram shows the regimes of operation of the system as a function of the ratio of the rates of flow of the non-Newtonian droplet phase and the continuous oil phase, and molecular weight of the polymer (Arratia et al., 2009). Reprinted with permission from New Journal of Physics, The effects of polymer molecular weight on filament thinning and drop breakup in microchannels by P.E. Arratia, L.A. Cramer, J.P. Gollub and D.J. Durian,* ***11****, 11, 115006 Copyright (2009) Institute of Physics. See Plate 5*

concentrations, they adsorb to the freshly created interface. This process requires a finite time to complete, as it is limited by the process of diffusion of molecules or aggregates (micelles) of surfactant to the interface. Further, surfactants already adsorbed to the interface, when subjected to a shear flow may relocate along the interface. This can lead to formation of a gradient of concentration of surfactants and, as a consequence, to the appearance of gradients of interfacial tension. Such gradients create tangential stresses (Marangoni stresses) and may affect the dynamics of flow around the interface.

Baret *et al.* (2009) studied the kinetics of adsorption of surfactants to the interface of droplets created in a microfluidic flow-focusing device. The report included an observation that varying the concentration of surfactant did not produce changes in the volume of the droplets created in the junction. The stability of the droplets depended on the time of their incubation prior to the test against coalescence. The interval after which the droplets became stable depended on the concentration of surfactant. These observations suggest that during formation of the droplets the interface was indeed free of surfactants, which only gradually adsorbed onto the interface via diffusion from the bulk of the liquid. Baret *et al.* (2009) have also shown that when the droplet travels through a microchannel, surfactants adsorb first at the rear of the droplet and that even late into the process of adsorption, the concentration of surfactant at the interface is always greater at the rear and lower at the front of the droplet.

The uneven distribution of surfactant along the interface can have pronounced effects on both the dynamics of formation of droplets and on their transport in microfluidic channels. Anna and Mayer (2006) have shown that at high concentrations (a few-fold higher than the value of the CMC) of surfactant in the inner, aqueous phase the planar flow-focusing system exhibits a breakup mode in which creation of the large primary droplet is followed by formation of a long and thin thread that subsequently breaks into tiny droplets (Figure 5.25). At very large ratios of the rate of flow of the continuous and discontinuous phases, that is, when the droplet phase is pushed into the system at a rate much smaller than the rate of flow of the continuous liquid, the system transits into tip-streaming. Tip-streaming was first discussed by Taylor (1934) for a droplet positioned in the stagnation point of a four-mill configuration. The extensional flow around the droplet shears the molecules of surfactant towards two opposite termini of the droplet. An increased concentration of surfactant decreases the value of interfacial tension up to a point at which it is so low, that the tip yields to the shear stresses exerted by the external flow and ejects a long and thin thread, which breaks into tiny drops. This thread and droplets carry away the surfactant that accumulated at the tip and caused the ejection. The interface is free again and the process starts all over again via slow adsorption of surfactant onto the interface and redistribution of surfactant along the interface towards the tips. The flow upstream of the orifice in a flow-focusing junction is very similar to one half of the four-mill configuration and indeed the tip of the droplet phase is observed

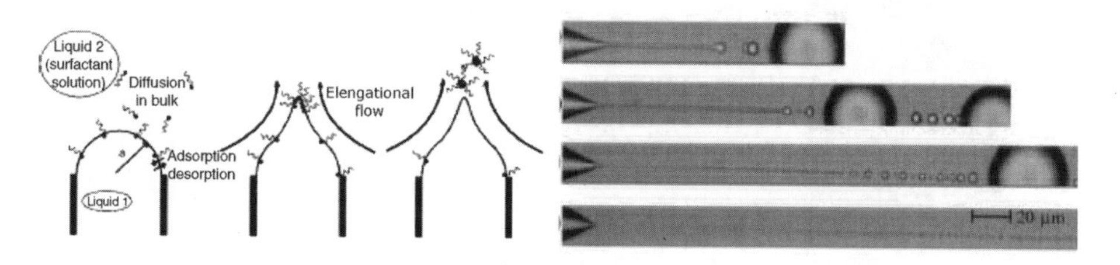

Figure 5.25 A schematic illustration of the process of adsorption of surfactant on the interface and simultaneous shearing of the surfactant molecules towards the tip of the droplet phase in microfluidic flow-focusing junction. Accumulation of surfactant at the tip causes periodic ejections of a thin thread. The micrographs in the right panel illustrate formation of a thread following an ejection of a large primary droplet. Reprinted from Anna, S. and Mayer, H., Microscale tipstreaming in a microfluidic flow focusing device, Physics of Fluids, 18, 121512. Copyright (2006) with permission from American Institute of Physics

to periodically emit a series of very small droplets (Anna and Mayer, 2006). Similar behaviors are also predicted numerically for a liquid emitted from a tube into a co-flowing outer fluid (Suryo and Basaran, 2006).

Hashimoto *et al.* (2008) have shown a rich variety of interfacial instabilities that also resemble tip streaming, but which occur during the flow of already pre-formed droplets in microchannels. When both the aqueous droplet and the continuous oil phase contained surfactants (Tween 20 and Span 80, respectively), the droplets traveling in the microfluidic channels undergo shear-driven interfacial instabilities (Figure 5.26). The shear exerted on the interface adjacent to the top and bottom walls of the microchannels pulled surfactant towards the rear of the droplet. The increased concentration of surfactant lowered the interfacial tension and allowed for thin sheets of liquid to be drawn from the droplets (see the schematic diagrams in Figure 5.26). These sheets, as they were not wetting the walls of the microchannel, subsequently became unstable and coiled into threads that broke up into droplets with diameters several times smaller than the height of the channel. These instabilities produced

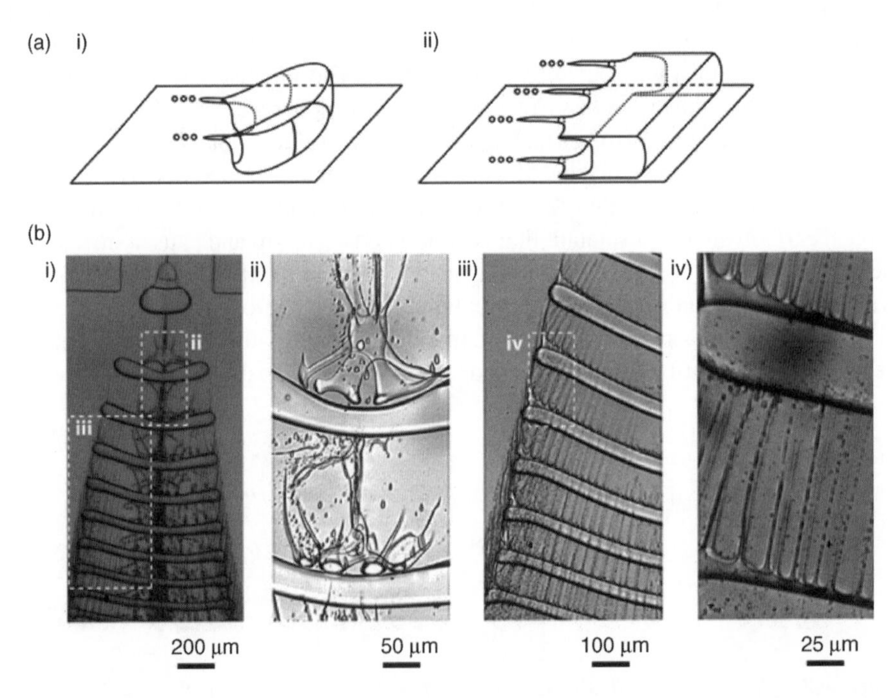

Figure 5.26 *Top illustrations (a) show schematically the shapes of the droplets translating in the microfluidic channel and undergoing a shear driven interfacial instability. The micrographs in the bottom row (b) show the instabilities (Hashimoto et al., 2008). Reprinted with permission from Soft Matter, Interfacial instabilities in a microfluidic Hele-Shaw cell by M. Hashimoto, P. Garstecki, H.A. Stone et al., **4**, 7, 1403–1413 Copyright (2008) Royal Society of Chemistry*

beautiful flow patterns in the microfluidic Hele–Shaw cell (Figure 5.26). Importantly for practical applications, if the tiny droplets are not of interest, particular combinations of surfactant may complicate the dynamics of flow and significantly alter the shape of the droplets. Care should be exercised in the choice of surfactants.

5.5.3 Wetting

As already indicated in the preceding chapter, the wetting properties of the walls of the channels with the two immiscible liquids are of utmost importance for guaranteeing stable formation of droplets in microfluidic devices. It is an absolute prerequisite for stable operation of planar microfluidic droplet generators that the droplet phase does not wet the walls of the channels. In other words, the continuous phase needs to wet the walls preferentially, so that the dispersed phase is always separated from the walls of the channels by at least a thin film of the matrix liquid. This requirement is specific to planar systems. In axi-symmetric devices it is possible to kinetically engulf the inner phase with the outer fluid and not to allow for contact of the droplet phase with the walls of the device with the sole use of hydrodynamics.

Even partial wetting of the walls of the device by the dispersed fluid completely alters the operation of the system. This is because patches of fluid wetting a solid substrate exhibit an increased stability against breakup and because motion of such wetting patches of fluid is typically erratic due to the pinning and de-pinning dynamics of the triple contact line. In 2003, Dreyfus, Tabeling, and Willaime (2003) studied formation of aqueous droplets in the continuous phase of tetradecane with addition of different amounts of surfactant (Span 80). At low concentrations of surfactant, water partially wetted the silicone walls of the microfluidic channels. Above a concentration of $\sim 10^{-5}$ w/w of Span 80 in tetradecane, the contact angle of a water droplet on silicone immersed in the continuous fluid becomes 180° and water does not wet the walls of the channels. Dreyfus, Tabeling, and Willaime (2003) reported that above this critical concentration of surfactant the system produced droplets (Figure 5.27). If the concentration of surfactant was too low, however, and water partially wetted the walls, the system produced erratic flow patterns (Figure 5.27).

Li *et al.* (2007) performed a screen of the operation of flow-focusing devices as a function of the surface energy of the walls of the channels. The differently modified surface chemistries of polydimethylosiloxane (PDMS) produced the following values of advancing contact angles of water (in air): 80, 92, 105, and 112°. The contact angles of the continuous phase (2 wt% solutions of Span 80 in mineral oil) on the same surfaces were equal to 34, 33, 32, and 30°. These values allowed for calculation of the contact angle of water on the same substrates in the presence of a surrounding continuous oil phase. This calculation (Li *et al.*, 2007) indicated that none of these substrates, at equilibrium, were wetted by water, that is, all of the tested surfaces were completely wetted by the continuous phase. Still, only the substrates characterized by the contact angle of water (in air) greater than 92° produced water in oil emulsions. The less hydrophobic channels (80 and 92°) produced erratic co-flow of water and oil. This observation pointed out that dynamic wetting effects and pinning are important and set

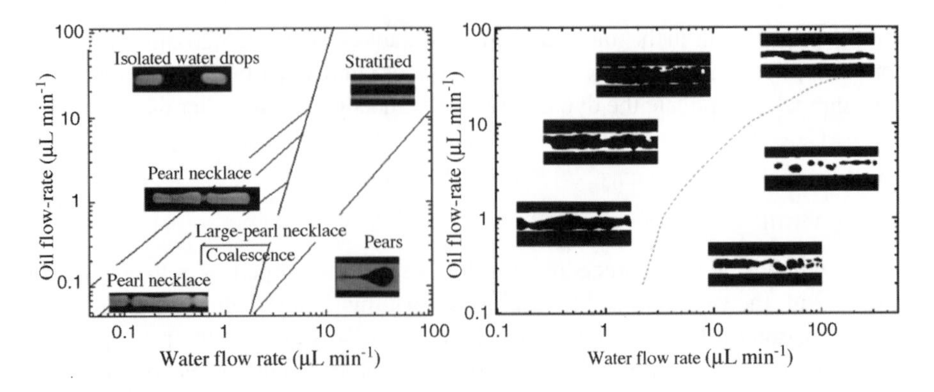

Figure 5.27 The diagrams of the modes of operation of a microfluidic droplet generator when the continuous fluid wets the walls preferentially (left graph) and when the droplet phase partially wets the channels (right graph) (Dreyfus, Tabeling, and Willaime, 2003). Reprinted with permission from Phys. Rev. Lett., Ordered and Disordered Patterns in Two-Phase Flows in Microchannels by R. Dreyfus, P. Tabeling and H. Willaime, **90**, 14, 144505 Copyright (2003) American Physical Society

more stringent requirements on the surface chemistry of the walls of the channels than the requirement of complete wetting of the substrate by the continuous phase at equilibrium.

Wetting of the substrate by the droplet phase can also be exploited in particular applications. For example, Fidalgo, Abell, and Huck (2007) reported a double T-junction system (Figure 5.28) in which the walls of the channels were hydrophobic and the system produced aqueous droplets in a continuous phase of oil in a stable manner. They modified a section of the floor of the outlet channel with hydrophilic poly (acrylic acid). The droplets passing over this hydrophilic patch wetted it. This process was successfully used to merge pairs of droplets: when two subsequent droplets wetted

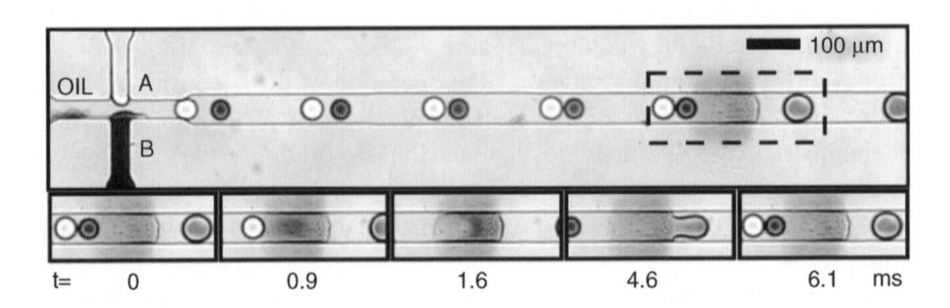

Figure 5.28 Micrographs of an all-hydrophobic system that had a small hydrophilic section on the floor of the outlet channel. Wetting of this section by aqueous droplets was successfully utilized to join two subsequent droplets and later eject them back into the shape of a droplet (Fidalgo, Bell, and Huck, 2007). Reprinted with permission from Lab on a Chip, Surface-induced droplet fusion in microfluidic devices by L.M. Fidalgo, C. Abell and W.T.S. Huck, 7, 8, 984–986 Copyright (2007) Royal Society of Chemistry

the hydrophilic section, the shear stress exerted by the continuous fluid pulled the combined aqueous solution from the patch to form a new, merged, droplet (Figure 5.28). Numerical work by Kusumaatmaja and Yeomans suggested that arrays of hydrophilic and hydrophobic patches could even be used for sorting droplets by their size (Kusumaatmaja and Yeomans, 2007).

5.5.4 Size of the Droplets

Microfluidic devices have been proven to be able to produce monodisperse droplets over a very wide range of volumes: from diameters of single micrometers (Xu and Nakajima, 2004) to millimeters (Engl *et al.*, 2007; Engl, Backov, and Panizza, 2008; Panizza *et al.*, 2008). Typically, the volume of the droplets can be tuned via an appropriate choice of the ratio of the rates of flow and the total rate of flow through the device. The diameters of the droplets are, however, strongly related to the diameter of the orifice, or channel at the junction and typically a range of dimensions between one tenth to ten times this characteristic size of the channel is achievable.

The throughput of the flow-focusing junctions can be estimated in a manner similar to the one that we proposed for T-junctions. The flow-focusing systems can produce monodisperse droplets in the squeezing and shearing modes, which produce highly monodisperse droplets up to slightly higher values of the capillary number than is the case for a T-junction. If the upper limit of the capillary number for formation of monodisperse droplets in a T-junction can be estimated as $Ca_c \approx 10^{-2}$, for the flow-focusing system it can be set to a higher value of $Ca_c \approx 10^{-1}$. This leads to the maximum throughput of the droplet phase $Q_d = Ca_c w^2 \sigma / \mu_c$ and maximum frequency $f_{max} = Ca_c \sigma / w \mu_c$. Obviously, these are only estimated limits of productivity, but can be used as a general guide for the choice of the scale of the device for production of droplets of desired volumes at the possibly high throughput. In the next chapter we will review parallel systems for formation of droplets that increase the throughput via parallelization of the drop-making junctions.

5.5.5 Supplying the Liquids

An important aspect in controlling the volume of the droplets is the choice of the strategy for feeding of fluids into the microfluidic chip. Ward *et al.* (2005) showed that there is a one-to-one correspondence between the dynamics of a flow-focusing junction controlled by the rates of flow of water and oil, and one controlled with the pressures applied to the inlets. However, the scaling of the size of the droplets is qualitatively different for each of the two cases. In practice, most syringe pumps that are commonly used to generate constant rates of flow introduce significant fluctuations or pulses that can severely modify the instantaneous rates of flow. As typically the volume of the droplets is very sensitive to the ratio of the rates of flow of the two immiscible phases, these pulses can generate large changes in the volume of the droplets and polydispersity of drops produced over long periods of time (e.g., minutes). This is probably the source of the typically significant scatter of data points on reported diagrams for scaling of the diameter of the droplets with various flow parameters. As the volume of the droplets is typically characterized in relation to the rate of flow of the phases, and as this volume explicitly depends on the ratio of the rates of flow and the total rate of flow via the squeezing or shearing mechanism of breakup, it is conceptually more straightforward to control the volume of the droplets via direct tuning of the rates of flow. An easy technique for generating constant rates of flow is based on the use of pressurized reservoirs connected to the chip via ducts of very high resistance to flow. If the hydraulic resistance of the connector is much higher than the resistance of the chip, the rate of flow will be, within a very good approximation, simply proportional to the pressure applied to the reservoir Korczyk *et al* (2010).

5.6 Designing Droplets

The superb control over flow that microfluidic systems offer can be exploited not only to form monodisperse droplets, but to literally design the architecture of the discrete elements of fluid. As the use of microfluidic systems for formation of particles and capsules is the main subject of this book, we will briefly overview the methods of obtaining structured droplets here, before the more detailed description of particular variants of these methods that will follow in subsequent chapters.

5.6.1 Control of the Interface of Homogeneous Droplets

Typically, the interfaces of droplets created in microfluidic systems are protected with surfactant adsorbed from the bulk of either the continuous or the droplet phase. Microfluidics, with the relative ease of design and fabrication of complicated systems of channels, allows for creation of droplets with tailored interfaces presenting a range of properties. For example, it is possible to dissolve reagents for interfacial polymerization in the droplet and the continuous phase and to form a polymeric (e.g., Nylon) membrane on the surface of the droplet, *in situ*, in the microfluidic channel (Quevedo, Steinbacher, and McQuade, 2005; Takeuchi *et al.*, 2005). Steinbacher *et al.* (2006) demonstrated a different interfacial reaction – an interfacial hydrolysis of silanes to obtain organosilicone shells on the surface of organic droplets in a continuous aqueous phase. Hettiarachi *et al.* (2007) showed formation of protective lipid coatings on the surfaces of bubbles of perfluorocarbon gas formed in a microfluidic system. These coatings, self-assembled from the continuous aqueous phase immediately after formation of individual bubbles, provided superb stability of micron-scale pockets of gas, in spite of the increased internal pressure due to the Laplace stresses.

It is also possible to coat the interface of a droplet (Nie *et al.*, 2008a) or bubble (Park *et al.*, 2009; Subramanian *et al.*, 2005) with colloidal particles. Nie *et al.* (2008a) showed a particularly interesting use of the control of the process that microfluidics offers. Namely, they dispersed the colloidal particles in the droplet phase. This allowed them to control the number of particles within a droplet and the resulting coverage of the interface. In later reports, Park *et al.* (2009), used the process of dissolution of carbon dioxide from the bubbles into the continuous aqueous phase to: (i) drastically decrease the volume of the bubbles, and (ii) change the pH of the continuous phase adjacent to the interface to a value that made the colloidal suspension unstable, and caused falling-out of the particles from the continuous phase onto the interface. The microfluidic approach to construction of Pickering emulsions and foams is interesting also because the confinement of the microchannels can be used to force the droplets or bubbles into non-spherical shells. Subramanian *et al.*, (2005) showed that adsorption of colloidal particles to the interface of such a deformed bubble can stabilize its non-spherical shape. These structures will be discussed in detail in Chapter 9.

5.6.2 Heterogeneous Droplets

The laminar character of the flow of liquids at the microscale can be used to form a compound stream of two miscible fluids and, furthermore, to break this stream into droplets composed of these two liquids. Nisisako, Torii, and Higuchi, (2004) were the first to report a microfluidic process for the formation of bi-colored (or, as they are often referred to Janus) droplets. They also showed that it is possible to polymerize these droplets into bi-colored particles. These same Authors later showed (Nisisako *et al.*, 2006) that these bi-colored particles can posses anisotropic electrostatic properties and can be used in electronic paper displays. Nie *et al.* (2006) took this technology a step further and demonstrated that even immiscible liquids (solutions of monomers) can be engineered to flow laminarly and break into bi- or tri-colored droplets. The resulting morphologies of the compound droplets can be then trapped in the unstable shapes via *in situ* photopolymerization. The use of microfluidics allows for further selective modification of the surface chemistry of only a portion of the interface of the particle (Nie *et al.*, 2006). Shepherd *et al.* (2006) demonstrated the formation of spherical and anisotropic assemblies of colloidal particles. These assemblies were obtained via extraction of solvent from droplets comprising suspensions of particles. Similarly to other reports on the formation of multi-colored particles, the droplet could be engineered to comprise sections of suspensions of differently colored particles. Formation of Janus particles will be discussed in detail in Chapter 10.

5.6.3 Multiple Emulsions

The most impressive use of the microfluidic control of flow in formation of droplets is exhibited in the methods of formation of multiple emulsions. Microfluidic networks of channels and the flow of fluids in these networks can be engineered to produce multiple droplets with control over practically all structural characteristics, including the number and character of the shells, and of the droplets contained within the core. The first report of the use of a microfluidic system for formation of multiple emulsions was published by Okushima *et al.* in 2004 (Okushima *et al.*, 2004). The technique used a sequence of two T-junctions and a sophisticated control of the surface chemistry of the channels. Figure 5.29 shows the schematics of the microfluidic system. The first T-junction is hydrophobic, including the inlet channel and the outlet from the junction. Hydrophobic character of the walls makes it possible to form aqueous droplets in a continuous oil phase. The network of channels was designed in such a way as to have the outlet of the

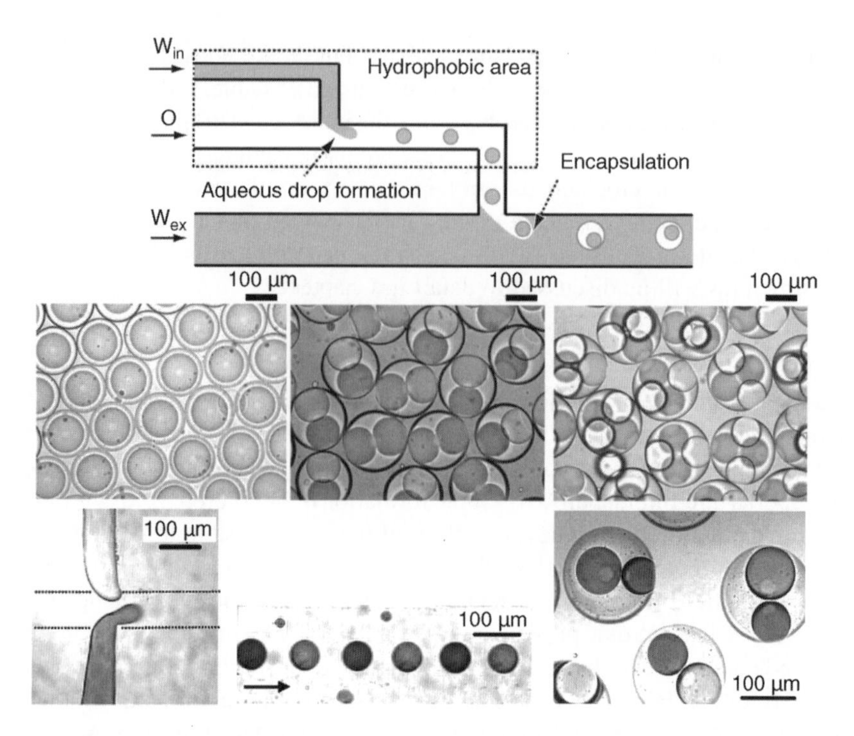

Figure 5.29 *The first demonstration of a microfluidic technique for formation of double emulsions. The scheme in the top panel shows the planar system of two consecutive T-junctions. The first junction is all hydrophobic to allow for stable formation of drops of water in oil. The main channel in the second junction is hydrophilic to facilitate breakup of the W–O emulsion into W–O–W drops (Nisisako, Okushima, and Torii, 2005). The middle row shows examples of double droplets with one, and four cores (Nisisako, Okushima, and Torii, 2005). Reprinted with permission from Soft Matter, Controlled formulation of monodisperse double emulsions in a multiple-phase microfluidic system by T. Nisisako, S. Okushima and T. Torii,* ***1****, 1, 23-27 Copyright (2005) Royal Society of Chemistry. Finally, the bottom panel shows micrographs of a similar system that allows for formation of double emulsions containing two different droplets in the core (Okushima et al., 2004). Reprinted with permission from Langmuir, Controlled production of monodisperse double emulsions by two-step droplet breakup in microfluidic devices by S. Okushima et al.,* ***20****, 23, 9905–9908 Copyright (2004) American Chemical Society. See Plate 6*

first T-junction become the inlet of the fluid to be dispersed in the second junction. In this second junction, the hydrophobic channel guided the emulsion of aqueous drops in an organic continuous fluid into a perpendicular, hydrophilic channel, which supplied the outermost continuous aqueous phase. In this configuration the stream of the W–O (water–oil) emulsion broke in the second junction to form water-in-oil-in-water (W–O–W) droplets. Okushima *et al.* (2004) and Nisisako, Okushima, and Torii, (2005) observed that it is possible to control the number of the aqueous droplets contained in the larger organic droplet simply by adjusting the frequency of breakup at the first and at the second junction. Similarly, tuning of the rates of flow of the each of the three phases allowed for some freedom in independent tuning of the volume fraction of the inner aqueous phase in the organic shell. Figure 5.29 shows three examples of monodisperse multiple droplets obtained this way, containing a single two or four inner aqueous droplets.

Seo *et al.* (2007) used a similar approach for formation of double emulsions in a planar flow-focusing system. Modification of the surface chemistry of sections of the device allowed for formation of W–O–W, O–O–W, and O–W–O droplets. Similarly to the double T-junction system (Nisisako, Okushima, and Torii, 2005; Okushima *et al.*, 2004), the control of the rates of flow and frequencies of formation of droplets in the consecutive flow-focusing junctions allowed for control of the volume fractions of the three phases and the number of inner droplets contained in each primary droplet. In a similar approach, Wan *et al.* (2008) demonstrated formation of droplets containing a controlled number of imbedded gaseous bubbles.

A number of Authors have also shown that, with clever tricks, it is possible to form multiple droplets in planar chips without selective modification of the surface chemistry (wettability) of the channels. For example, Arakawa, Yamamoto, and Shoji (2008) showed that a modified T-junction can form bubbles of gas wrapped by a thin membrane of an organic phase. The device was fabricated in silicone and glass. The walls of the channels were all hydrophilic and preferentially wetted by the outer, continuous, aqueous phase. The bubbles probably formed via the squeezing mechanism of breakup as they formed slugs squeezed between the side walls of the channel. As the water–gas interface is characterized by the highest value of interfacial tension, the structure of the gaseous core in an organic shell was stabilized by the gradation of interfacial tensions between gas–oil and oil–water. Saeki *et al.* (2008) used an all-hydrophobic system to form W–O–W emulsions. The system comprised a T-junction that formed the W–O droplets. The outlet of this T-junction lead to a wider and deeper channel that also possessed additional side inputs for the aqueous phase. The droplets flowing into this larger channel left most of the oil on the hydrophobic walls, while preserving a thin organic shell before detachment into the bulk, continuous aqueous phase (Saeki *et al.*, 2008).

In 2005, Nie *et al.* (2005) reported a detailed study of formation of multiple droplets in a planar flow focusing system, which included the central channel that delivered the inner (core) liquid, two, adjacent to the inner one, channels that delivered the liquid to form the shell around the core, and a set of two outermost channels guiding the continuous liquid. The inlet channels all terminated at the same distance upstream of the orifice. The device was fabricated in polyurethane, which favored wetting by the outermost, continuous, aqueous phase. The inner phase comprised silicone oil that was immiscible with monomer solutions constituting the shell (intermediate) fluid. These Authors found that in this simple geometry, the hydrodynamic forces were sufficient to engulf the inner silicone oil in the solution of monomer. These two liquids formed a concentric stream that, in the orifice, broke into multiple droplets. Nie *et al.* (2005) constructed a detailed diagram of the architectures of the droplets, attainable via appropriate tuning of the rates of flow of the three immiscible phases

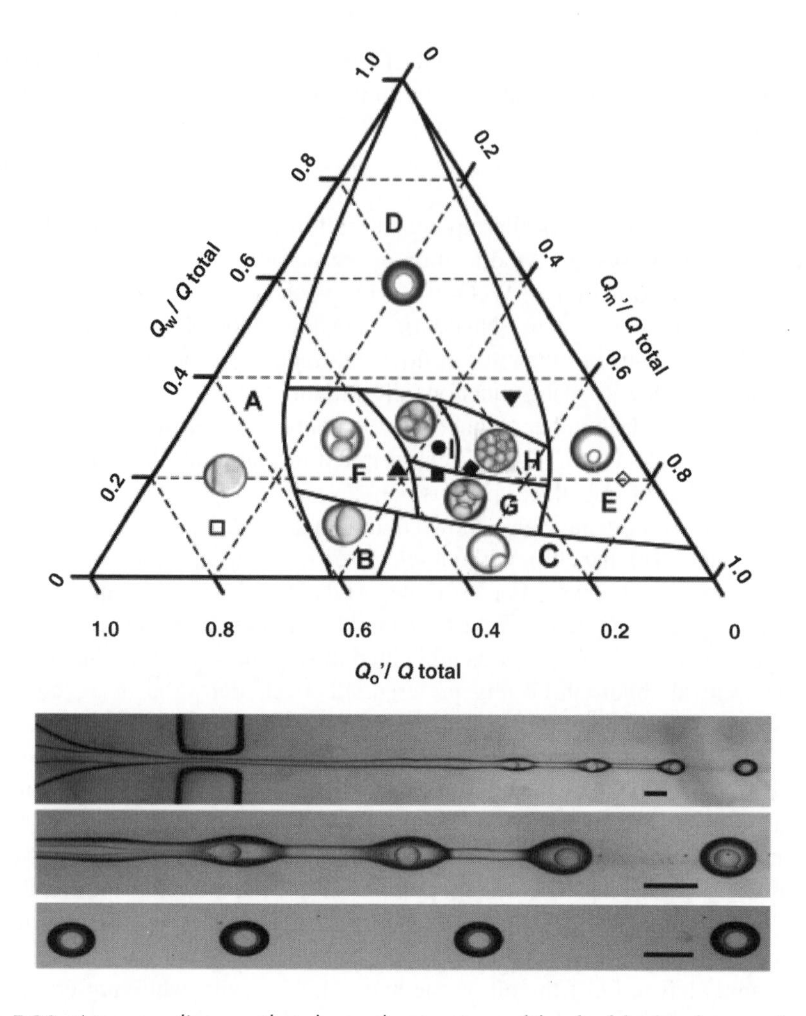

Figure 5.30 A ternary diagram that shows the structure of the double droplets as a function of the rates of flow of the three phases [of the inner oil phase (Q_o], the intermediate monomer phase (Q_m), and the outer continuous aqueous phase (Q_w)]. The micrographs illustrate the operation of the system during formation of double droplets and their flow in the outlet channel (Nie et al., 2005). Reprinted with permission from Journal of the American Chemical Society, Polymer Particles with Various Shapes and Morphologies Produced in Continuous Microfluidic Reactors by Z. Nie et al., **127**, 22, 8058–8063 Copyright (2005) American Chemical Society

(Figure 5.30). The diagram can be rationalized via the observations made that: (i) the diameter of each of the jets (inner and outer) can be very well approximated via conservation of mass, and (ii) that these jets breakup via the Rayleigh–Plateau instability with the diameter of the droplets being closely related to the wavelength of the fastest mode of collapse of an unstable column of liquid. The conservation of mass can be applied via an approximation that both jets flow at the speed of the continuous liquid. This former speed can be easily estimated from the rate of and the cross-section of the channel. Nie *et al.* (2005) showed that tuning of the diameter of the jet (by modulating the appropriate rate of flow) can be used to set the diameters of the droplets and, by appropriate adjustment of the diameters of each (inner and outer) jet, to set the number of the inner droplets.

Similar observations were reported simultaneously by Utada *et al.* (2005), which used an axi-symmetric system for formation of multiple emulsions. This system was also capable of formation of double droplets, providing detailed control over the inner and outer radii. These Authors reported two modes of breakup of the composite stream: dripping and jetting. In the dripping regime the compound stream broke in the orifice (upper panel in Figure 5.31), while in the jetting regime, the composite thread extended

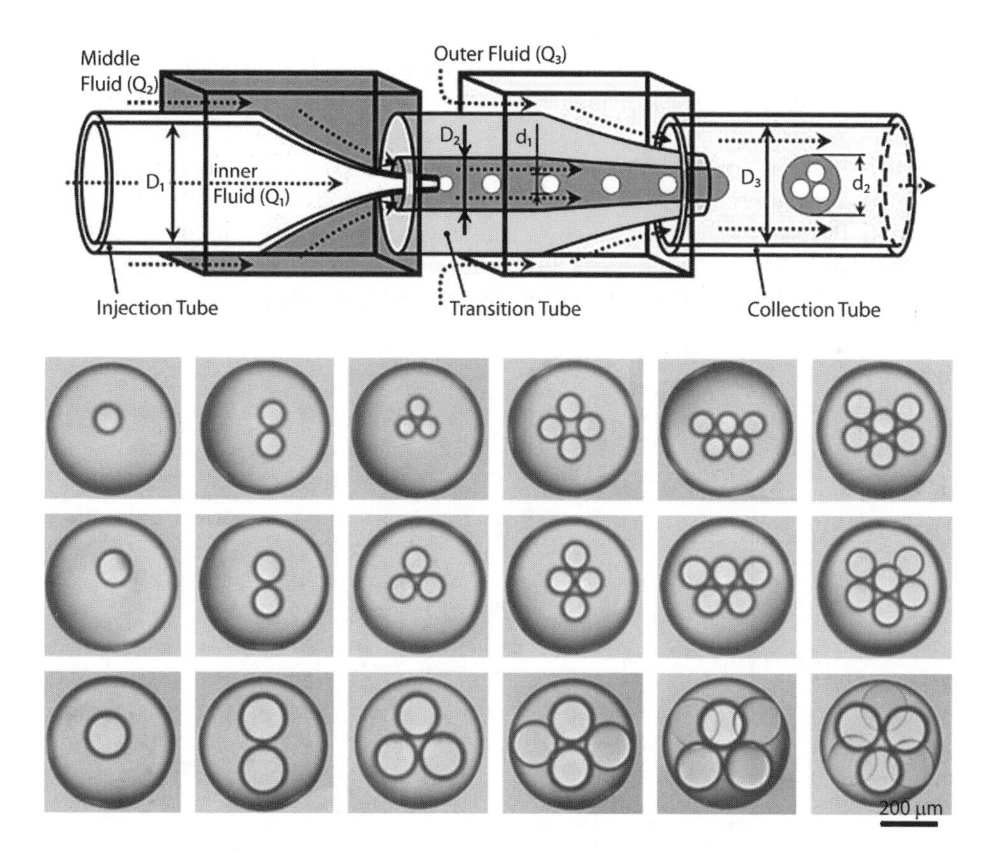

Figure 5.31 *Micrographs of the microcapillary device proposed by Utada et al. (2005) for formation of double emulsions. The dripping regime (upper picture) allows for excellent control of the diameters of the droplets while the jetting mode (bottom picture) is more difficult to control (Utada et al., 2005). Reproduced from Chu, L.Y, Utada, A.S, Shah R.K, et al., Controllable monodisperse multiple emulsions, Angewandte Chemie-International Edition, 46, 47, 8970–8974. Copyright (2007) with permission from Wiley-VCH*

deep downstream into the outlet capillary. While the dripping regime produced monodisperse droplets, the jetting regime effectively prohibited fine control of the distribution of diameters of the drops. Again, the simple analysis based on the Rayleigh–Plateau instability of a liquid column allowed a correct estimate of the speed of flow at which the system crossed over from the dripping to the jetting regime, and the dimensions of the double droplets. The cross-over between the regimes can be understood through a comparison of the time needed for the perturbation of the width of the jet to grow and break-off a droplet (pinching time) with the time (convection time) needed for the compound stream to elongate into a morphology unstable with respect to Rayleigh–Plateau instability. If the pinch time is greater than the convection time, then the thread grows before it is broken off and the system crosses-over to the jetting regime. While the total rate of flow controls the breakup mode, the ratio of the rates of flow controls the volume fraction of the inner phase and the sheeting liquid, allowing fine tuning of the thickness of the shell.

In summary, microfluidic devices can be designed and constructed to successfully form multiple droplets. The stoichiometry of the droplets (number and the type of core droplets contained in the outermost drop) can be effectively controlled. Most of the reported techniques of formation of multiple emulsions (Loscertales *et al.*, 2002; Nie *et al.*, 2005; Utada *et al.*, 2005) indicate formation of a coaxial jet that breaks via the Rayleigh–Plateau instability. It is known that flow can modify the stability criteria of a column of fluid. The instabilities can be divided into absolute ones, which are localized in space with respect to the walls of the device, and convective ones, which can be

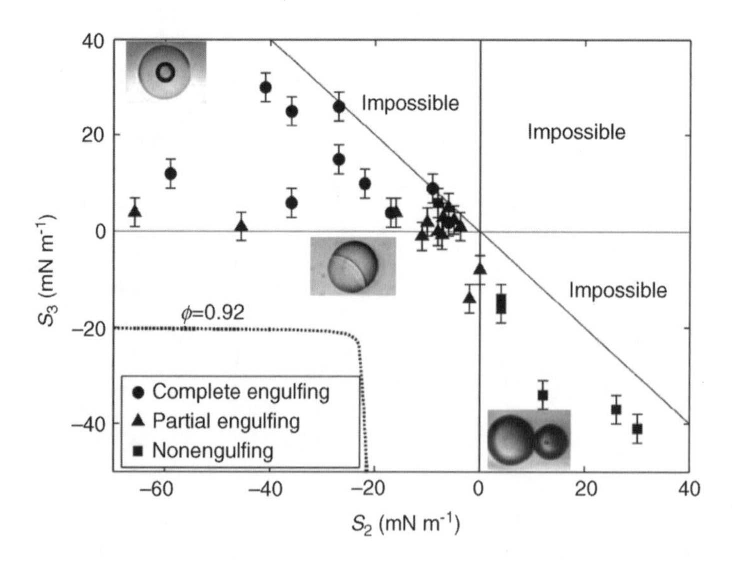

Figure 5.32 *A diagram illustrating the equilibrium states of double emulsions (Pannacci et al., 2008). Reprinted with permission from Phys. Rev. Lett., Equilibrium and Nonequilibrium States in Microfluidic Double Emulsions by N. Pannacci, H. Bruus, D. Bartolo et al.,* ***101****, 16, 164502 Copyright (2008) American Physical Society*

washed away from the orifice by the flow of either of the immiscible fluids. The subject of stability of jets of liquid subjected to both flow and confinement are an active area of study. For example, Guillot and coworkers (Guillot *et al.*, 2007; Guillot, Colin, and Ajdari, 2008) derived the equations that predict the transition between dripping and laminar co-flow of concentrically flowing immiscible liquids in the confinement of a circular capillary. Utada and coworkers (Utada *et al.*, 2007; Utada *et al.*, 2008) pointed out that this transition can be controlled by the viscous effects in the outer fluid and inertial ones in the inner liquid. As the subject is relatively new and certainly complex, by virtue of including a large number of parameters and the necessity to understand the interplay of viscous, inertial, and interfacial forces, there are no practical guidelines available to date.

An important aspect in formation of multiple emulsions using microfluidics is not only the precision with which the double droplet can be prepared, but also the opportunity to trap the droplets in non-equilibrium morphologies. Pannacci *et al.* (2008) analyzed the stability of a double droplet that can either be stable in the completely engulfed morphology (the inner droplet being completely surrounded by the shell), or in a partially engulfed shape (with the two droplets having a surface of contact). The third possibility is that the droplets are completely separated at equilibrium. Which of these scenarios yields the equilibrium morphology depends on the values of the three interfacial tensions σ_{12}, σ_{23}, and σ_{13}, where the indices 1, 2, and 3 correspond to the three liquids (counting from the innermost to the outermost, continuous liquid). In Figure 5.32 one can see the regions of stability of these three morphologies on a map spanned by the spreading parameters $S_2 = \sigma_{13} - (\sigma_{12} + \sigma_{23})$ and $S_3 = \sigma_{23} - (\sigma_{13} + \sigma_{23})$. This diagram can guide the choice of liquids and surfactants for obtaining stable double emulsions of the desired type.

Pannacci *et al.* (2008), in their analysis, also measured the time that a non-equilibrium morphology of a completely engulfed droplet needs to acquire the equilibrium Janus configuration. This time is linearly proportional to the radius of the droplet and for diameters of several hundreds of micrometers can be as long as a large fraction of a second, giving enough time for, for example, photopolymerization performed *in situ* to arrest the non-equilibrium shape.

5.7 Conclusions

This chapter reviewed the microfluidic methods of formation of droplets and bubbles. The popular microfluidic systems – T-junctions and flow-focusing generators – provide control over all aspects of the produced emulsions, from the volume of individual droplets, through the distribution of their sizes, the volume fraction of the dispersed phase, and the architecture of the droplets. In the next chapter we will describe the challenges and latest developments associated with the application of microfluidic techniques to high-throughput formation of droplets, a requirement in many practical applications.

References

Abate, A.R., Poitzsch, A., Hwang, Y., Lee, J., Czerwinska, J., and Weitz, D.A. (2009) *Phys. Rev. E*, **80**, 026310.

Adamson, D.N., Mustafi, D., Zhang, J.X.J., Zheng, B., and Ismagilov, R.F. (2006) *Lab Chip*, **6**, 1178–1186.

Anna, S.L., Bontoux, N., and Stone, H.A. (2003) *Appl. Phys. Lett.*, **82**, 364–366.

Anna, S.L. and Mayer, H.C. (2006) *Phys. Fluids*, **18**, 121512.

Arakawa, T., Yamamoto, T., and Shoji, S. (2008) *Sens. Actuators A-Phys.*, **143**, 58–63.

Arratia, P.E., Gollub, J.P., and Durian, D.J. (2008) *Phys. Rev. E*, **77**, 036309.

Arratia, P.E., Cramer, L.A., Gollub, J.P., and Durian, D.J. (2009) *New J. Phys.*, **11**, 115006.

Baret, J.C., Kleinschmidt, F., El Harrak, A., and Griffiths, A.D. (2009) *Langmuir*, **25**, 6088–6093.

Christopher, G.F., Noharuddin, N.N., Taylor, J.A., and Anna, S.L. (2008) *Phys. Rev. E*, **78**, 036317.

Chu, L.Y, Utada, A.S, Shah R.K, et al., (2007) Controllable monodisperse multiple emulsions, Angewandte Chemie-International Edition, 46, 47, 8970–8974.

Cubaud, T. and Ho, C. M. (2004) *Phys. Fluids*, **16**, 4575–4585.

Cubaud, T., Tatineni, M., Zhong, X.L., and Ho, C.M. (2005) *Phys. Rev. E*, **72**, 037302.

Cubaud, T. and Mason, T.G. (2008) *Phys. Fluids*, **20**, 053302.

De Menech, M. (2006) *Phys. Rev. E*, **73**, 031505.

De Menech, M., Garstecki, P., Jousse, F., and Stone, H.A. (2008) *J. Fluid Mechan.*, **595**, 141–161.

Dollet, B., van Hoeve, W., Raven, J.P., Marmottant, P., and Versluis, M. (2008) *Phys. Rev. Lett.*, **100**, 034504.

Dreyfus, R., Tabeling, P., and Willaime, H. (2003) *Phys. Rev. Lett.*, **90**.

Duffy, D.C., McDonald, J.C., Schueller, O.J.A., and Whitesides, G.M. (1998) *Anal.Chem.*, **70**, 4974–4984.

Engl, W., Tachibana, M., Panizza, P., and Backov, R. (2007) *Int. J. Multiphase Flow*, **33**, 897–903.

Engl, W., Backov, R., and Panizza, P. (2008) *Curr. Opin. Colloid Interface Sci.*, **13**, 206–216.

Fidalgo, L.M., Abell, C., and Huck, W.T.S. (2007) *Lab Chip*, **7**, 984–986.

Fu, T.T., Ma, Y.G., Funfschilling, D., and Li, H.Z. (2009) *Chem. Eng. Sci.*, **64**, 2392–2400.

Fuerstman, M.J., Lai, A., Thurlow, M.E., Shevkoplyas, S.S., Stone, H.A., and Whitesides, G.M. (2007) *Lab Chip*, **7**, 1479–1489.

Funfschilling, D., Debas, H., Li, H.Z., and Mason, T.G. (2009) *Phys. Rev. E*, **80**, 015301.

Ganan-Calvo, A.M. (1998) *Phys. Rev. Lett.*, **80**, 285–288.

Ganan-Calvo, A. and Gordillo, J.M. (2001) *Phys. Rev. Lett.*, **87**, 274501.

Ganan-Calvo, A. (2004) *Phys. Rev. E*, **69**, 027301.

Garstecki, P., Gitlin, I., DiLuzio, W., Whitesides, G.M., Kumacheva, E., and Stone, H.A. (2004) *Appl. Phys. Lett.*, **85**, 2649–2651.

Garstecki, P., Stone, H.A., and Whitesides, G.M. (2005) *Phys. Rev. Lett.*, **94**, 164501.

Garstecki, P., Fuerstman, M.J., Stone, H.A., and Whitesides, G.M. (2006) *Lab Chip*, **6**, 437–446.

Guillot, P. and Colin, A. (2005) *Phys. Rev. E*, **72**, 066301.

Guillot, P., Colin, A., Utada, A.S., and Ajdari, A. (2007) *Phys. Rev. Lett.*, **99**, 104502.

Guillot, P., Colin, A. and Ajdari, A. (2008) *Phys. Rev. E*, **78**, 016307.

Gupta, A., Murshed, S.M.S., and Kumar, R. (2009) *Appl. Phys. Lett.*, **94**, 164107.

Hallmark, B., Parmar, C., Walker, D., Hornung, C.H., Mackley, M.R., and Davidson, J.F. (2009) *Chem. Eng. Sci.*, **64**, 4758–4764.

Hashimoto, M., Garstecki, P., Stone, H.A., and Whitesides, G.M. (2008) *Soft Matt.*, **4**, 1403–1413.

Hettiarachchi, K., Talu, E., Longo, M.L., Dayton, P.A., and Lee, A.P. (2007) *Lab Chip*, **7**, 463–468.

Husny, J. and Cooper-White, J.J. (2006) *J. Non-Newtonian Fluid Mech.*, **137**, 121–136.
Jensen, M.J., Stone, H.A., and Bruus, H. (2006) *Phys. Fluids*, **18**, 077103.
Jullien, M.C., Ching, M., Cohen, C., Menetrier, L., and Tabeling, P. (2009) *Phys. Fluids*, **21**, 072001.
Korczyk, P.M., Cybulski, O., Makulska, S., and Garstecki, P. *Lab Chip*, **11**, 173 (2010).
Kusumaatmaja, H. and Yeomans, J.M. (2007) *Langmuir* **23**, 956–959.
Lee, W., Walker, L.M., and Anna, S.L. (2009) *Phys. Fluids*, **21**, 032103.
Li, W., Nie, Z.H., Zhang, H., Paquet, C., Seo, M., Garstecki, P., and Kumacheva, E. (2007) *Langmuir*, **23**, 8010–8014.
Link, D.R., Anna, S.L., Weitz, D.A., and Stone, H.A. (2004) *Phys. Rev. Lett.*, **92**, 054503.
Liu, H.H. and Zhang, Y.H. (2009) *J. Appl. Phys.*, **106**, 034906.
Lorenceau, E., Sang, Y.Y.C., Hohler, R., and Cohen-Addad, S. (2006) *Phys. Fluids*, **18**, 097103.
Loscertales, I.G., Barrero, A., Guerrero, I., Cortijo, R., Marquez, M., and Ganan-Calvo, A. (2002) *Science*, **295**, 1695–1698.
Marre, S., Aymonier, C., Subra, P., and Mignard, E. (2009) *Appl. Phys. Lett.*, **95**, 134105.
Nie, Z.H., Xu, S.Q., Seo, M., Lewis, P.C., and Kumacheva, E. (2005) *J. Am. Chem. Soc.*, **127**, 8058–8063.
Nie, Z.H., Li, W., Seo, M., Xu, S.Q., and Kumacheva, E. (2006) *J. Am. Chem. Soc.*, **128**, 9408–9412.
Nie, Z.H., Il Park, J., Li, W., Bon, S.A.F., and Kumacheva, E. (2008a) *J. Am. Chem. Soc.*, **130**, 16508–16509.
Nie, Z.H., Seo, M.S., Xu, S.Q., Lewis, P.C., Mok, M., Kumacheva, E., Whitesides, G.M., Garstecki, P., and Stone, H.A. (2008b) *Microfl. Nanofluidics*, **5**, 585–594.
Nisisako, T., Torii, T., and Higuchi, T. (2002) *Lab Chip*, **2**, 24–26.
Nisisako, T., Torii, T., and Higuchi, T. (2004) *Chem. Eng. J.*, **101**, 23–29.
Nisisako, T., Okushima, S., and Torii, T. (2005) *Soft Mat.*, **1**, 23–27.
Nisisako, T., Torii, T., Takahashi, T., and Takizawa, Y. (2006) *Adv. Mater.*, **18**, 1152–1156.
Okushima, S., Nisisako, T., Torii, T., and Higuchi, T. (2004) *Langmuir*, **20**, 9905–9908.
Ong, W.L., Hua, J.S., Zhang, B.L., Teo, T.Y., Zhuo, J.L., Nguyen, N.T., Ranganathan, N., and Yobas, L. (2007) *Sens. Actuators A-Phys.*, **138**, 203–212.
Panizza, P., Engl, W., Hany, C., and Backov, R. (2008) *Colloids Surf. A-Physicochem. Eng. Aspects*, **312**, 24–31.
Pannacci, N., Bruus, H., Bartolo, D., Etchart, I., Lockhart, T., Hennequin, Y., Willaime, H., and Tabeling, P. (2008) *Phys. Rev. Lett.*, **101**, 164502.
Park, J.I., Nie, Z., Kumachev, A., Abdelrahman, A.I., Binks, B.R., Stone, H.A., and Kumacheva, E. (2009) *Angew. Chem. Int. Ed.*, **48**, 5300–5304.
Pipe, C.J. and McKinley, G.H. (2009) *Mech. Res. Commun.*, **36**, 110–120.
Quevedo, E., Steinbacher, J., and McQuade, D.T. (2005) *J. Am. Chem. Soc.*, **127**, 10498–10499.
Saeki, D., Sugiura, S., Kanamori, T., Sato, S., Mukataka, S., and Ichikawa, S. (2008) *Langmuir*, **24**, 13809–13813.
Schroeder, V., Behrend, O., and Schubert, H. (1998) *J. Colloid Interface Sci.*, **202**, 334–340.
Seo, M., Paquet, C., Nie, Z.H., Xu, S.Q., and Kumacheva, E. (2007) *Soft Mat.*, **3**, 986–992.
Shepherd, R.F., Conrad, J.C., Rhodes, S.K., Link, D.R., Marquez, M., Weitz, D.A., and Lewis, J. A. (2006) *Langmuir*, **22**, 8618–8622.
Shui, L.L., van den Berg, A., and Eijkel, J.C.T. (2009) *J. Appl. Phys.*, **106**, 124305.
Song, H., Bringer, M.R., Tice, J.D., Gerdts, C.J., and Ismagilov, R.F. (2003a) *Appl. Phys. Lett.*, **83**, 4664–4666.
Song, H. and Ismagilov, R.F. (2003b) *J. Am. Chem. Soc.*, **125**, 14613–14619.
Song, H., Tice, J.D., and Ismagilov, R.F. (2003c) *Angew. Chem. Int. Ed.*, **42**, 768–772.
Steegmans, M.L.J., Schroen, C., and Boom, R.M. (2009a) *Chem. Eng. Sci.*, **64**, 3042–3050.
Steegmans, M.L.J., Schroen, K., and Boom, R.M. (2009b) *Langmuir*, **25**, 3396–3401.

Steinbacher, J.L., Moy, R.W.Y., Price, K.E., Cummings, M.A., Roychowdhury, C., Buffy, J.J., Olbricht, W.L., Haaf, M., and McQuade, D.T. (2006) *J. Am. Chem. Soc.*, **128**, 9442–9447.
Steinhaus, B., Shen, A.Q., and Sureshkumar, R. (2007) *Phys. Fluids*, **19**, 073103.
Subramanian, A.B., Abkarian, M., Mahadevan, L., and Stone, H.A. (2005) *Nature*, **438**, 930–930.
Sugiura, S., Nakajima, M., Tong, J., Nabetani, H., and Seki, M. (2000) *J. Colloid Interface Sci.*, **227**, 95–103.
Sugiura, S., Nakajima, M., and Seki, M. (2002) *Langmuir*, **18**, 3854–3859.
Suryo, R. and Basaran, O.A. (2006) *Phys. Fluids*, **18**, 082102.
Takeuchi, S., Garstecki, P., Weibel, D.B., and Whitesides, G.M. (2005) *Adv. Mater.*, **17**, 1067–1072.
Tan, Y.C., Cristini, V., and Lee, A.P. (2006) *Sens. Actuators B-Chem.*, **114**, 350–356.
Taylor, G.I. (1934) *Proc. R. Soc., London, Ser. A*, **146**, 501.
Thorsen, T., Roberts, R.W., Arnold, F.H., and Quake, S.R. (2001) *Phys. Rev. Lett.*, **86**, 4163–4166.
Tice, J.D., Song, H., Lyon, A.D., and Ismagilov, R.F. (2003) *Langmuir*, **19**, 9127–9133.
Tice, J.D., Lyon, A.D., and Ismagilov, R.F. (2004) *Anal. Chim. Acta*, **507**, 73–77.
Utada, A.S., Lorenceau, E., Link, D.R., Kaplan, P.D., Stone, H.A., and Weitz, D.A. (2005) *Science*, **308**, 537–541.
Utada, A.S., Fernandez-Nieves, A., Stone, H.A., and Weitz, D.A. (2007) *Phys. Rev. Lett.*, **99**, 094502.
Utada, A.S., Fernandez-Nieves, A., Gordillo, J.M., and Weitz, D.A. (2008) *Phys. Rev. Lett.*, **100**, 014502.
van der Graaf, S., Steegmans, M.L.J., van der Sman, R.G.M., Schroen, C., and Boom, R.M. (2005) *Colloids Surf. A-Physicochem. Eng. Aspects*, **266**, 106–116.
van der Graaf, S., Nisisako, T., Schroen, C., van der Sman, R.G.M., and Boom, R.M. (2006) *Langmuir*, **22**, 4144–4152.
van Steijn, V., Kreutzer, M.T., and Kleijn, C.R. (2007) *Chem. Eng. Sci.*, **62**, 7505–7514.
van Steijn, V., Kleijn, C.R., and Kreutzer, M.T. (2009) *Phys. Rev. Lett.*, **103**, 214501.
Wan, J., Bick, A., Sullivan, M., and Stone, H.A. (2008) *Adv. Mater.*, **20**, 3314–3318.
Ward, T., Faivre, M., Abkarian, M., and Stone, H.A. (2005) *Electrophoresis*, **26**, 3716–3724.
Weber, M. and Shandas, R. (2007) *Microfl. Nanofluidics*, **3**, 195–206.
Wong, H., Radke, C.J., and Morris, S. (1995a) *J. Fluid Mechan.*, **292**, 95–110.
Wong, H., Radke, C.J., and Morris, S. (1995b) *J. Fluid Mechan.*, **292**, 71–94.
Xu, J.H., Li, S.W., Tan, J., Wang, Y.J., and Luo, G.S. (2006a) *Langmuir*, **22**, 7943–7946.
Xu, J.H., Li, S.W., Wang, Y.J., and Luo, G.S. (2006b) *Appl. Phys. Lett.*, **88**, 133506.
Xu, J.H., Li, S.W., Tan, J., and Luo, G.S. (2008) *Microfl. Nanofluidics*, **5**, 711–717.
Xu, Q.Y. and Nakajima, M. (2004) *Appl. Phys. Lett.*, **85**, 3726–3728.
Yobas, L., Martens, S., Ong, W.L., and Ranganathan, N. (2006) *Lab Chip*, **6**, 1073–1079.
Zhao, Y.C., Chen, G.W., and Yuan, Q. (2006) *AICHE J.*, **52**, 4052–4060.
Zheng, B., Roach, L.S., and Ismagilov, R.F. (2003) *J. Am. Chem. Soc.*, **125**, 11170–11171.

6

High-Throughput Microfluidic Systems for Formation of Droplets

CHAPTER OVERVIEW

6.1 Introduction

In Chapter 5 we reviewed microfluidic techniques for formation of droplets and the hydrodynamics of breakup in T-junctions and flow-focusing devices. These systems are capable of producing highly monodisperse emulsions and provide effective control over the volumes of the droplets. As will be discussed in detail in Chapters 7–10, this ability to form monodisperse droplets of well-defined volumes can be used to produce polymeric particles and capsules of potential use in research and in the pharmaceutical, cosmetic, and food industries. Interest in such applications, even if motivated by the better control over the characteristics of the particles than those available via classical synthetic routes, is subject to the requirements of large throughput and of being cost-effective. Because microfluidic devices operate with flows in microchannels and with small amounts of liquids, these requirements have prompted interest in construction of

Microfluidic Reactors for Polymer Particles. Eugenia Kumacheva and Piotr Garstecki.

systems that combine the control offered by microfluidics with largely increased throughput.

The exquisite control that microfluidic systems offer over the formation of droplets depends critically on the effects of confinement. This, together with the mechanism of breakup in microfluidic junctions, sets the limits on productivity of a single-droplet generator. The only vista to increasing throughput is thus via parallelization, that is, via use a multiplicity of microjunctions. The volumes of the droplets are a function of the geometry of the junction and of the rates of flow of the liquids through the microfluidic droplet generator (and in general can also be functions of the pressure drops through the junctions). Thus, fairly obviously, it is not enough to fabricate identical microstructures, it is also necessary to supply each of them with the same rates of flow of the fluids.

Conceptually the simplest route is parallel multiplication of the sources of fluids – be it syringe pumps or independent pressurized containers. This, however, is highly impractical and expensive. Thus, the given rate of flow of the liquid supplied from an external source needs to be distributed into an array of droplet generators. This is a straightforward task for simple liquids. As the distribution of flow at low Reynolds numbers is governed by equations analogous to Kirchhoff's and Ohm's Laws, it is not complicated to design a network of channels that guarantees equal rate of flow through a set of junctions. Two simple solutions to this are a fractal structure of the distribution network (Luque *et al.*, 2007), or a set of identical channels guiding the liquid from a common pressurized reservoir to each of the droplet generators (Luque *et al.*, 2009). Unfortunately, for multiphase flow the situation complicates because the blocking–squeezing mechanism of breakup that dominates generation of droplets in microfluidic devices modifies the resistance of the junction and leads to oscillations of pressure upstream of the junction during the process of formation of a droplet. The already-formed droplets flowing in the outlets also increase the resistance to flow in these channels and, as a consequence, can also modify the distribution of flow between the junctions.

For bubbles the effects of coupling between the resistance generated by the bubbles flowing in the outlet channel and the dynamics of breakup at the junction can lead to oscillations of the volume of the bubbles of large amplitude and large period (Raven and Marmottant, 2006; Sullivan and Stone, 2008). For incompressible liquids these effects have smaller amplitude, but the coupling between the resistance of the outlet and the dynamics at the junction can still lead to behaviors characteristic of non-linear systems, with oscillations and chaotic dynamics.

Obviously one way of eliminating these problems is to design an adequately large hydrodynamic resistance of the inlets for each of the immiscible phases for each of the junctions separately. These resistances should be sufficiently large that the fluctuations of resistance at the junction and downstream of the junction could not significantly modify the rate of flow through it. This approach, however, could lead to a stark contrast with the idea of miniaturization, because providing large resistances to the inlets requires designing long channels that occupy space on the chip. Another vista is to connect each of the droplet generators directly to reservoirs of pressurized liquids and, likewise, to a wide common outlet that also presents a reservoir of pressure. In this architecture fluctuations of resistance at any particular junction should not affect the flow through any other junction, because it is supplied from and drained into a large reservoir that cannot be affected by small changes at any of the junctions. This approach is also not free of challenges, mainly in the fabrication and in optimization of packing.

Finding the optimal compromise between the different strategies and an effective design of parallelized systems of microfluidic droplet generators requires detailed characterization of the effects of hydrodynamic coupling between the junctions and between the junction and its outlet, and subsequent optimization of the network that distributes the liquids. At the moment, knowledge about these effects is not complete, and new observations and designs are being reported each year. In the following, we will introduce the effects that modify the pressure distribution in the microfluidic networks, show evident examples of coupling of the dynamics of neighboring droplet generators, and a few examples of working, highly parallel systems. Finally, we will discuss the few available examples of microfluidic systems that allow for parallel formation of droplets of different volumes and of varying chemical content.

6.2 Effects that Modify the Pressure Distribution

A simple fluid without any intrusions, when it flows through a capillary or a microchannel, develops a parabolic profile of speed of flow: the fluid flows the fastest in the center of the channel, and rests at the walls of the capillary. Thus, if we choose two small volumes of liquid that are originally at the same position along the capillary, but their locations in the cross-section are different, the distance between them will grow linearly with the time of flow. A droplet or a bubble introduced into this flow modifies the velocity field of flow. The simplest way to see the reason for this modification is to observe that the droplet (bubble) in the flow separates the continuous liquid behind it from the liquid in front of it. Even if this separation is not perfect, that is, even if a fraction of the continuous liquid bypasses the droplet, the parabolic profile that develops without the flowing obstacle can no longer be sustained. As a result, close to the caps of the droplet the continuous fluid recirculates and additional eddies are created. There is also a circulation of the liquid inside the droplet, and some flow along the droplet (Figure 6.1). All of these additional flows increase the viscous dissipation in the carrier (continuous) fluid and in the liquid inside the droplet. As a result, it demands a higher pressure drop along the capillary to maintain the same average speed of flow as without the bubble or droplet inside of it, that is, for the flow of the carrier liquid itself. Correspondingly, for a constant pressure drop along the capillary an introduction of a droplet decreases the speed of flow. This is equivalent to an increased resistance to flow in that capillary: a droplet carries an additional charge of hydrodynamic resistance.

Modeling the details of the velocity field around a droplet flowing in a capillary is difficult. Taylor (1961) and Bretherton (1961) put forward the foundations for these studies. Taylor proposed a relationship for the thickness (proportional to the square root of the capillary number, Ca) of the film deposited on a capillary after passage of a

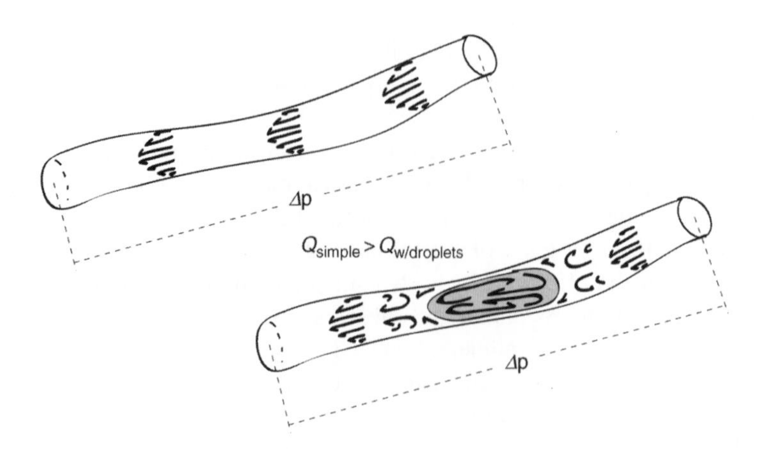

Figure 6.1 The flow of a simple fluid through a microchannel develops a parabolic profile of speed of flow, which minimizes the viscous dissipation for a given pressure drop Δp along the capillary. A droplet introduced into the flow modifies the flow field and increases dissipation. For a constant pressure drop this results in reduction of the volumetric rate of flow through that capillary

semi-infinite bubble, while Bretherton proposed the scaling for the pressure drop (proportional to $Ca^{2/3}$) along a finite bubble. The explosion of interest in the subject associated with the introduction of microfluidic techniques in recent years produced new experimental data on the hydrodynamic resistance contributed by the droplets and bubbles (Fuerstman *et al.*, 2007; Hodges, Jensen, and Rallison, 2004; Labrot *et al.*, 2009; Sessoms *et al.*, 2009). However, as the flow around a bubble (droplet) depends critically on a large number of parameters (volume of the immiscible segment, interfacial tension between the fluids, surface coverage with surfactant, viscosities of the two fluids, the speed of flow) a unified picture (or equations for the resistance introduced by bubbles and droplets in capillaries) is not yet available. Importantly, the droplets flowing in outlet capillaries of microfluidic droplet generators do increase the resistance to flow and do modify the rate of flow through the junctions at which they were created. As will be evident in the examples given below, the effects of increased resistance of channels guiding droplets and the blocking of the junction by a growing droplet can introduce strong hydrodynamic coupling between proximate droplet (bubble) generators.

6.3 Hydrodynamic Coupling

Perhaps the most striking example of the effects of hydrodynamic coupling on microfluidic emulsification was shown in 2005 by Garstecki, Fuerstman, and Whitesides (2005) in a system comprising five flow-focusing junctions connected in series and fed from a common source of the continuous liquid (Figure 6.2). For a given, fixed total rate of flow of the continuous liquid, at low pressures applied to the stream of gas, the tip

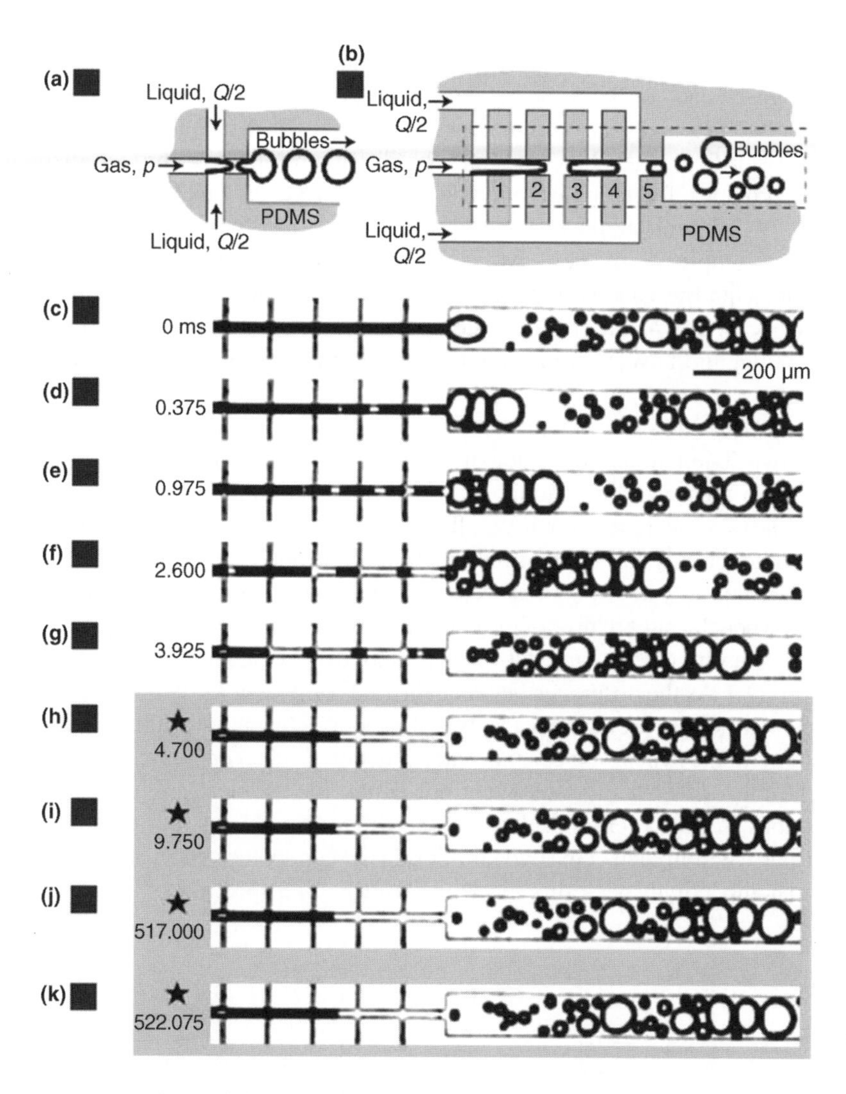

Figure 6.2 *Insets (a) and (b) show schematically a single flow-focusing junction and the system of five flow-focusing junctions connected in series with a common supply of the continuous liquid. Insets (c–h) show micrographs of the system at different instants within a single period of operation of the system. Insets (h–k) show snapshots separated by an integer multiple of the period, illustrating the reproducibility of the periodic dynamics of the system (Garstecki, Fuerstman, and Whitesides, 2005). Reprinted with permission from Nature Physics, Oscillations with uniquely long periods in a microfluidic bubble generator by P. Garstecki, M.J. Fuerstman and G.W. Whitesides,* ***1****, 3, 168–171 (2005) Macmillan Publishers Ltd*

of this stream approaches only the first orifice and the system produces monodisperse bubbles. Thus, the system operates in a period-1 mode. As the pressure applied to the stream of gas is increased, the tip of the gaseous phase enters the array of the orifices. Blocking of a number of orifices by the stream of gas modifies the distribution of the continuous liquid between the orifices, which in turn affects the rate of breakup of the stream of gas at each of the orifices. The rate of breakup, in turn, affects the times of release of bubbles and their volumes, which affects the distribution of the bubbles between the orifices, their blocking, and the distribution of flow through them. This feedback provides for complex dynamics of the system, which can produce, for example, 29 different bubbles in one limit cycle (or one "period"). Interestingly, in spite of the complexity of these cyclic dynamics, the system is stable against small perturbations in flow. Clearly hydrodynamic "cross-talk" between the orifices dramatically changes the dynamics of the system, from period-1 formation of monodisperse bubbles to long limit cycles producing non-trivial, multimodal distribution of volumes of bubbles.

Barbier *et al.* (2006) systematically explored the effects of hydrodynamic coupling between two T-junctions connected in parallel (Figure 6.3). Both junctions were supplied with the continuous liquid drawn from a common source. In addition, the outlets of the two T-junctions ran in parallel over a short distance to later recombine into a common outlet channel. Each of the junctions was independently supplied with the aqueous streams (Q_w and Q'_w) that were to be broken up into droplets. Barbier *et al.* found that when the rates of flow of the two water streams were equal, the system formed droplets synchronously (in- and out-of-phase modes) over a wide range of the rates of flow of oil (Q_o). As the difference between the rates of flow of the two aqueous streams increased, the range over which the two T-junctions were synchronized decreased (Figure 6.3). In the synchronized mode each junction produced monodisperse droplets. When the rates of flow of water were not equal, the droplets produced at each of the junctions in the synchronized mode had different volumes, but continued to be monodisperse within each of the two families. In the non-synchronized mode the system produced highly polydisperse droplets with a wide Fourier spectrum of the times of their emission.

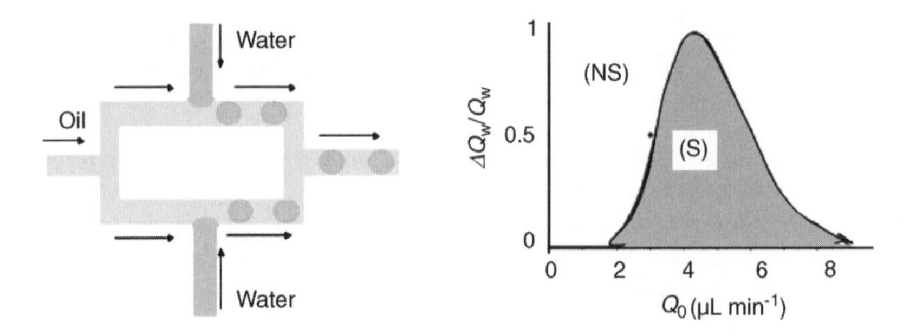

Figure 6.3 A schematic representation of the system of two T-junctions connected in parallel. The graph shows the regimes of operation of the system [synchronized (S) and non-synchronized (NS) formation of droplets at the two junctions] as a function of the rate of flow of oil and the relative difference in the two rates of flow of water ($\Delta Q_w = Q_w - Q'_w$). Reproduced with permission from (Barbier et al., 2006). Reprinted with permission from Physical Review E, Producing droplets in parallel microfluidic systems by V. Barbier, H. Willaime, P. Tabeling and F. Jousse, ***74****, 4, 046306 Copyright (2006) American Physical Society*

Barbier *et al.* varied the lengths of parallel sections of the channels, both upstream of the junctions and downstream of them, and found that increasing either of them resulted in a decrease in the range of rates of flow that led to synchronization of formation of the droplets. They hypothesized, and provided a simple physical model that backed up this supposition, that the shorter the parallel sections, the stronger the coupling between the two droplet generators. In the case of this system, this coupling was favorable in view of the formation of monodisperse droplets, as synchronization stabilized this mode of operation of the system.

Frenz *et al.* (2008) explored a slightly different system (Figure 6.4), which also had two generators of droplets fed from a common source of the continuous (oil) phase. The drops were emitted into short, parallel channels that merged downstream to a common, wider outlet. Formation of droplets at the two junctions synchronized when droplets in each of the parallel arms where large enough to fill their cross-section – thus the resistance to flow introduced by the droplets was critical to observing the hydrodynamic coupling between the two generators. Frenz *et al.* observed that it is possible to synchronize the two droplet generators over a fairly large range of values of the ratio of the rates of flow of the two droplet phases (Q_x/Q_y between 1:5 and 5:1). Symmetric feeding of the two generators ($Q_x/Q_y = 1$) resulted in the (synchronized) frequency of formation of droplets following an empirical scaling of $f_x = \alpha Q_o Q_x^{\beta}$, where Q_o is the rate of flow of oil, and the exponent $\beta = 0.82$. These Authors presented an insightful analysis that assumed that even when the feeding of the two droplet phases is asymmetric ($Q_x/Q_y \neq 1$), each half of the double droplet generator behaves as a half of a symmetrically fed generator, but with a different rate of flow of oil. As the asymmetrically fed generator was observed to synchronize formation of droplets at each of the two junctions, and because the relationships for the frequencies of

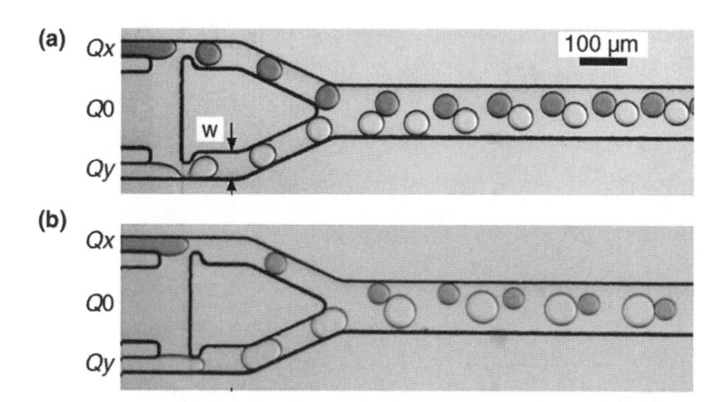

Figure 6.4 *Micrographs of the microfluidic system of two parallel droplet generators coupled via a common source of the continuous liquid. Inset (a) shows the operation of the system when the rates of flow of the aqueous streams (Q_x and Q_y) are equal (28 nL s^{-1} each). Inset (b) illustrates synchronization of an asymmetric case, when $Q_x = 16$ nL s^{-1} and $Q_y = 50$ nL s^{-1} (Frenz et al., 2008). Reprinted with permission from Langmuir, Microfluidic Production of Droplet Pairs by L. Frenz, J. Blouwolff, A.D. Griffiths and J.-C. Baret,* ***24****, 20, 12073–12076 Copyright (2008) American Chemical Society*

a symmetrically fed system have an explicit dependence on the rates of flow, they can be used to retrieve the effective division of Q_o between the two sides. This, in turn allows construction of an equation that yields the synchronization frequency as a function of the rate of flow of oil and of the two droplet phases, even if $Q_x/Q_y \neq 1$.

Importantly, this system enables formation of pairs of droplets of different chemical composition with the ability to tune the relative volumes of the droplets. This ability is important in the use of microdroplets as reaction chambers because it makes it possible to create pairs of drops comprising different solutions. Merging of such pairs forms reaction mixtures that can be processed on chip. Hung *et al.* (2006) used a system of two T-junctions facing each other and emitting droplets into a common main channel that carries the continuous liquid. They demonstrated that in such a system it is possible to form droplets of two different solutions alternately. Further, they showed that by adjusting the ratio of the rates of flow of the two aqueous streams it was possible to obtain a sequence of a different number of droplets of one type separated by single droplets of the second composition. Subsequent merging of such packets of droplets allowed for on-chip synthesis of cadmium–selenide nanoparticles. Chen *et al.* (2007) used this effect to introduce bubbles of gas in between droplets carrying reagents for an enzymatic reaction. In both of these reports it was the effect of blocking of the main channel by the growing (or passing) droplet that introduced the oscillations of pressure at the inlet of the second phase and led to synchronization.

Synchronization of formation of droplets at neighboring generators can also be used to form emulsions comprising repeating units of different numbers of droplets of distinct chemical composition. Chokkalingam, Herminghaus, and Seemann (2008) reported a system similar to that of Barbier *et al.* (2006) for synchronizing formation of droplets of two different solutions in all three regimes of breakup that they observed in their device (blocking–squeezing, step–emulsification, and jetting). This allowed for formation of ordered, flowing lattices of these heterogeneous droplets. Hashimoto, Garstecki, and Whitesides (2007) used a set of three flow-focusing generators emitting droplets (or bubbles) to a common outlet channel to form "composite" emulsions comprising bubbles and droplets of different fluids, arranged into periodic structures with the ability to control the stoichiometry of the unit cell (Figure 6.5).

6.4 Integrated Systems

Examples given above clearly show the effects of hydrodynamic coupling on the formation of droplets and bubbles. These effects can either interfere with the regular and predictable dynamics of an individual generators, thus hindering the ability to produce monodisperse emulsions, or can stabilize such regular dynamics. They can also be used to engineer sequences of droplets of distinct chemical identity, both for lab-on-chip, and, potentially, preparatory applications. The systems discussed above, however, do not present the high level of parallelization needed for the high-throughput formation of droplets. Massive integration requires microfabrication of a dense array of identical generators. In general, this task does not present significant challenges. The readily available methods of microfabrication allow for fabrication of microfluidic junctions that can be optimally packed on chips. However, as every fabrication method has its limits of fidelity, the junctions are never truly "identical". The small differences in dimensions of orifices or junctions can translate into differences in the volumes of droplets either directly, that is, two junctions of different dimensions but supplied with identical rates of flow can form differently sized droplets, and indirectly, as the differences in dimensions of the junctions can modify the rates of flow through them.

There are a number of ways to design the distribution network for liquids to supply the junctions. One is to interface the junctions directly with reservoirs of pressurized liquids – minimizing the hydraulic resistances of the inlets to and outlets from the junctions. A direct contact with a large reservoir of pressure can minimize the hydrodynamic coupling between the junctions: variations of the resistance to flow through the particular junction that arise during the process of formation of a single droplet are too small to significantly alter the pressure in the large reservoir that connects to a large number of droplet generators. Membrane emulsification draws directly from this idea: a porous membrane separates a large reservoir of pressurized liquid that flows through the pores and breaks into droplets on the other side of the membrane. The droplets are carried away by a slow flow of the continuous phase. The membranes can be microfabricated to have almost ideally reproducible diameters of the pores (Kobayashi, Mukataka, and Nakajima, 2005). These systems can form almost perfectly monodisperse emulsions, however, the throughput is strongly limited by clogging of the pores (van Dijke *et al.*, 2009). An alternative is to use a long and narrow slit

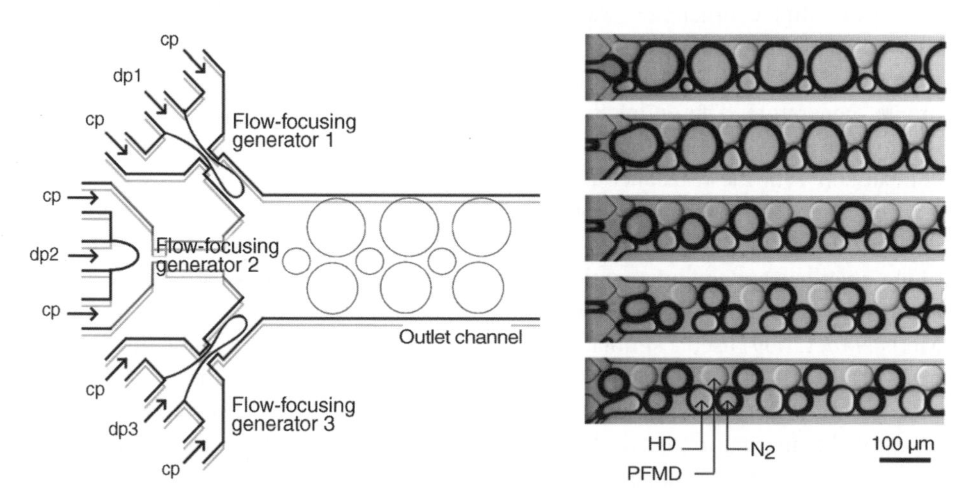

Figure 6.5 *A schematic representation of the system of three flow-focusing units generating droplets (or bubbles) into a common outlet channel and micrographs of the composite emulsions formed in this system (Hashimoto, Garstecki, and Whitesides, 2007). Reprinted with permission from Small, Synthesis of Composite Emulsions and Complex Foams with the use of Microfluidic Flow-Focusing Devices by M. Hashimoto, P. Garstecki and G.M. Whitesides,* ***3****, 10, 1792–1802 (2007). Copyright Wiley-VCH Verlag GmbH & Co. KGaA. Reproduced with permission.*

(van Dijke *et al.*, 2009) instead of the pores. These systems are more robust and also produce monodisperse droplets, but the throughput of even an integrated system comprising a multitude of slits is very limited (e.g., to less than 1 mL h^{-1}).

Hashimoto *et al.* (2008) studied formation of bubbles and droplets in model integrated systems comprising two and four flow-focusing junctions in parallel. The junctions were closely connected directly upstream of the orifices and downstream of them, as they all emitted droplets into a common and wide outlet channel. This study showed a pronounced difference in the dynamics of the systems that produced bubbles of gas and those forming droplets of liquid. In the former case, the parallel flow-focusing junctions synchronized for both in-phase, and out-of-phase formation of bubbles over wide ranges of the rates of flow of the continuous liquid and pressure applied to gas. This coupling resulted in formation of both monodisperse and bi-disperse bubbles. In contrast, formation of droplets of liquids did not show these effects. Over almost the entire range of rates of flow (per junction) that yield formation of droplets in a single junction, the two- and four-orifice systems produced monodisperse droplets of the same volume and did not show any significant effects of hydrodynamic coupling.

A conceptually similar approach of connecting the junctions to inlet and outlet reservoirs was used by Tetradis-Meris *et al.* (2009) to construct a truly integrated system of 180 flow-focusing generators on a $77 \times 108\,mm^2$ chip. The flow-focusing junctions had uniform, square ($20 \times 20\,\mu m^2$) cross-sections for all the four channels forming the junction. The parallel system successfully formed droplets of a mean diameter of 21.14 μm with a coefficient of variation of 4.74% of the mean. In their system, these workers used a distribution network of the liquids, which contained channels of a cross-section (e.g., $500 \times 1500\,\mu m^2$), that was much larger than the cross-section of the droplet generators. The outlets had similarly large cross-sections. This ensured that the hydrodynamic resistance to flow in the inlets and outlets was negligible in comparison with the resistance to flow in the junctions, and effectively realized the concept of contacting the junctions directly with a reservoir of pressurized liquid.

However, even for such a large – several orders of magnitude – ratio of the resistances of the junctions and the inlet–outlet network, the topology of the distribution network can still have an impact on the quality of the emulsion produced in the system. Tetradis-Meris *et al.* (2009) used a simulation of flow of a simple fluid through the distribution network to analyze how do small variations in the dimensions of the flow-focusing junctions translate into variations of rates of flow through them. They compared two types of the distribution networks: a symmetrical tree, which has the inlet of liquid split into two identical channels, each subsequently splitting into two, and so on, to form a network feeding $N = 2^n$ junctions, and a "ladder" network, which connected each junction to the reservoir of liquid directly. The results showed large differences: the tree amplified small differences in dimensions of the junctions into differences in rates of flow much stronger than the ladder design.

A different approach to equal distribution of the rates of flow of the junctions uses hydrodynamic resistance of either inlets-to or outlets-from the junctions to regulate the flow through them. This approach has the advantage that it is less prone to differences in resistance of the ducts associated with a particular junction due to imperfections of fabrication. This is because the junctions themselves have small lengths of the channels and the variations of the widths of the orifices or channels translate into large relative differences of resistance of the junction. On the other hand, in long microfluidic channels, the small erratic variations in their width should average out over their length and the relative variations in their resistance should be much smaller (e.g., proportional to the ratio of the width of the channel to its length).

To date there is no systematic knowledge of the relationship between the quality of the emulsion and the placing of the increased hydrodynamic resistance (either upstream, downstream of the orifices, or both) nor on the impact of the ratio of the resistance of inlets/outlets to the resistance of the droplet generators. Still, this approach has been shown to serve to produce monodisperse emulsions in a high-throughput

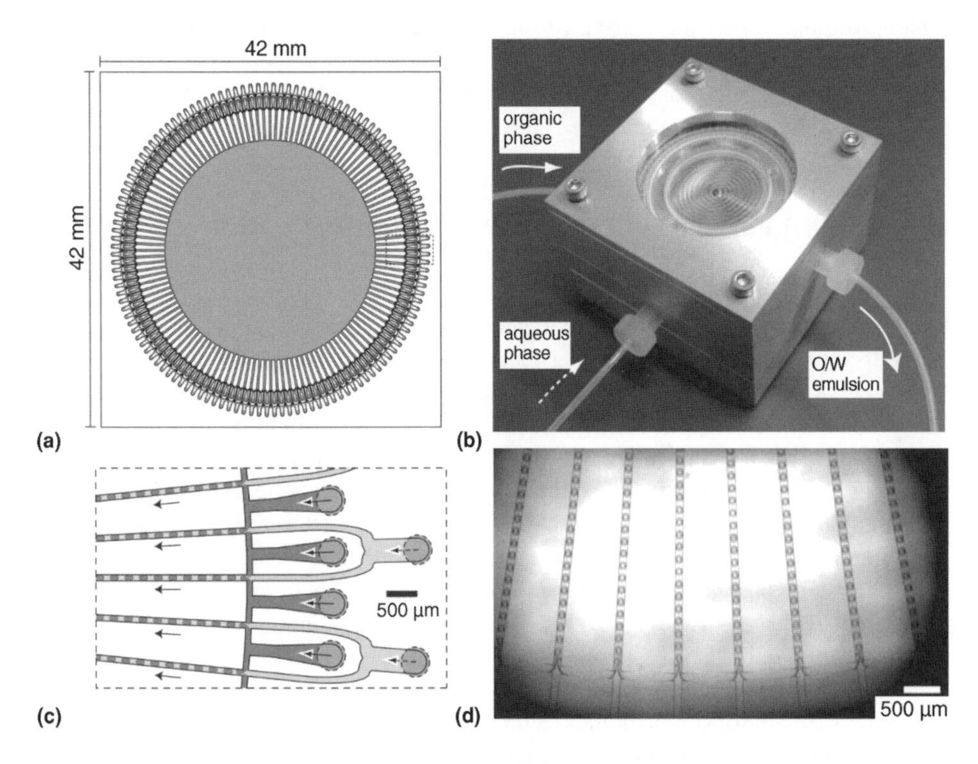

Figure 6.6 *A schematic representation (a) of the layer of microfluidic channels comprising 128 flow-focusing junctions for formation of droplets and a photograph (b) of the multilayer microfluidic device for high-throughput formation of droplets. Inset (c) shows a magnification of the geometry of the parallel network of droplet generators and inset (d) presents a micrograph of a fragment of the system during formation of monodisperse droplets (Nisisako and Torii, 2008). Reprinted with permission from Lab on a Chip, Microfluidic large-scale integration on a chip for mass production of monodisperse droplets and particles by T. Nisisako and T. Torii,* ***8****, 2, 287–293 Copyright (2008) Royal Society of Chemistry*

manner. For example, Nisisako and Tori (2008) reported a system that had a circular arrangement of parallel flow focusing junctions that emitted droplets into radially converging microchannels of a square ($100 \times 100\,\mu m^2$) cross-section (Figure 6.6). The junctions were supplied with liquids via large diameter (0.4 mm) vertical channels that connected to pressurized reservoirs located in the bottom layers of the device. The resistance of the radial outlet channels was approximately 100-times larger than the hydraulic resistance of the inlets. The system produced tightly monodisperse emulsions of simple and bi-colored (Janus) droplets with diameters of approximately 100 μm at a total rate of flow of the dispersed phase of $0.3\,L\,h^{-1}$ with a coefficient of variation as low as 1.3% of the mean.

In contrast, Li *et al.* (2009) placed the hydrodynamic resistance upstream of the flow focusing junctions in a set of 8 chips, each comprising 16 droplet generators. The junctions produced droplets into less resistant outlet channels that provided time of flow and space to illuminate the droplets of monomer, to polymerize them into particles. These Authors varied the resistance on the inlets for the droplet phase and found that increasing this resistance produced more tightly monodisperse droplets and particles. The system was capable of formation of approximately 140 μm droplets with a coefficient of variation well below 5%, at a rate of $50\,g\,h^{-1}$.

6.5 Parallel Formation of Droplets of Distinct Properties

The control that microfluidic systems offer over the flow of simple and immiscible fluids can be extended far beyond the sole idea of increasing the throughput of formation of monodisperse emulsions comprising identical (or nearly identical) droplets. Microfluidic networks can be used to distribute liquids in precisely designed proportions, so that each of the orifices could be supplied with a *different* rate of flow of the dispersed or continuous phase to produce droplets of different predesigned diameters at each of the junctions. This could lead to, for example, formulation of heterogeneous suspensions of droplets or particles with the distribution of their diameters determining (Xu *et al.*, 2009) the temporal profile of release of active substances imbedded in them. Li *et al.* (2008) used a system of four flow-focusing junctions operating in parallel, each having a different width of the orifice to produce quadra-modal emulsions with four well-resolved and monodisperse populations of droplets.

Microfluidic systems can readily form gradients of concentration of chemistry in continuous streams of simple liquids (Jeon *et al.*, 2000). The mechanism of operation of these systems relies on the idea of: (i) introducing two miscible solutions of different concentration onto the chip, (ii) successive splitting of each of the streams and merging with the sub-streams of the other solution, (iii) allowing for each of the set of merged streams to mix, and (iv) spliting and recombining the mixed streams again to obtain a larger number of streams of different concentrations. In this way it is possible to form N streams of different compositions from the two inputs. The gradation of concentration of each of the original solutions in the N streams can be designed via appropriate choice of the lengths of the branches in the network. This – gradient generating – system of

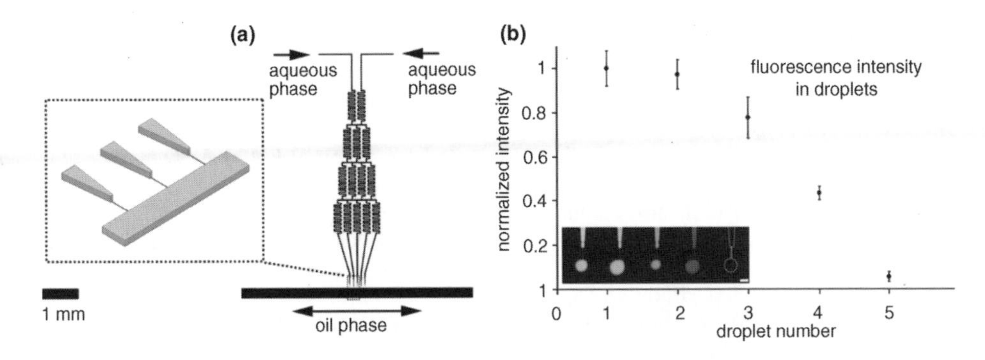

Figure 6.7 *Schematic representation of the gradient forming network coupled to five T-junction droplet generators (a) and a graph illustrating the intensity of fluorescence from droplets created at the five junctions supplied with different concentrations of fluorescein (b) (Lorenz et al., 2008). Reprinted with permission from Analytica Chicmica Acta, Simultaneous generation of multiple aqueous droplets in a microfludic device by R.M. Lorenz, G.S. Fiorini, G.D.M. Jeffries et al.,* ***630****, 2, 124–130 Copyright (2008) Elsevier Ltd*

channels can be used to feed the inlets of junctions for the formation of droplets, to create an emulsion comprising droplets of different chemical composition. Lorenz *et al.* (2008) demonstrated coupling of a gradient forming network with five outlets to five T-junctions emitting droplets into a common outlet channel. This system produced a predesigned distribution of concentration of fluorescein in the droplets (Figure 6.7).

Damean *et al.* (2009) used the same concept to construct an analytical system that tested the same enzymatic reaction within droplets in four parallel lines, each line supporting a reaction at different concentrations of the substrate. In general, these systems offer a unique opportunity for formulating suspensions of droplets or particles with all the characteristics under precise control, including arbitrary distribution of the diameters of individual droplets and of their chemical composition.

6.6 Conclusions

In summary, there is a number of experimental realizations of parallel microfluidic systems for formation of droplets. The exact effect of using hydrodynamic resistance and its placing upstream or downstream of the junctions has not been systematically explored, but the existing reports certify that parallelization is possible. Judging by the dimensions of individual junctions, a higher degree of integration should be possible. For example, for junctions of typical cross-sections of $100 \times 100\,\mu m^2$ it should be possible to fit at least one or two junctions per 1 mm^2 of the smallest facet of the device (100–200 droplet generators per cm^2). The available levels of integration are much lower, yielding $\sim$10–20 junctions per cm^2 (Li *et al.*, 2009, Nisisako and Torri, 2008). Thus, further progress in the area of parallelization of microfluidic droplet generators and increasing throughput of microfluidic emulsification can be expected.

References

Barbier, V., Willaime, H., Tabeling, P., and Jousse, F. (2006) *Phys. Rev. E*, **74**, 046306.
Bretherton, F.P. (1961) *J. Fluid Mechan.*, **10**, 166–188.
Chen, D.L.L., Li, L., Reyes, S., Adamson, D.N., and Ismagilov, R.F. (2007) *Langmuir*, **23**, 2255–2260.
Chokkalingam, V., Herminghaus, S., and Seemann, R. (2008) *Appl. Phys. Lett.*, **93**, 254101.
Damean, N., Olguin, L.F., Hollfelder, F., Abell, C., and Huck, W.T.S. (2009) *Lab Chip*, **9**, 1707–1713.
Frenz, L., Blouwolff, J., Griffiths, A.D., and Baret, J.C. (2008) *Langmuir*, **24**, 12073–12076.
Fuerstman, M.J., Lai, A., Thurlow, M.E., Shevkoplyas, S.S., Stone, H.A., and Whitesides, G.M. (2007) *Lab Chip*, **7**, 1479–1489.
Garstecki, P., Fuerstman, M.J., and Whitesides, G.M. (2005) *Nat. Phys.*, **1**, 168–171.
Hashimoto, M., Garstecki, P., and Whitesides, G.M. (2007) *Small*, **3**, 1792–1802.
Hashimoto, M., Shevkoplyas, S.S., Zasonska, B., Szymborski, T., Garstecki, P., and Whitesides, G.M. (2008) *Small*, **4**, 1795–1805.
Hodges, S.R., Jensen, O.E., and Rallison, J.M. (2004) *J. Fluid Mech.*, **501**, 279–301.
Hung, L.H., Choi, K.M., Tseng, W.Y., Tan, Y.C., Shea, K.J., and Lee, A.P. (2006) *Lab Chip*, **6**, 174–178.
Jeon, N.L., Dertinger, S.K.W., Chiu, D.T., Choi, I.S., Stroock, A.D., and Whitesides, G.M. (2000) *Langmuir*, **16**, 8311–8316.
Kobayashi, I., Mukataka, S., and Nakajima, M. (2005) *Langmuir*, **21**, 7629–7632.
Labrot, V., Schindler, M., Guillot, P., Colin, A., and Joanicot, M. (2009) *Biomicrofluidics*, **3**, 012804.
Li, W., Young, E.W.K., Seo, M., Nie, Z., Garstecki, P., Simmons, C.A., and Kumacheva, E. (2008) *Soft Mat.*, **4**, 258–262.
Li, W., Greener, J., Voicu, D., and Kumacheva, E. (2009) *Lab Chip*, **9**, 2715–2721.
Lorenz, R.M., Fiorini, G.S., Jeffries, G.D.M., Lim, D.S.W., He, M.Y., and Chiu, D.T. (2008) *Anal. Chim. Acta*, **630**, 124–130.
Luque, A., Perdigones, F.A., Esteve, J., Montserrat, J., Ganan-Calvo, A.M., and Quero, J.M. (2007) *J. Microelectromech. Syst.*, **16**, 1201–1208.
Luque, A., Perdigones, F., Esteve, J., Montserrat, J., Ganan-Calvo, A., and Quero, J.M. (2009) *J. Micromech. Microeng.*, **19**, 045029.
Nisisako, T. and Torii, T. (2008) *Lab Chip*, **8**, 287–293.
Raven, J.P. and Marmottant, P. (2006) *Phys. Rev. Lett.*, **97**, 154501.
Sessoms, D.A., Belloul, M., Engl, W., Roche, M., Courbin, L., and Panizza, P. (2009) *Phys. Rev. E*, **80**, 016317.
Sullivan, M.T. and Stone, H.A. (2008) *Philos. Trans. R. Soc. London A-Math. Phys. Eng. Sci.*, **366**, 2131–2143.
Taylor, G.I. (1961) *J. Fluid Mech.*, **10**, 161–165.
Tetradis-Meris, G., Rossetti, D., de Torres, C.P., Cao, R., Lian, G.P., and Janes, R. (2009) *Ind. Eng. Chem. Res.*, **48**, 8881–8889.
van Dijke, K., Veldhuis, G., Schroen, K., and Boom, R. (2009) *Lab Chip*, **9**, 2824–2830.
Xu, Q.B., Hashimoto, M., Dang, T.T., Hoare, T., Kohane, D.S., Whitesides, G.M., Langer, R., and Anderson, D.G. (2009) *Small*, **5**, 1575–1581.

7

Synthesis of Polymer Particles in Microfluidic Reactors

CHAPTER OVERVIEW

Microfluidic Reactors for Polymer Particles. Eugenia Kumacheva and Piotr Garstecki.

7.1 Introduction

Microfluidic generation of polymer particles can be classified by the number of phases in the original, prior-to-synthesis system; by the mechanism of solidification of "precursor" droplets in multiphase systems; and by the type of polymerization process, which may include "on-chip" or "off-chip" polymerization conducted in the continuous or batch mode.

In a single-phase microfluidic synthesis, the entire microchannel is filled with a mixture of the monomer (or a pre-polymer), an initiator, and if required, a catalyst and a solvent. The monomer undergoes polymerization as it travels through the microchannels (Wu *et al.*, 2004; Serra *et al.*, 2005; Iwasaki and Yoshida, 2005; Honda *et al.*, 2005). For example, atom transfer radical polymerization of 2-hydroxypropyl methacrylate was carried out in the microfluidic reactor (Wu, 2004). Particle synthesis is achieved by *site-specific* polymerization of the monomer, which is achieved through localized photoinitiated polymerization by irradiating the liquid through a mask, so that the features of the mask determine the shapes of the resulting polymer particles.

Multiphase flows are formed in microchannels when two of more liquids that are immiscible, at least on the time scale of an experiment, are introduced in the microchannels. The resulting flows can take forms of suspended droplets, plugs, and wall-wetting films (Günther and Jensen, 2007). Polymer particles are generated in a two-step procedure that includes microfluidic emulsification and subsequent solidification of the resulting droplets or plugs.

In one-phase systems, synthesis of polymer particles is achieved by photoinitiated polymerization. In two-phase systems, precursor droplets may compartmentalize a pure monomer or a liquid reactive prepolymer, or a solution of a monomer or a polymer. The solidification of droplets occurs by polymerization, crosslinking gelation, or by the removal of the solvent.

In one-phase systems, particles are always produced in the continuous mode *in situ*, that is, in the microfluidic reactor. Solidification of droplets in two-phase synthesis can be achieved in the continuous mode, or by combining *in situ* synthesis (or in more general terms, *in situ* solidification) with batch post-polymerization. Sometimes, microfluidic devices are only used for the emulsification and the collected droplets are solidified in a batch process. Strictly speaking, this process cannot be called "microfluidic synthesis," as it only exploits microfluidic droplet generators.

The present chapter summarizes current progress in the microfluidic synthesis of rigid polymer particles by means of polymerization. Currently, two polymerization mechanisms are used to synthesize rigid polymer particles in microfluidic reactors, namely, free-radical polymerization and polycondensation. (Redox polymerization has been successfully used for the synthesis of gel microbeads and will be discussed in Chapter 8). Free-radical polymerization is the most frequently used route to generating polymer particles with a uniform and core-shell morphology, whereas polycondensation has been used for interfacial polymerization and it yielded polymer microcapsules.

In this chapter we will focus on the microfluidic synthesis of rigid spherical polymer particles with uniform structures. The synthesis of hydrogel particles, generation of polymer capsules and the synthesis of particles with non-spherical shapes and complex morphologies are discussed in Chapters 8–10.

7.2 Particles Synthesized by Free-Radical Polymerization

7.2.1 Polymerization in Multi-Phase Flow

In 2001–2002 Sugiura and coworkers showed that microfluidic emulsification is a useful route to producing highly monodispersed droplets of monomers, which upon polymerization can yield polymer particles with a narrow distribution of sizes (Sugiura *et al.*, 2001; Sugiura, Nakajima, and Seki, 2002). Free-radical polymerization of monomers emulsified in microfluidic devices was demonstrated for the styrenic monomer. Droplets of divinyl benzene mixed with a thermal initiator benzoyl peroxide were generated by breaking up the monomer stream on the terraces of microchannels. The droplets were dispersed in the aqueous continuous phase comprising an anionic surfactant sodium dodecyl sulfate.

A microfluidic emulsification was achieved (Sugiura *et al.*, 2001; Sugiura, Nakajima, and Seki, 2002) by pressurizing divinyl benzene into the continuous aqueous phase through precisely fabricated microchannels. In order to increase the efficiency of emulsification, each plate had a large number of tiny channels along each terrace line. The stream of the monomer acquired the shape of the microchannel and was cut off by the stress imposed by the flow of the continuous phase. The resulting droplets gained a spherical shape, in order to minimize their surface energy. The process of droplet formation resembled membrane emulsification (Nakashima, Shimizu, and Kukizaki, 1991), and in principle, it could be applied to the preparation of both oil-*in*-water and water-*in*-oil emulsions. The size of the droplets was controlled by the value of interfacial tension between the monomer phase and the continuous aqueous phase, the dimensions of the microchannels, and the flow rates of the liquids. Polydispersity of the droplets was characterized as the coefficient of variation, defined as $CV = (\sigma/d)/100$, where CV is the coefficient of variation (%), d is the number-average diameter, and σ is the standard deviation of the diameter of the droplets. Microfluidic emulsification produced divinyl benzene droplets with a polydispersity of below 3%.

The emulsion of divinyl benzene mixed with initiator was stabilized with 2 wt% of poly(vinyl alcohol) and heated to 90 °C, in order to initiate polymerization of the monomer. The diameter of crosslinked polymer particles was determined by the size of the "precursor" droplets of divinyl benzene. The polydispersity of the microbeads increased in comparison with precursor droplets, but overall it did not exceed 8%.

In this work, the approach to *continuous* microfluidic reactors for synthesis of polymer particles has not been implemented. Overall, particle production resembled conventional suspension polymerization. The use of the microfluidic device was limited only to the production of precursor monomer droplets, whereas the polymerization was conducted in a batch mode. The exploitation of two distinct stages – the microfluidic emulsification of the monomer and its batch polymerization – ensured that the microchannels were not clogged by the solidified particles. The broadening in the distribution of sizes of the particles, as compared with the "precursor" droplets, was the result of the coalescence between the monomer droplets during their transportation and/or during batch polymerization.

A microfluidic approach to the synthesis of polymer particles was further developed by Nisisako, Torii, and Higuchi (2004). This group generated polymer particles by photoinitiated or thermally initiated polymerization of microfluidically emulsified acrylic monomer, poly(1,6-hexanediol diacrylate), in an aqueous solution of poly(vinyl alcohol). Both T-junction and Y-shaped microfluidic droplet generators fabricated in quartz glass were used for the preparation of droplets. Figure 7.1 shows the generation of the monomer droplets in the T-junction device at varying flow rates of the continuous phase (Q_c) at a constant flow rate of the disperse phase. The size of the droplets and the breakup frequency of the monomer thread were precisely controlled in a particular range of Q_c (or in the well-defined range of the flow-rate ratios of the monomer and continuous aqueous phases). The size of monomer droplets was varied from 30 to 120 μm by changing the ratio of flow rates of the liquids. At insufficiently high values of

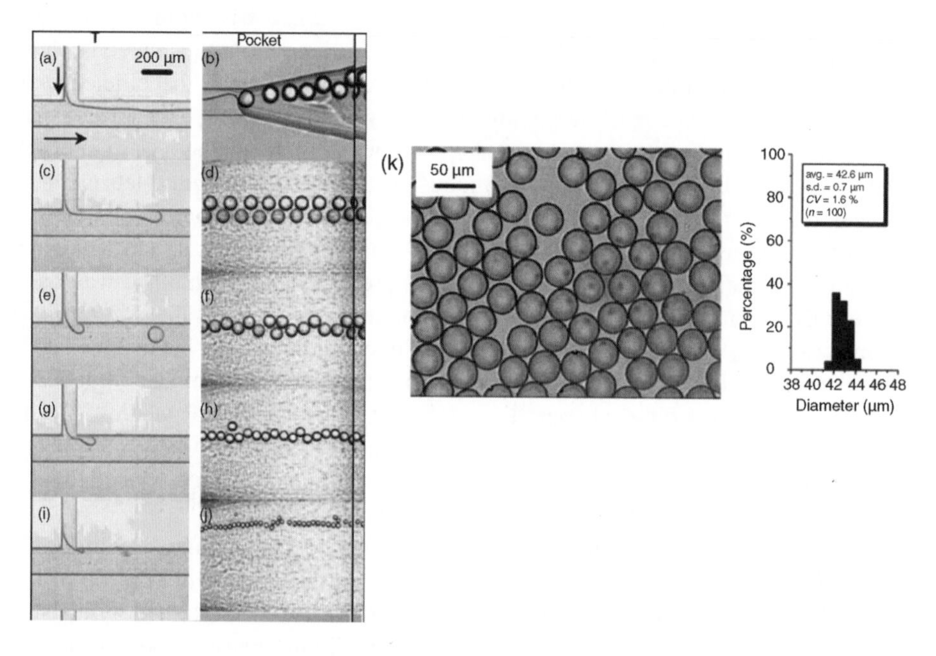

__Figure 7.1__ Microfluidic emulsification in the T-junction/pocket at a fixed monomer flow rate of 0.1 mL h^{-1} and the flow rate of the continuous phase of (a, b) Q_c = 0.5 mL h^{-1}; (c, d) Q_c = 1.0 mL h^{-1}; (e, f) Q_c = 2.0 mL h^{-1}; (g, h) Q_c = 4.0 mL h^{-1}; (i, j) Q_c = 18.0 mL h^{-1}; (k, l) Q_c = 22.0 mL h^{-1}. All images were recorded at 10 000 fps. (k) Microspheres of poly(1,6-hexanediol diacrylate) cured by photoinitiated polymerization and size distribution of the microbeads with mean diameter of 43 μm and CV = 1.6% (Nisisako, Torii, and Higuchi, 2004). Reprinted with permission from Chemical Engineering Journal, Novel microreactors for functional polymer beads by T. Nisisako, T. Torii and T. Higuchi, __101__, 1–3, 23–29 Copyright (2004) Elsevier Ltd

Q_c, the monomer stream spread over the wall of the microchannels and the formation of droplets occurred outside the T-junction. Above the threshold value of Q_c, the breakup point on the monomer thread moved downstream, which resulted in the formation of polydisperse droplets before the pocket area. The authors admitted the formation of small satellite droplets with diameter $<5\,\mu m$, however the main population of large droplets had a very narrow size distribution with polydispersity below 2%. It was suggested that the satellites could be easily removed by conventional filtration techniques.

Following the emulsification step, conversion of droplets into particles was conducted in the batch process. The resulting droplets were collected in a beaker and cured for 1–2 min either by exposing the droplets to UV light radiation, or by heating them in a hot-water bath to 70–80 °C for 4–5 min.

In the conceptually similar process, off-chip photopolymerization was conducted for core-shell droplets generated by breaking up a coaxial jet of water (an inner liquid) and DuPont prepolymer Somos 6120 (an outer liquid) (Loscertales *et al.*, 2002). The compound jet with a diameter from tens of nanometers to tens of micrometers was generated by the action of electro-hydrodynamic forces. The core-shell droplets were collected on a plate and the shell comprising Somos 6120 was photopolymerized.

In the processes described above, the role of microfluidics was limited to the emulsification of the monomer or reactive prepolymer, and therefore the concept of the continuous microfluidic synthesis of polymer particles was not fully implemented. In 2005, Kumacheva and Whitesides and coworkers reported *continuous* microfluidic synthesis of polymer particles (Xu *et al.*, 2005). Figure 7.2 shows a schematic drawing of the continuous microfluidic reactor for polymer particles (Xu *et al.*, 2005). Two components – the emulsification and the polymerization compartments – were integrated into the planar microfluidic reactor. Monomer droplets with dimensions from about 20 to 150 μm were generated in a microfluidic flow-focusing device (MFFD)

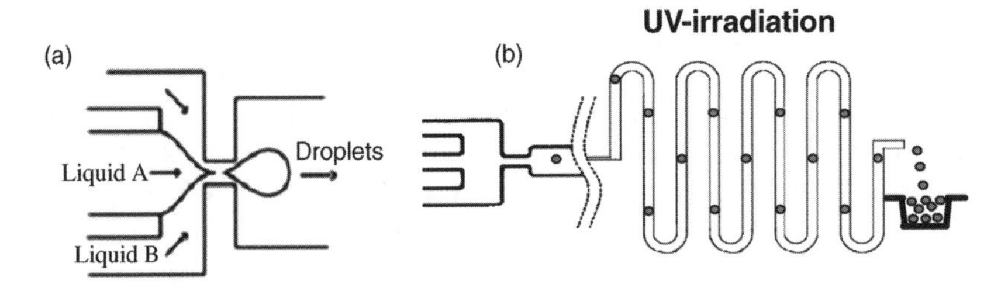

Figure 7.2 *Schematics of the experimental setup for the continuous synthesis of polymer particles in microfluidic reactors: (a) emulsification of a non-polar monomer (liquid A) in an aqueous continuous phase (liquid B) in the flow-focusing device. Immiscible liquids A and B are forced into the narrow orifice where the inner liquid thread breaks to release monodisperse droplets into the downstream channel (Xu et al., 2005). (b) Monomer droplets flow in the extension wavy channel where they undergo photoinitiated polymerization. Reprinted with permission from Langmuir, Controlled synthesis of nonspherical microparticles using microfluidics by D. Denukuri K. Tsoi, T. A. Hatton and P. S. Doyle,* ***21****, 6, 2113–2116 Copyright (2005) American Chemical Society*

(Anna, Bontoux, and Stone, 2003). A pressure gradient applied along the long axis of the MFFD forced two immiscible liquids (a monomer, liquid A) and an aqueous continuous phase (liquid B) through a narrow orifice. Various multifunctional monomers such as tripropylene glycol diacrylate (TPGDA), dimethacrylate oxypropyldimethylsiloxane (DMOS), ethylene glycol diacrylate, divinylbenzene, and pentaerythritol triacrylate were introduced into the MFFD as liquid A. Liquid B was typically an aqueous solution of the surfactant sodium dodecyl sulfate (SDS) or a polymer stabilizer poly(vinyl alcohol) (PVA).

In the orifice the continuous phase surrounded the monomer thread, which became unstable and broke up in a periodic manner, releasing droplets into the downstream channel. Subsequent to the emulsification, monomer droplets were forced into the serpentine extension channel (a polymerization compartment) where they were exposed to UV-irradiation and polymerized *in situ* by photoinitiated free-radical polymerization (Xu *et al.*, 2005). The role of the extension channel was to provide sufficient time for the continuous *on-chip* polymerization. Solid crosslinked polymer microspheres were collected at the exit of the microfluidic reactor. For a particular type of the microfluidic droplet generator, control of particle size was achieved by changing the flow rates of the monomer and continuous phases to produce droplets with different dimensions. The narrow size distribution of the droplets was well-preserved during polymerization and the resulting particles had coefficients of variation, *CV*, as low as 1.5%. In addition to spherical polymer particles, non-spherical microbeads were produced (see Chapter 10).

In a similar work, photoinitiated free-radical synthesis was used for the continuous polymerization of plugs generated by breaking off a stream of the reactive UV-sensitive prepolymer Norland Optical Adhesive 60 (NOA60, Norland Products) (Dendukuri *et al.*, 2005). Figure 7.3 illustrates the design of the experimental setup used for the synthesis of polymer particles. Emulsification at the T-junction produced large plugs, which upon polymerization yielded non-spherical particles. This confinement-based route to microfluidic production of non-spherical particles is discussed in greater detail in Chapter 10. Here we point out that polymerization of the liquid photopolymer occurred *in situ*, that is, on a microfluidic chip, and in the continuous process. Following the break off of the stream of NOA60 at a T-junction (Thorsen *et al.*, 2001), the plugs moved to the downstream channel where they were exposed to UV-irradiation and polymerized (Figure 7.3). *In situ* polymerization enabled precise control of the distribution of sizes of the polymer particles.

Both research groups (Xu *et al.*, 2005; Dendukuri *et al.*, 2005) produced "one polymer particle at a time" by polymerizing individual "precursor" droplets. Whereas in general this approach raised the question about the productivity of microfluidic reactors for the synthesis of polymer particles, it had several important advantages. One of them was the ability to preserve an extremely narrow size distribution of droplets generated by the microfluidic emulsification. Because of the highly periodic nature of breakup of the monomer thread, a plug of the continuous phase existing between the droplets produced a well-defined spacing between them, as shown schematically in Figure 7.2b. This plug prevented collisions and coalescence between the droplets prior to their

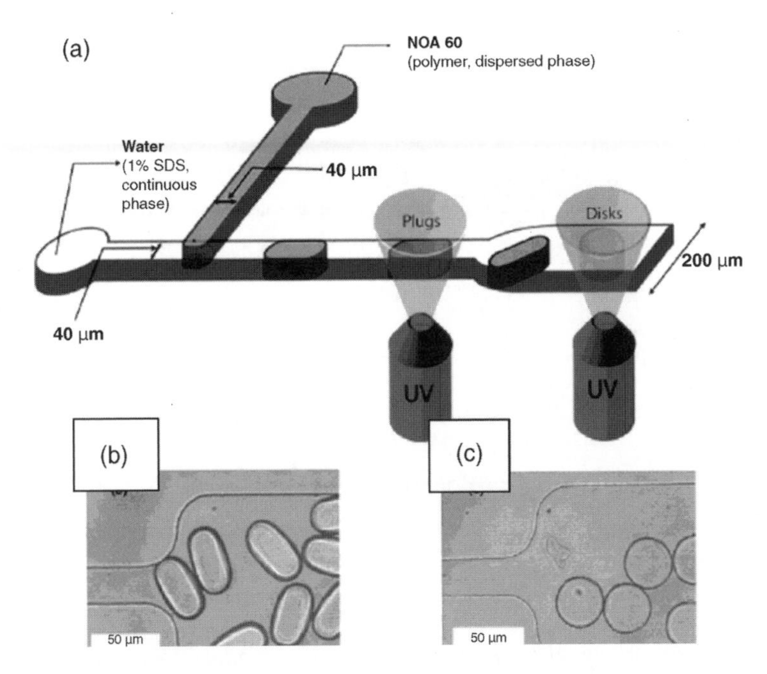

Figure 7.3 Microchannel geometry used to create plugs and disks: (a) schematic of channel with plug and disk creation zones marked; (b) polymerized plugs in the 200 μm section of the channel, 38 μm height; and (c) polymerized disks in the 200 μm section of the channel, 16 μm height (Dendukuri et al., 2005). Reprinted with permission from Langmuir, Continuous Fabrication of Biocatalyst Immobilized Microparticels using Photopolymerization and Immiscible Liquids in Microfluidic Systems by W. J. Jeong, J. Y. Kim, J. Choo et al., **21**, 9, 3738–3741 Copyright (2005) American Chemical Society

solidification. The spacing between the droplets, determined by the length of the plug, could be readily controlled by tuning the frequency of break off of the monomer thread via the variation in the flow rates of the droplet and continuous phases.

Complementary to the synthesis of polymer particles, continuous microfluidic polymerization was also reported for the synthesis of microfibers and microtubes (Jeong *et al.*, 2005). The approach employed the formation of a multi-liquid stream and its "on-the-fly" *in situ* photopolymerization, as shown in Figure 7.4. The microfluidic device was fabricated by incorporating a pulled glass micropipette into a preformed hole in a PDMS substrate comprising a microchannel network. When the polymerizable liquid monomer 4-hydroxybutyl acrylate and non-polymerizable sheath fluid (50 vol% poly(vinyl alcohol) were introduced into the microfluidic device, a 3D-coaxial sheath flow stream around the monomer stream flow was formed at the merging position (labeled as X) of both flows. The outlet tube was exposed to UV-irradiation, in order to trigger the photoinitiated polymerization of the stream of 4-hydroxybutyl acrylate. The polymerized thread moved along the direction of flow without touching the inner walls of the device and emerging from the outlet as a fiber.

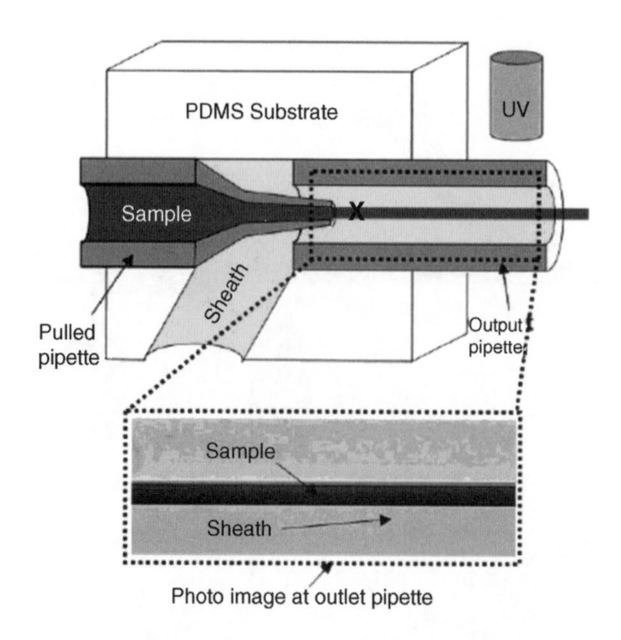

Figure 7.4 The experimental setup for the microfluidic synthesis of microfibers. Adapted with permission from Lab on a Chip, Hydrodynamic microfabrication via "on the fly" photopolymerization of microscale fibers and tubes by W. Jeong et al., 4, 576–580. Copyright (2004) Royal Society of Chemistry

7.2.1.1 Emulsification of Polymerizable Liquids

General trends in microfluidic emulsification are discussed in Chapter 4. Here we will only describe the essentials of the emulsification of monomers with different compositions and properties and the relationships between the emulsification and polymerization stages, which are pertinent to the control of the ultimate size and size distribution of the resulting polymer particles.

The ability to produce polymer particles by the microfluidic synthesis raised the question about the types of monomers that produce droplets with a narrow size distribution. In particular, the question was whether microfluidic emulsification can be used to generate highly monodisperse droplets from multifunctional monomers or reactive prepolymers with high viscosities and/or non-Newtonian properties.

A study of the emulsification of multifunctional monomers in a MFFD was conducted for four liquid acrylate monomers: ethylene glycol dimethacrylate (EGDMA), tri(propylene glycol) diacrylate, pentaerythritol triacrylate (PETA-3), and pentaerythritol tetraacrylate (PETA-4) (Seo *et al.*, 2005). The properties of these monomers are given in Table 7.1. The density and interfacial tensions of the monomers with the continuous aqueous phase – an aqueous solution of SDS – showed a moderate change from EGDMA to PETA-4, whereas viscosity increased more than 500-fold. Prior to the emulsification experiments it was ensured that all monomers had Newtonian properties. Figure 7.5a shows that viscosity of all monomers did not change with increasing shear rate.

Table 7.1 *Properties of monomers emulsified in MFFD (Seo et al., 2005)*

Monomer	Density ρ (g cm^3)	Viscosity μ (cP)	Interfacial tension γ (dyn cm^{-1})
EGDMA	1.05	3.5	1.0
TPGDA	1.03	14	2.8
PETA-3	1.18	586	3.1
PETA-4	1.19	1813	3.4

Source: Reprinted with permission from *Langmuir*, Continuous Microfluidic Reactors for Polymer Particles by M. Seo, Z. Nie, S. Xu *et al.*, **21**, 25, 11614–11622 Copyright (2005) American Chemical Society

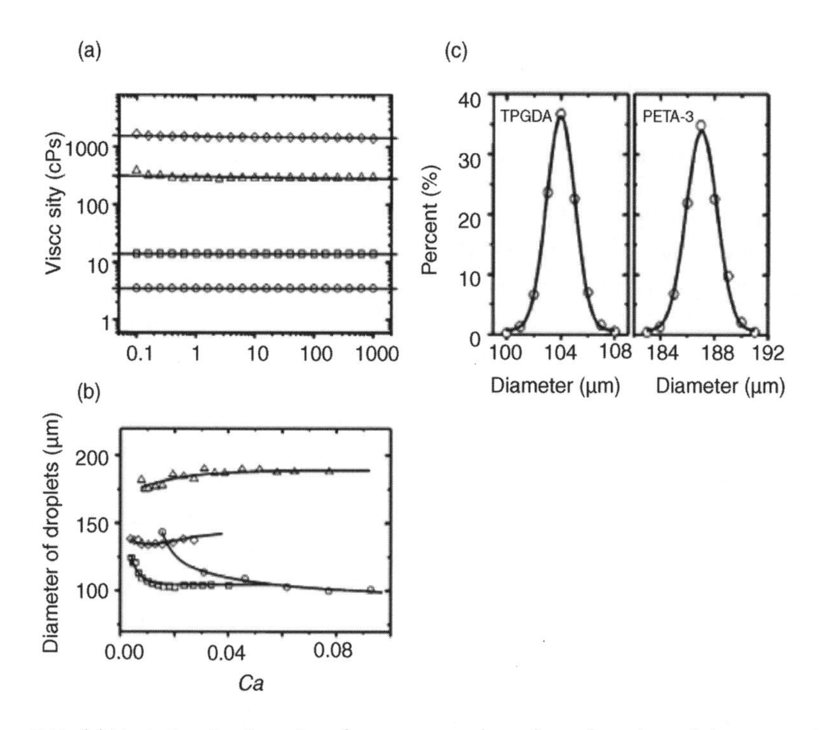

Figure 7.5 *(a) Variation in viscosity of monomers plotted as a function of shear rate. (a and b): PETA-4 (△), PETA-3 (◇), TPGDA (□), and EGDMA (○). (b) Variation in droplet diameter with increasing capillary number, Ca, of the continuous aqueous phase for a constant water-to-monomer flow-rate ratio, Q_w/Q_m of 60. For Ca > 0.028, a thread of PETA-4 did not break up into droplets. (c) Distribution of diameters of TPGDA droplets (average diameter 103.9 μm, CV = 0.94%, Q_m = 0.035 mL h^{-1}, Q_w = 2.1 mL h^{-1}) and PETA-3 droplets (average diameter 187.5 μm, CV = 0.9%, Q_m = 0.06 mL h^{-1}, Q_w = 3.6 mL h^{-1}) (Seo et al., 2005). Reprinted with permission from Langmuir, Continuous Microfluidic Reactors for Polymer Particles by M. Seo, Z. Nie, S. Xu et al., **21**, 25, 11614–11622 Copyright (2005) American Chemical Society*

Emulsification was conducted in the microfluidic flow-focusing device (Anna, Bontoux, and Stone, 2003), however, it can be expected that the trends observed in the formation of monomer droplets are applicable to other mechanisms of microfluidic emulsification, such as to the breakup of liquid threads at T-junction (Thorsen *et al.*, 2001) or terraces in microchannels (Sugiura *et al.*, 2001; Sugiura, Nakajima, and Seki, 2002). Figure 7.5b shows the variation in the diameter of monomer droplets plotted as a function of the capillary number, *Ca*, of the continuous phase. The size of droplets was determined by the properties of the monomer liquid and the flow rates of the continuous and droplet phases. At low values of *Ca* all monomers produced relatively large droplets. With increasing *Ca*, the diameters of droplets of low-viscous monomers EGDMA and TPGDA decreased until the size of droplets became invariant. The stabilized dimensions of droplets of EGDMA and TPGDA at high values of *Ca* were close to approximately 100 μm.

By contrast, for the intermediate and large values of *Ca*, a different trend was observed for tri- and tetra-functional monomers PETA-3 and PETA-4, that is, the monomers with high viscosity. With increasing values of *Ca* the dimensions of droplets formed by PETA-3 first slightly increased and then remained almost invariant. For PETA-4 with increasing *Ca*, the size of droplets slightly increased and for $Ca > 0.028$, a thread of PETA-4 did not break up into droplets.

It was concluded that the difference in the formation of droplets from the monomers listed in Table 7.1 was caused by the difference in their viscosity (Table 7.1). The results were in agreement with the reported effect of viscosity of Newtonian liquids on their microfluidic emulsification (Nie *et al.*, 2008). In particular, the unusual variation in the dimensions of droplets formed from PETA-3 and PETA-4 was ascribed to the high viscosity of these fluids: in contrast to the low viscosity monomers, the breakup of the liquid threads of PETA-3 and PETA-4 in droplets occurred with the formation of a long, narrow neck on a monomer thread. The "tail" resulting from the necking was acquired by the droplet and contributed to the increase in droplet volume.

The role of viscosity of the droplet phase in microfluidic emulsification is described in greater detail in Chapter 4, however here it should be emphasized that although the size and the distribution of the sizes of droplets depended on the type of the liquid, for each monomer listed in Table 7.1 a window of flow rates of the continuous and droplet phases existed in which the droplets had a narrow distribution of sizes. Generally, the coefficient of variation of the droplets was below 2% in the broad range of emulsification conditions. As an example, Figure 7.5c shows the distribution in dimensions of the TPGDA and PETA-3 droplets generated under optimized conditions. Despite their high viscosity, multifunctional monomers PETA-3 and in particular, PETA-4 were successfully emulsified in droplets with narrow size distributions, although these droplets had a substantially larger size than those generated from low-viscous monomers EGDMA and TPGDA.

In general, the ultimate size of polymer particles produced by microfluidic synthesis is determined by several factors. Firstly, it is the size of "precursor" droplets, which depends on the geometry of the microfluidic droplet generator, the flow rates of the continuous and droplet phases, and the viscosity and interfacial tension of the droplet and continuous phases. Following emulsification and prior to polymerization, the dimensions of droplets can be changed due to their coalescence or partial dissolution of the monomer in the continuous phase. (Non-controlled coalescence also leads to the broadening of the distribution of sizes of the droplets and the corresponding polymer particles.) Therefore, both coalescence and dissolution of droplets are considered to be undesired effects, as they counteract the ability to generate particles with a predetermined size.

Microfluidic emulsification is typically conducted in microfluidic droplet generators with feature sizes of the order of tens or hundreds of micrometers and hence it produces precursor droplets with comparable dimensions. Smaller droplets (and corresponding particles) can be generated in several ways. Firstly, it is the emulsification in microfluidic droplet generators with smaller dimensions of microchannels, as was demonstrated for the generation of several micrometer-size bubbles (Hettiarachchi *et al.*, 2007). In this case, however, the productivity of such microfluidic reactors is significantly reduced. Emulsification at high flow-rate ratios of the droplet-to-continuous phases dramatically reduces the size of droplets (Anna and Mayer, 2006), however it produces droplets with broadened distribution of sizes. The use of electrohydrodynamic jetting (Basaran, 2002; Loscertales *et al.*, 2002) or emulsification in the presence of a large amount of surfactant (Anna and Mayer, 2006) can also be used for the production of smaller droplets, which have a broad size distribution, so that the particles derived from these droplets require further fractionation.

7.2.2 Synthesis in Single-Phase Flow

Microfluidic synthesis of polymer particles described above employed multiphase flow of two liquids: a monomer or a prepolymer and a liquid forming a continuous phase. The method required the use of fluids that are largely immiscible, at least, on the time scale of the emulsification and polymerization processes. In addition, the efficiency of the microfluidic reactor was restricted to the production of "one particle at a time", although the productivity of the reactor can be increased by scaling out the microfluidics synthesis by conducting it in integrated multiple parallel channels.

These limitations were overcome by using a different approach to the continuous microfluidic synthesis of polymer particles, that is, including a series of methods employing microfluidic synthesis in one-phase flow. In 2006 Dendukuri *et al.* used projection photolithography to photopolymerize a diacrylate oligomer continuously flowing through a planar microfluidic device (Dendukuri *et al.*, 2006). Figure 7.6 illustrates these experiments. A stream of a polymerizable liquid (a low molecular weight polyethyleneglycole diacrylate) comprising a photoinitiator is passed through a rectangular microchannel of the PDMS microfluidic device. The liquid is exposed through a mask to controlled pulses of UV light using an inverted microscope. Polymerization is initiated in the continuous stream of the prepolymer by a mask-defined UV light beam emanating from the objective, and the shape of polymerized particles is determined by the pattern of the mask. Because the time of polymerization of the prepolymer is faster than 100 ms, the resulting particles do not move appreciably

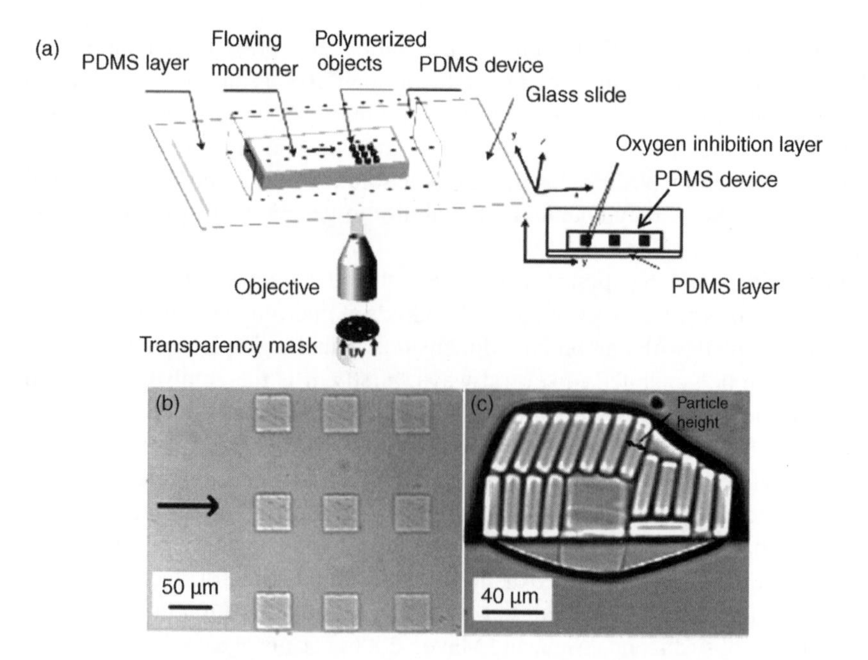

Figure 7.6 (a) Schematic illustrating the experimental setup used in the continuous synthesis of polymer particles. A mask containing the desired features is inserted in the field-stop plane of the microscope. The monomer stream flows through the microfluidic device fabricated in PDMS in the direction of the horizontal arrow. Particles are polymerized, by a mask-defined UV light beam emanating from the objective, and then continue to flow within the unpolymerized monomer stream. The inset shows the side view of the polymerized particles and the monomer layer that is not polymerized due to oxygen inhibition. (b) A bright-field microscopy image (xy plane) of an array of cuboids moving through the unpolymerized monomer. (c) A cross-sectional view of the cuboids seen in (b) upon collection in a droplet that has turned most particles on their sides. (Adapted, with permission, from Dendukuri et al., 2005). Reprinted with permission from Langmuir, Controlled Synthesis of Nonspherical Microparticles Using Microfluidics by D. Denukuri K. Tsoi, T.A. Hatton and P.S. Doyle, **21**, 6, 2113–2116 Copyright (2005) American Chemical Society

during the polymerization process and the distortion of particle shapes is minimized. Permeability of PDMS to molecular oxygen results in the diffusion of oxygen through the walls of the microchannel and the formation of chain-terminating peroxide radicals (Decker, 1985). The existence of an oxygen inhibition layer adjacent to the top and the bottom PDMS surfaces allows for the formation of a liquid lubrication layer and the flow of particles within the non-polymerized oligomer stream along the microchannel. Following polymerization, the resulting polymer particles are collected in the reservoir.

Figure 7.6b and c shows the array of particles synthesized from poly(ethylene glycol diacrylate) by using a mask with square features. The height of the particles with a range of shapes was determined by the height of the microchannel and the thickness of the oxygen inhibition layer of about 2.5 μm. (A more detailed discussion of the microfluidic synthesis of polymer particles with non-spherical shapes is provided in Chapter 10.)

The size of the smallest polymer particle that could be synthesized by using projection photolithography was determined by the optical resolution of the microscope – the smallest distinguishable feature that can be discerned. The length over which the sidewalls of the particles were straight was determined by the depth of field of the microscope objective (the length over which the beam of light had a constant diameter). The authors pointed out that higher resolution (and hence the smaller size of particles) comes at the cost of decreased depth of field. Other important practical considerations in particle synthesis included finite polymerization times and the smallest printable feature size on a transparency mask of approximately 10 μm.

In the continuous flow lithography, particle throughput could not be increased without compromising resolution: at high flow rates of the liquid, which is used to increase the productivity of the synthesis, smearing and deformation of particle features occurred. Enhanced resolution of the continuous projection lithography technique was achieved by implementing a modified method, namely stop-flow lithography (SFL) (Dendukuri *et al.*, 2007). A new setup used compressed air-driven flows in preference to syringe pumps exploited in the projection lithography method. A flowing stream of oligomer poly(ethylene glycol) diacrylate was stopped prior to polymerizing an array of polymer particles, thereby providing for enhanced resolution in the particle synthesis. The polymerized particles were then forced out and the cycle of stop-flow polymerization was repeated, again. Particles synthesized by SFL had dimensions down to 1 μm, whereas the productivity of the microfluidic synthesis was orders of magnitude higher compared with continuous flow lithography.

Recently, a new variant of one-phase microfluidic synthesis, namely, lock and release lithography, utilized the deformation of microfluidic devices fabricated in PDMS devices under external pressure (Bong, Pregibon, and Doyle, 2008). This process was used to synthesize three-dimensional and multifunctional particles. Three-dimensional microfluidic devices fabricated in PDMS contained positive relief structures protruding from the ceiling. The polymerized particles were "locked" in a flow and then forced out of the microfluidic device by using a high-pressure pulse, which deformed the PDMS device and released the particles. Composite particles were synthesized by flowing the first polymerizable liquid, polymerizing, and locking a particle in place using one mask. Then the second polymerizable liquid was introduced to form an overlapped region around the first particle using a different mask.

7.3 Polymer Particles Synthesized by Polycondensation

Whereas currently microfluidic synthesis of polymer particles by free radical polymerization is a well-established procedure, polycondensation reactions leading to the production of polymer particles have been reported by only several groups. These reports describe the preparation of polymer capsules (particles with a rigid shell engulfing a liquid core) using polymerization at the interface between the droplets and the continuous phase. (A more detailed discussion of the microfluidic preparation of capsules is provided in Chapter 9.)

In 2005 Whitesides and coworkers demonstrated continuous microfluidic synthesis of Nylon-6,6 capsules (Takeuchi *et al.*, 2005). This work was not only the first demonstration of the microfluidic synthesis of polymer particles by polycondensation but it also showed the ability to solve the problems associated with the wetting of the walls of the microfluidic reactor with the droplet phase. Nylon-6,6 capsules were produced in a microfluidic axi-symmetric flow-focusing device (Figure 7.7). Emulsification of aqueous 1,6-diaminohexane in hexadecane was carried out in a cylindrical glass tube with a narrow cross-section half-way down its length, which served as the orifice of the flow-focusing device. In this geometry, the droplet phase did not come in contact with the surface of the tube and problems associated with wetting of the device with aqueous 1,6-diaminohexane were avoided. The glass tube was incorporated in

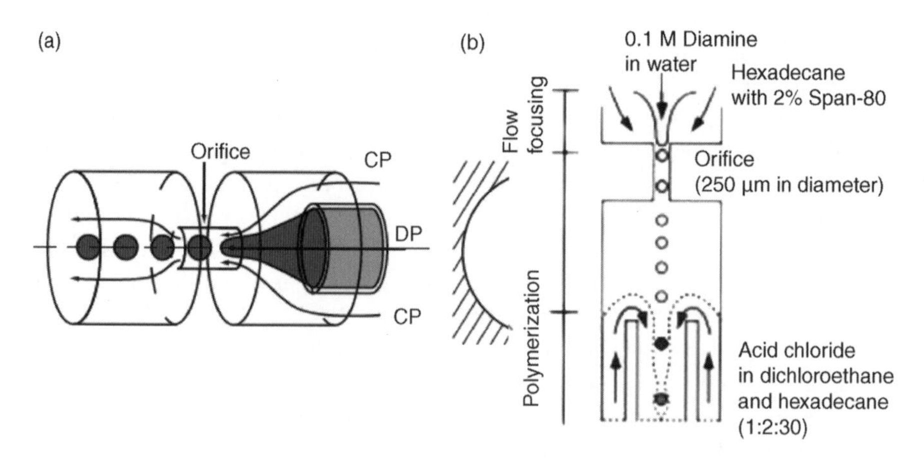

Figure 7.7 *(a) An axysymmetric (3D) flow-focusing channel. The channel is composed of a cylindrical tube with a narrow cross-section halfway down its length. The narrow region serves as the orifice where fluid is focused and breaks into aqueous droplets. In this geometry, wetting problems are avoided as the aqueous phase does not make contact with the walls. (b) A schematic of the microfluidic reactor for the synthesis of Nylon-6,6 aqueous capsules. An aqueous solution of 1,6-diaminohexane (0.1 M) was introduced into the channel at the first inlet. The continuous phase, Span-80 in hexadecane (2% v/v) was introduced into the channel from a second set of inlets. A solution of adipoyl in dichloroethane and hexadecane (1 : 2 : 30) was introduced into the channel from a third inlet. Polymerization between adipoyl chloride and 1,6-diaminohexane occurred at the surface of droplets. A solution of dodecanol in hexadecane was used to quench the polymerization reaction. (Adapted, with permission, from Takeuchi et al., 2005). Reprinted from Advanced Materials, An Axisymmetric Flow-focusing Microfluidic Device by S. Takeuchi, P. Garstecki, D. B. Weibel, and G. M. Whitesides,* ***17****, 1067–1072 (2005). Copyright Wiley-VCH Verlag GmbH & Co. KGaA. Reproduced with permission*

a PDMS slab. A solution of the second reagent – adipoyl chloride in a mixture of dichloroethane and dioxane – was introduced through a third inlet, as shown in Figure 7.7.

Rapid interfacial polymerization reaction between adipoyl chloride (dissolved in the continuous phase) and 1,6-diaminohexane (dissolved in aqueous droplets) led to the continuous formation of capsules of Nylon-6,6. The particles were collected in a solution of dodecane-1-ol in hexadecane, which quenched unreacted adipoyl chloride and terminated the polymerization reaction. Without quenching, the polymerization proceeded until all 1,6-diaminohexane diffused out of the droplets and the particles collected at the outlet of the microfluidic reactors were becoming crosslinked. The productivity of the microfluidic reactor was approximately 500 particles per minute and the reactor worked without clogging microchannels for 6 h. We note that a similar polycondensation reaction was employed for the microfluidic synthesis of a nylon membrane at the interface of the aqueous solution of 1,6-diaminohexane and the solution of adipoyl chloride in xylenes (Zhao *et al.*, 2002).

Interfacial polycondensation conducted in the microfluidic format was also reported by Quevedo, Steinbacher, and McQuade (2005). Droplets were generated from a solution of sebacoyl chloride and 1,3,5-benzene tricarboxylic acid chloride in a chloroform–cyclohexane mixture. The continuous aqueous phase contained polyethyleneimine. The authors used a simplified analogue of the microfluidic T-junction: a small-gauge needle inserted in the orthogonal way through the wall of the poly(vinyl chloride) tubing and situated in the middle of the channel. Despite substantial simplification in comparison with traditional two-dimensional and three-dimensional microfluidic droplet generators, the emulsification in the tube-and-needle device produced droplets (and the resulting polyamide capsules) with polydispersity lower than 9%. The diameter of the polyamide capsules was tuned in the range of from 300 to 900 μm by varying the flow rate ratios of the continuous-to-droplet phases. These dimensions were larger than typical dimensions of particles produced in planar microfluidic reactors (see, e.g., Xu *et al.*, 2005), however they could be reduced by using needles with a smaller diameter. The use of the tube-and-needle microfluidic device had an advantage of easy cleaning and replacement, if the tube was clogged. No information was provided about the productivity of this simplified microreactor.

7.4 Combination of Free-Radical Polymerization and Polycondensation Reactions

Under optimized conditions, free radical polymerization and step-growth polymerization produce interpenetrating polymer networks (IPNs) (Sperling, 1981). Polymer particles with the IPN structures have a broad range of applications which include the production of sound and vibration damping materials, automotive bumpers, insulators for under-the-hood automotive wiring, and particles acting as ion-exchange resins. These applications set a number of requirements to the structure and properties of the particles, such as a certain degree of flexibility, controlled size of domains in phase separated polymer structure, or a high energy absorption capability (Sperling and Mishra, 1996).

Generally, IPN particles are synthesized by suspension, mini-emulsion, or emulsion polymerization. Microfluidic synthesis of polymer particles with an IPN structure has been demonstrated by conducting, in sequence, two reactions: photoinitiated free-radical polymerization of TPGDA (Reaction 1) and condensation of poly(propylene glycole) tolylene 2,4-diisocyanate (PU-pre) with diethanolamine (DEA) (Reaction 2) (Li *et al.*, 2008). The schematics of these reactions are shown in Figure 7.8. Whereas free radical polymerization reactions of acrylates are fast (reaction rate constants are of the order of $\sim 10^2$–$10^4\,L\,mol^{-1}\,s^{-1}$ (Moore, 1977; Kaczmarek and Decker, 1994), the condensation between the –NCO and –OH groups at room temperature occurs at a low rate. The challenge in the microfluidic synthesis of IPN particles was to reach a high extent of Reaction 2 on short time scales. The authors hypothesized and experimentally established that heat generated in an exothermic free-radical polymerization of an acrylate monomer in Reaction 1 triggers the polycondensation of the urethane oligomer in Reaction 2. Indeed, the rate of condensation between the –NCO and –OH groups increases at elevated temperatures (the activation energy of these reactions is in the range of from about 50 to $140\,kJ\,mol^{-1}$) (Sultan and Busnel, 2006; Li *et al.*, 2008). On the other hand, free radical polymerization reactions of acrylates are exothermic: the enthalpies of these reactions are in the range of from 55 to $86\,kJ\,mol^{-1}$ (Miyazaki and Horibe, 1988; Anseth, Wang, and Bowman, 1994; Brandrup *et al.*, 1999).

Figure 7.9 shows the approach to a two-step microfluidic polymerization of polymer particles with the IPN structure. A droplet of a mixture of TPGDA and a photoinitiator 2-diethoxyacetophenone (DEAP) (reagents for Reaction 1) and PU-pre, DEA, and a catalyst dibutyltin dilaurate (DBTDL) (reagents for Reaction 2) is exposed to UV-irradiation. Photoinitiated free-radical polymerization yields a polyTPGDA network and generates heat, which activates the polycondensation of PU-pre. When both polymerization reactions proceed to high conversion and when macroscopic phase separation of the polymers in the particle is suppressed, the two-step synthesis yields polyTPGDA–polyurethane particles with an IPN structure.

High conversion of PU-pre within the time of residence of the droplets in the microfluidic reactor was achieved by controlling the amount of heat generated in Reaction 1 and accounting for the amount of heat dissipated in the continuous phase. Control of the amount of heat produced in Reaction 1 was realized by varying the relative fractions of the reagents in the droplet. The stoichiometric ratios between the –NCO groups of PU-pre and –OH groups of DEA were ensured by controlling the rates of flow of the corresponding liquid monomers supplied to the microfluidic reactor.

By determining the value of activation energy, ΔE_a, of Reaction 2 and establishing the relationship between the fraction of TPGDA and conversion of the –NCO groups of PU-pre, the authors optimized the composition of particles and the conditions of microfluidic polymerization, in order to achieve high conversion of PU-pre in particles with an IPN structure (Figure 7.10).

The examination of the internal structure of the microbeads using confocal fluorescence microscopy (CFM) suggested that no macroscopic phase separation occurred in

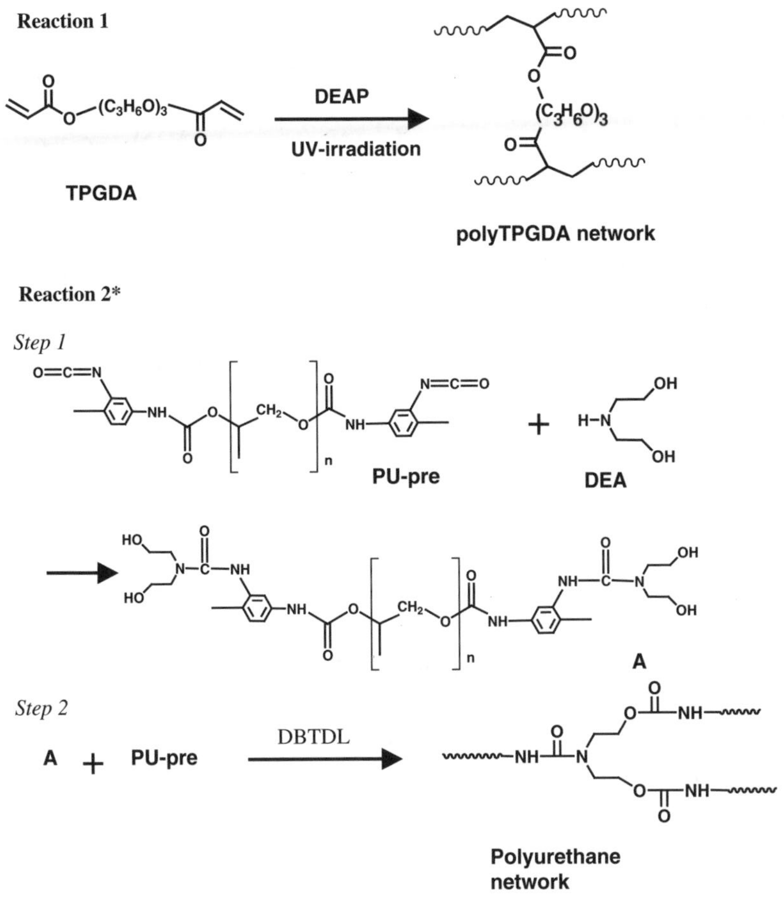

Figure 7.8 *Polyaddition of TPGDA (Reaction 1) and polycondensation of PU-pre (Reaction 2). In Step 1 of Reaction 2 the –NCO groups of PU-pre rapidly react with the –NH groups of DEA (no catalyst is required for this step). In Step 2 the reaction between the –OH groups of DEA and the –NCO groups of PU-pre occurs at a low rate and requires a catalyst dibutyltin dilaurate (DBTDL). We proposed that the heat generated in Reaction 1 will accelerate the second step of Reaction 2. (Adapted, with permission, from Li et al., 2008). Reprinted with permission from Journal of the American Chemical Society, Multi-step Microfluidic Polymerization Reactions Conducted in Droplets: The Internal Trigger Approach by W. Li, H.H. Pham, Z. Nie et al.,* ***130****, 30, 9935–9941 Copyright (2008) American Chemical Society*

the polymer particles prepared from the reaction mixture at the optimized ratio of monomers required in Reactions 1 and 2. Furthermore, to verify the role of microfluidic synthesis in controlling the morphology of the microbeads, polymer particles were synthesized by conventional suspension polymerization, using the same recipe as in

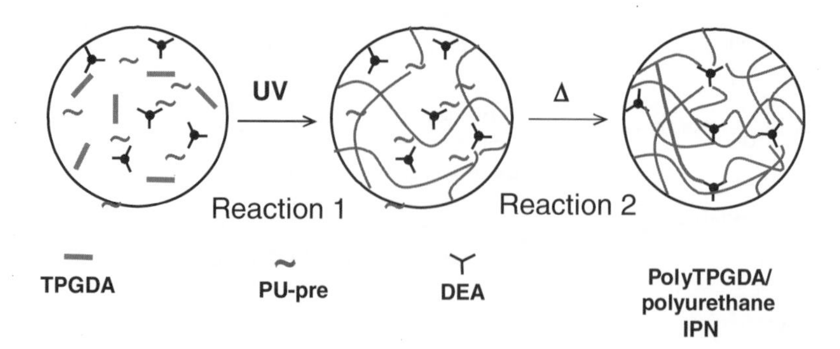

Figure 7.9 *Schematics of multi-step polyaddition and polycondensation reactions conducted in a droplet comprising a mixture of TPGDA monomer (—), PU-pre (~), and PU-crosslinker DEA (⤙). The molecules of the photoinitiator and catalyst are not shown in order to simplify the schematics. (Adapted, with permission, from Li et al., 2008). Reprinted with permission from Journal of the American Chemical Society, Multi-step Microfluidic Polymerization Reactions Conducted in Droplets: The Internal Trigger Approach by W. Li, H.H. Pham, Z. Nie et al., **130**, 30, 9935–9941 Copyright (2008) American Chemical Society. See Plate 7*

the microfluidic synthesis. As expected, the resulting particles featured a broad size distribution, due to the coalescence between the droplets of monomer mixtures. More importantly, macroscopic phase separation occurred in the microbeads produced by suspension polymerization. A more homogeneous structure of the particles obtained by microfluidic synthesis was ascribed to a better control of Reaction 1, initiated by the more uniform UV-irradiation of every droplet moving through the microchannel, in addition to more uniform heat dissipation in the continuous aqueous phase.

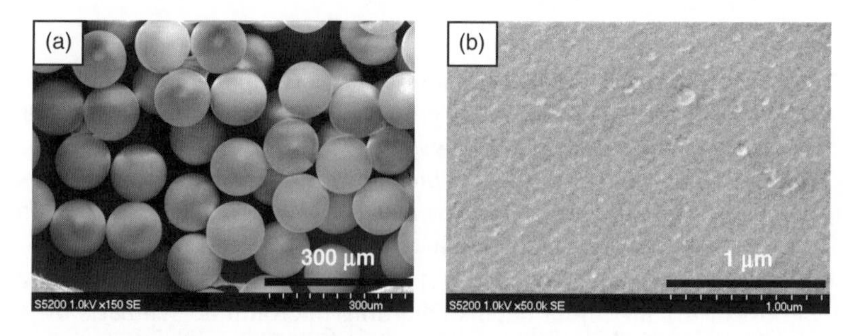

Figure 7.10 *SEM images of the polymer particles (a) and the surface of polymer particles (b) synthesized by combining free-radical polymerization (Reaction 1) and condensation polymerization (Reaction 2) compartmentalized in droplets. The ratio between the monomer mixtures required in Reaction 1 and 2 (see text) is 0.98. (Adapted, with permission, from Li et al., 2008). Reprinted with permission from Journal of the American Chemical Society, Multi-step Microfluidic Polymerization Reactions Conducted in Droplets: The Internal Trigger Approach by W. Li, H.H. Pham, Z. Nie et al., **130**, 30, 9935–9941 Copyright (2008) American Chemical Society*

7.5 General Considerations on the Use of Other Polymerization Mechanisms

Currently, photoinitiated free-radical polymerization and polycondensation dominate the microfluidic synthesis of rigid polymer particles. For various reasons, other polymerization reactions, such as living radical polymerizations or ionic polymerizations, have not been used for the continuous microfluidic synthesis of polymer particles. Partly, this is explained by the lack of motivation in polymerization reactions, which require relatively long times to achieve reasonably high monomer-to-polymer conversions. In order to achieve sufficient polymerization times which are directly related to the time of residence of droplets in the microfluidic reactor, the lengths of microchannels would have to be dramatically increased. This would lead to an increase in the dimensions of the microfluidic reactor. In addition, reduction of pressure along extended microchannels would increase the possibility of clogging of the reactor with the solidified particles.

On the other hand, fast polymerization reactions, such as ionic polymerizations, are currently difficult to implement in microfluidic reactors, due to the stringent requirements to the purity of monomers, the inert atmosphere, or specific temperatures. In principle, these requirements can be addressed in integrated microreactors, however; currently, the benefits of microfluidic synthesis of particle synthesis under such conditions are questionable.

To summarize, fast polymerization reactions that are activated by the external triggers and are not highly demanding for specific reaction conditions such as low temperature or inert atmosphere, hold higher promise in the microfluidic synthesis of polymer particles. To date, polymerization reactions for the microfluidic synthesis of rigid polymer particles were either thermoinitiated, or photoinitiated by using UV irradiation, however other sources of energy can be used to trigger polymerization, such as γ-irradiation, microwaves, near-IR irradiation, or their combination.

7.6 Important Aspects of Microfluidic Polymerization of Polymer Particles

Currently, most of the important details regarding microfluidic synthesis of polymer particles are collected for the multiphase free-radical polymerization; however the majority of the effects influencing microfluidic synthesis are applicable to polycondensation reactions and a single-phase free-radical polymerization. These effects are summarized below, in order to provide guidance for the microfluidic synthesis of polymer colloids. The criteria for the successful synthesis included high conversion of monomer to polymer, a narrow particle size distribution, and the ability to control the size of particles by varying the flow rates of the continuous and droplet phases.

7.6.1 Modes of Microfluidic Polymerization

One of the greatest challenges in microfluidic synthesis of polymer particles is achieving high conversion of monomer into polymer during the time of the microfluidic polymerization, that is, during the time of residence of particles in the microchannels. The lower the rate of polymerization reaction the longer should be the time of residence of the particles in the reactor. The time of residence is given by $t \approx l/v$ where l is the length of microchannel and v is the velocity of particles in the microchannels. For slowly polymerizing polymers, an increase in polymerization time can be achieved by increasing the length of the polymerization compartment of the microfluidic reactor and/or by reducing the velocity of the particles. Both factors are not desirable. Decrease in v complicates control of the dimensions of the particles, which are determined by the flow rates of the continuous and droplet phases, and may lead to the clogging of microchannels with the particles. In addition, the overall productivity of the microfluidic synthesis (the number of particles produced per unit time) reduces. On the other hand, with increasing length of the polymerization compartment the dimensions of the microfluidic reactor increase.

Depending on the polymerization kinetics, in order to achieve high conversion the synthesis of particles in microfluidic reactors can be performed in several ways. Firstly, both emulsification of a monomer mixture and polymerization can be carried out in the continuous mode solely in a microfluidic reactor (or *in situ*), as shown in Figure 7.11a for the photoinitiated synthesis of polymer particles. Droplets generated in the microfluidic droplet generator flow through the polymerization compartment – the downstream channel typically having a serpentine shape – where they are exposed to UV irradiation. Rapid conversion of monomer into polymer leads to droplet solidification in the serpentine channel, so that no post-polymerization is needed for the particles collected at the outlet of the reactor. Typically, this mode of polymerization is used for the microfluidic synthesis of rapidly polymerizing monomers, such as acrylic monomers. In another limit, following emulsification and "prepolymerization" of the monomer mixture in the microfluidic reactor, partly polymerized particles are transferred into a batch reactor and post-polymerized using, for example, thermoinitiated polymerization (Figure 7.11b). This polymerization process can be used for the synthesis of, for example, polystyrene particles.

Two other polymerization modes include polymerization in the microfluidic reactor followed by continuous polymerization in the external extension channel (Figure 7.11c), or a combination of the modes shown in Figure 7.11c, followed by batch polymerization, as shown in Figure 7.11d. Strictly speaking, the extension channel can be considered as a part of the microfluidics reactor. The external extension channel can, in principle, collect particles prepolymerized in multiple microfluidic reactors, thereby increasing the productivity of microfluidics synthesis.

In Figure 7.11 all polymerization processes include polymerization *in situ* or "on-chip", that is, in the microfluidic reactor. This feature is critical: in order to fully implement the advantages of continuous microfluidics synthesis it is important to

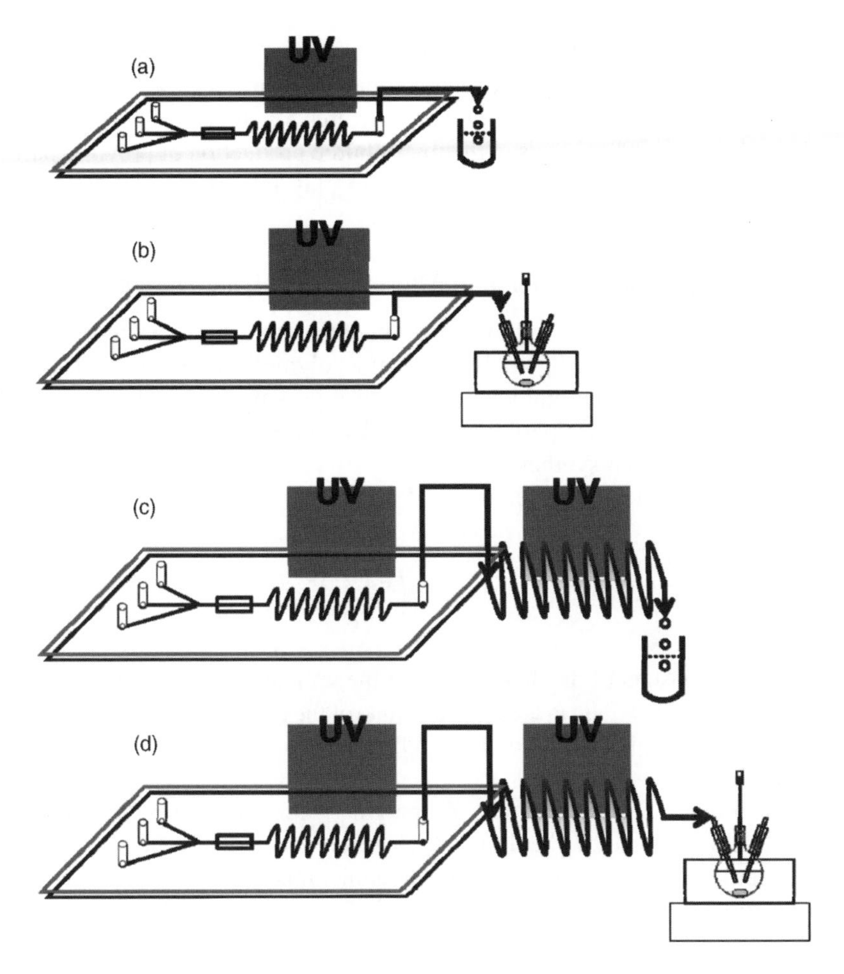

Figure 7.11 *Polymerization processes used for the synthesis of polymer particles: (a) continuous "on-chip" microfluidic synthesis; (b) combination of "on-chip" and batch polymerization; (c) continuous synthesis combining polymerization "on-chip" and in the extension microchannel; and (d) combination of polymerization processes shown in (a–c). (Adapted, with permission, from Seo, 2008). Reproduced with permission from PhD Thesis of M. Seo Copyright (2008) M. Seo*

prepolymerize particles in a microfluidic reactor. The synthesis of polymer particles with a narrow size distribution requires that emulsification is followed by the polymerization of "one particle at a time", in order to protect particles against coalescence and preserve their narrow size distribution in the subsequent continuous or batch polymerizations. Polymerization of polymer particles with non-conventional shapes also depends on the ability to "pre-shape" them in the polymerization compartment of the microfluidic reactor.

There are also indications that *in situ* polymerization helps to control the internal structure and morphology of particles, due to the controlled amount of energy obtained by every droplet flowing through the microchannels. This feature is particularly important when polymerization and phase separation processes are responsible for the formation of particles with porous structures or IPN structures (Dubinsky *et al.*, 2008; Li *et al.*, 2008). Thus in the microfluidic production of polymer particles, both emulsification *and* on-chip polymerization play important roles in producing polymer particles.

7.6.2 Achieving High Conversion in Microfluidic Polymerization

In the view of the factors listed above, it is imperative to achieve as high as possible a conversion of the monomer into polymer during polymerization *in situ*, that is, in the microfluidic reactor. As photoinitiated free-radical polymerization dominates microfluidic synthesis of polymer particles, below we summarize the main factors that influence the rate of such synthesis.

The rate of free radical polymerization, R_p, is:

$$R_p = k_p \left(\frac{fk_d[I]}{k_t} \right)^{\frac{1}{2}} [M] \qquad (7.1)$$

where k_d is rate constant of the dissociation of the initiator, f is the fraction of radicals that start chain growth, k_p and k_t are rate constants of the propagation and termination of the polymer chain, respectively, and $[I]$ and $[M]$ are the concentrations of the initiator and the monomer.

Based on Equation (7.1), the rate of polymerization increases with increasing concentration of monomer in the droplets. Therefore when monomers are mixed with solvents, in order to reduce viscosity of the monomer mixture or to synthesize particles with a macroporous structure (Dubinsky *et al.*, 2008), the rate of polymerization is reduced.

Using highly efficient photoinitiators is vital to the successful microfluidic synthesis. For a particular photoinitiator, the change in its concentration is an alternative approach to increasing the rate of polymerization. In a photoinitiated free-radical polymerization, the rate of chain propagation (R_p) depends on the concentration of photoinitiator, c_{in}, as

$$R_p \propto [1 - \exp(-\varepsilon l c_{in})]^{0.5} \qquad (7.2)$$

where l is the sample thickness and ε is the absorptivity of the photoinitiator (Decker, 1998; Decker, 2002). Thus with increasing c_{in} the rate of polymerization increases, although more gradually than with increasing the concentration of the monomer. An increase in the concentration of the initiator results in a decreasing molecular weight of the polymer, however this effect becomes less important for crosslinked polymers synthesized from multifunctional polymers.

In practice however, the concentration of the photoinitiator has to be optimized in order to control the molecular weight of the polymer and to avoid overheating of the droplets (and a microreactor, in general). Figure 7.12 illustrates the effect of the

Figure 7.12 *Typical SEM images of polyTPGDA polymer particles produced by continuous polymerization in a microfluidic reactor at concentrations of photoinitiator HPCK,* c_{in}: *(a) 2 wt%; (b) 4 wt%; and (c) 6 wt%. Corresponding droplets were obtained at* $Q_w = 4\,\text{mL}\,\text{h}^{-1}$, $Q_m = 0.1\,\text{mL}\,\text{h}^{-1}$. *Scale bar is* $100\,\mu\text{m}$. *(Acquired, with permission, from Seo et al., 2005). Reprinted with permission from Langmuir, Continuous Microfluidic Reactors for Polymer Particles by M. Seo, Z. Nie, S. Xu et al.,* ***21****, 25, 11614–11622 Copyright (2005) American Chemical Society*

concentration of the photoinitiator 1-hydroxycyclohexylphenyl ketone (HPCK) on the polymerization of polymer particles from TPGDA. The concentration $c_{\text{in}} = 2$ wt% was insufficient to provide the formation of fully solidified particles during the time of residence of particles in the microfluidic reactor of up to 120 s. Under such conditions, the particles that were collected at the exit of the microfluidics reactor had a rigid polymer "skin" and a liquid monomer core and when imaged using SEM, they collapsed under vacuum (Figure 7.12a).

The optimized concentration of HPCK was from 3.5 to 4.5 wt%: at $c_{\text{in}} = 4.0$ wt% conversion of TPGDA into polymer was 95–97%. The resulting polyTPGDA particles had a well-defined spherical shape and a smooth surface (Figure 7.12b). We note that relatively high concentrations of an initiator are typical for photoinitiated free radical polymerization, in comparison with thermoinitiated polymerization. For a higher content of photoinitiator exceeding 6 wt%, fast polymerization of TPGDA led to particle "explosion," due to the large amount of heat released during the polymerization reaction (Figure 7.12c).

Energy flux is another important factor in optimizing the time of polymerization of polymer particles in microfluidic synthesis. High energy flux of the light source allows the use of shorter times of irradiation of droplets in the microfluidic reactor and, hence, it reduces the time of polymerization. For example, for solidification of Norland Optical Adhesive 60, flux provided by a 100 W UV lamp allowed a very short exposure time of ($<1\,\mu$s) to polymerize the liquid plugs, given that an energy input of only $3\,\text{J}\,\text{cm}^{-2}$ was required for a full cure of the prepolymer (Dendukuri *et al.*, 2005).

One of the most important factors in photoinitiated microfluidic synthesis is a rationalized selection of the photoinitiators, which is based on the analysis of spectra of the irradiation source, the absorption spectrum of the initiator, and the absorption spectrum of the material of the microfluidic device. Firstly, the selected photoinitiator

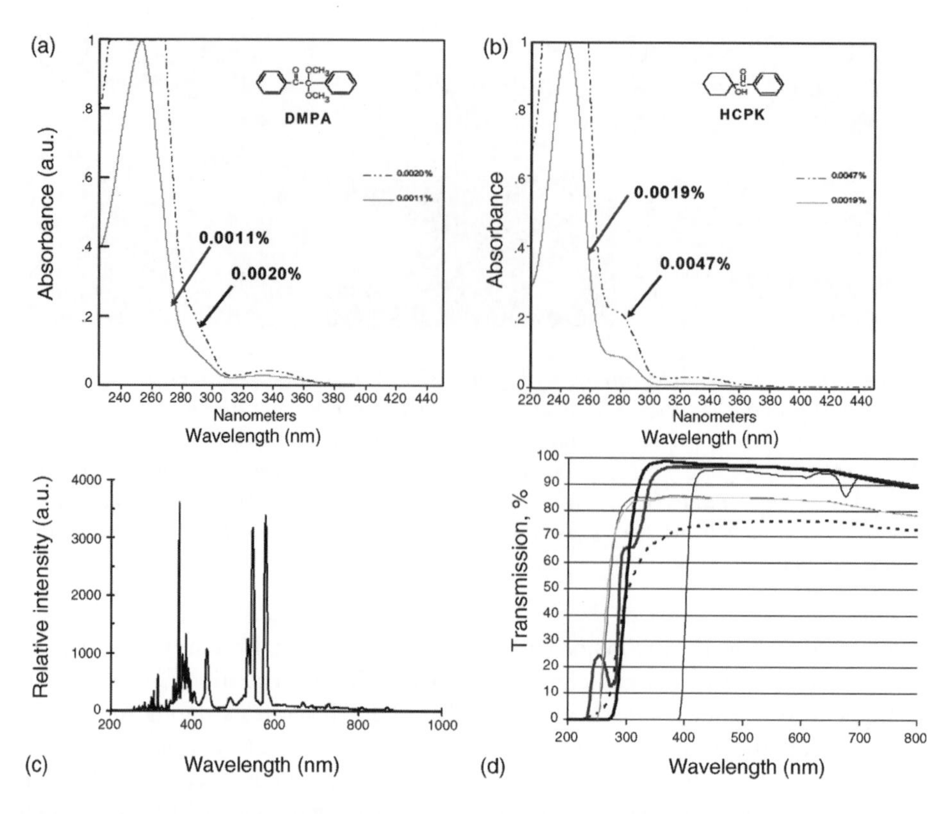

Figure 7.13 *A rationale for the selection of photoinitiators. (a, b) Absorption spectra of photoinitiators 2,2-dimethoxy-2-phenyl acetophenone (a) and 1-hydroxycyclohexylphenylketone (b) acquired in methanol at two different weight concentrations. (c) Absorption spectrum of the UV-irradiation source (Dr. Hönle UVA Print 40C, F-lamp, 400 W). (d) Absorption spectra of the typical materials used for the fabrication of microfluidic devices: polycycloolefine (—), Schott B270 super white crown glass (—), Acrylite-OP-1 (—), Acrylite-OP-2 (—), cured poly(dimethyl siloxane) (···), and polycarbonate (—). See Plate 8*

has to absorb efficiently in the spectral range of the light source used for the irradiation. Secondly, it is important to ensure that this wavelength range does not overlap with absorption of light by the material of the microfluidic reactor. As an example, Figure 7.13 shows absorption spectra of photoinitiators, an absorption spectrum of PDMS (frequently used for the fabrication of microfluidic reactors), and the spectral variation of intensity of the UV-irradiation source (Dr. Hönle UVA Print 40C, F-lamp, 400 W). Figure 7.13a shows that the maximum intensity of UV-light is generated in the spectral ranges of 350–450 and 520–600 nm. No significant absorption occurs in this spectral range by PDMS (Figure 7.13b). Figure 7.13a,b shows absorption spectra of two typical photoinitiators, 1-hydroxycyclohexylphenyl ketone (HCPK) and 2,2-dimethoxy-2-phenyl acetophenone (DMPA), respectively, which were measured at different photoinitiator concentrations in methanol. In both cases, strong and moderate absorption occurs in the spectral ranges of 240–260 and 310–370 nm, respectively, however only the second band corresponds to the high intensity peaks of the light source (Figure 7.13c), suggesting that the use of these two photoinitiators and the irradiation source is not optimized.

Furthermore, the selection of the material of the reactor based on its optical properties requires the measurement of the material's transmission spectrum. It should be understood that absorption of light generated by the irradiation source will have two important implications. Firstly, photopolymerization conducted in the microreactor will be less efficient and the additional post-polymerization may be required to achieve high conversion of monomer into polymer. Secondly, exposure to irradiation will lead to the heating of the microfluidic reactor. This unwanted effect may result in distortion of microchannels and their clogging with partly polymerized particles. Figure 7.13d features transmission spectra of several materials that are typically used for the fabrication of microfluidic devices. Polycarbonate strongly absorbs in the spectral range of up to about 400 nm, and therefore it should be ruled out as a potential material for the fabrication of the microfluidic reactor for the photo-initiated synthesis of polymer particles using the light source and photoinitiators discussed above. Two Acrylite samples and PDMS show reasonably good transmission of up to 65–85% in the spectral range 310–370 nm. These polymers can be used as the materials of the microfluidic reactors. Polycycloolefin and Shott B270 Superwhite Crown Glass show the best transmission of up to 90–95% in the designated spectral range and thus, based on their optical properties, can be recommended for the fabrication of reactors.

7.6.3 *In Situ* Polymerization of Monomer Droplets

During *on-chip* polymerization each monomer droplet has to be polymerized individually. In this process, two challenges have to be overcome: the coalescence of droplets and the clogging of microchannels by polymerized particles. In order to synthesize particles with a narrow size distribution it is important to avoid collisions between the droplets in the polymerization compartment, that is, in the extension downstream channel and, as much as possible, to avoid adhesive contacts between the droplets and the walls of the downstream channel. Utilization of the materials of microfluidic reactors that have low affinity for monomer droplets and, if needed, an appropriate modification of the surface of microfluidic reactors, proved useful in suppressing the wetting of the channel walls with liquid monomers.

When the cross-section of the extension channel is larger than that of the preceding downstream channel, the velocity of the dispersion decreases and the droplets begin to collide and coalesce. Even if the cross-sectional areas of the microchannels are maintained constant, the drop in pressure along the downstream channel leads to a reduction in droplet velocity and, potentially, to collisions of droplets. It should be noted, however, that contacts between precursor droplets do not necessarily lead to their coalescence; nevertheless, they may affect particle size distribution. For example, sometimes droplets form doublets and triplets moving together through the extension microchannel. Polymerization of such assemblies yields polymer aggregates. In addition, aggregation of partly polymerized droplets can lead to the clogging of the polymerization microchannel.

The process of clogging of microchannels is complicated and is not well understood. Clogging occurs even when the diameter of the particles is significantly smaller than the effective diameter of the polymerization microchannel. Recently, the dynamics of clogging processes have been addressed under experimental conditions that differ from those used for the microfluidic synthesis of polymer particles (Wyss *et al.*, 2006). Thus, here we provide only general considerations that are based on observations made in polymerization experiments, which may be useful when designing microfluidic reactors for the synthesis of polymer particles.

Clogging occurs when individual particles or aggregates of particles stick to the walls of the extension polymerization channel. Both factors change the velocity of other particles moving through the microchannel, which ultimately form a large cluster of microbeads around the primary particle and eventually, lead to complete blockage of the channel.

In order to suppress clogging of the extension polymerization channel with partly or completely polymerized polymer particles, the dimensions, the shape, and the surface chemistry of the polymerization compartment, and the flow rates of liquids have to be optimized. As one of the major reasons for droplet coalescence or particle aggregation is the decrease in their velocity in the polymerization microchannel, one of the ways to suppress this effect is to adjust the dimensions of the downstream channel and the polymerization channel. Generally, these channels have the same height; therefore gradual reduction in the width of the polymerization channel along its length may prove useful. Another way to compensate for the reduction in droplet/particle velocity is to add a carrier liquid to the polymerization microchannel, which would help to compensate for the drop in pressure. Overall, using long *on-chip* polymerization extension channels is not helpful, and for slowly polymerizing monomers it may be beneficial to use the design of the reactor shown in Figure 7.11b–d, so that in-chip, *in situ* polymerization is only used for monomer prepolymerization.

The surface chemistry and topography of the extension polymerization channel are important factors to be considered. Smooth motion of droplets and particles through the polmerization channel can be counteracted by the physical entrapment of particles at surface asperieties of the channel and/or by the adherance of droplets or particles to the microchannel walls. Both effects slow down the particles flowing through the microchannel and, potentially, they may lead to the blockage of the polymerization compartment. Therefore, the surface of the microchannel has to be smooth and adhesion between the droplets (and particles) and the surface of the polymerization section has to be suppressed. This requirement is equally important for microfluidic emulsification and *in situ* polymerization.

An important factor in polymerization of droplets generated in the emulsification compartment of the microfluidic reactor is the relationship between the size of droplets and the dimensions of extension microchannel. Polymerization of monomer plugs that are touching the surface of the microchannels will be discussed in Chapter 10. Here we stress that problems associated with microchannel roughness and particle adhesion to the walls, in addition to friction between the particles and the walls of the microchannel, are amplified for large droplets touching the walls of microchannels. Some of the problems can be minimized by producing droplets with a diameter that is substantially smaller than the height and the width of the microchannel. Although for some materials shrinkage during polymerization helps to overcome problems with adhesion to the microchannel walls, it is advisable to generate droplets with a diameter that is, at least, 20–30 μm smaller than the dimensions of the polymerization microchannel.

Increasing the temperature in the polymerization compartment during microfluidic synthesis is another important factor to consider. Most of polymerization reactions are exothermic; thus a substantial amount of heat is dissipated in the continuous phase. This heat alone, in addition to the potential absorption of light by the material of the microfluidic reactor in photoinitiated reactions may cause an increase in the temperature of the microreactor. Possible deformation of the extension polymerization channel can also lead to clogging. Therefore it is advisable to maintain the temperature of the reactor constant by mounting a cooling jacket around the polymerization compartment.

7.7 Synthesis of Composite Particles

Microfluidic synthesis of composite particles includes one additional step: prior to the microfluidic emulsification, the host monomer or liquid prepolymer has to be is mixed with an organic or inorganic additive(s). The additive can be introduced into the host monomer in a liquid or a solid state to form a solution or a dispersion. Several important requirements have to be fulfilled in the preparation of a mixture for the microfluidic synthesis of polymer particles. The first group of requirements is characteristic of the production of composite materials, regardless of their methods of preparation, and they will not be discussed in great detail. Briefly, in order to synthesize polymer particles that are compositionally and structurally uniform it is important to achieve good compatibility of a host monomer with an additive. This applies to the preparation of macroporous particles from monomers dissolved in a porogen solvent (Dubinsky *et al.*, 2008) or the synthesis of copolymer particles (Lewis *et al.*, 2005) when two co-monomers form a one-phase system. In the synthesis of hybrid polymer–inorganic particles, for example, polymer microspheres loaded with semiconductor or magnetic nanoparticles, it is important to ensure that the nanoparticles do not aggregate in the host monomer and do not change their properties when mixed with the host monomer.

Features that are specific for the continuous microfluidic synthesis of polymer particles include pre-mixing or the supply of the individual ingredients of the mixture to the microfluidic reactor, mixing realized in the microfluidic reactor, and polymerization of the monomer or liquid prepolymer.

Mixing of the host monomer and additive(s) is an important step in the preparation of composite particles by microfluidic synthesis. Mixing is simple and straightforward when the additive is miscible with the host monomer [e.g., mixtures of TPGDA with porogen solvents (Dubinsky *et al.*, 2008), with acrylic acid (Lewis *et al.*, 2005), and with liquid crystal 5CB (Xu *et al.*, 2005). Typically, a preformed mixture is placed in the reservoir from which it is continuously fed to the microfluidic reactor. When the additive is reactive, the reaction can occur in the supply tubing, and ultimately block it. In this case, mixing is realized in the mixing compartment of the microfluidic reactor, just before the emulsification (Li *et al.*, 2008). Such a process is beneficial for the optimization of the formulation for the chemical reactions, as the variation in the composition of the mixture can be readily achieved by supplying reactants at varying flow rates (Li *et al.*, 2008).

We note that mixing between two liquids that is realized in laminar flow may be a challenge: the typical values of convective-to-diffusion time scales, quantified in terms of the Pèclet number, Uh/D, are between 10^1 and 10^5 where U is the velocity of liquid in the direction of flow, h is a characteristic dimension of the cross-section of the microchannel, and D is the molecular diffusion coefficients (Ottino and Wiggins, 2004; Stone, Strook, and Ajdari, 2004). Thus convection is significantly faster than molecular diffusion. Efficient mixing in microfluidics relies on the ability to create a large contact interface between the two fluids flowing through the microchannel, and it is typically achieved by using chaotic advection in microfluidic devices with transversal components of the flow that stretch and fold volumes of fluid over the cross-section of the channel. Furthermore, focusing of multiple liquid streams during the formation of droplets favors good mixing between the liquids.

When an additive is immiscible with the host monomer or prepolymer and has a density higher than the host monomer, demixing of the components may occur even before the ingredients are supplied to the microfluidic reactor. Precipitation of the heavier component can also lead to the clogging of the supply tubing. The segregation may lead to the enrichment of composite particles with a lighter component and the dependence of particle composition on time. Adherence of the additive to the walls of the tubing may cause the same effect as the difference in density. In such situations, stirring of the mixture before its entrance to the microfluidics reactor or in the microfluidic channels may prove beneficial.

The introduction of additives in the host monomer may also change the macroscopic properties of the droplet phase such as viscosity, interfacial tension with the continuous phase, and the wettability of the walls of microchannels. The change in these properties may strongly affect microfluidic emulsification of the monomer mixture. The effect of viscosity of the droplet phase is illustrated in Figure 7.5. Whereas the change in viscosity results in the change of the dimensions of droplets and requires optimization of the formation of droplets, microfluidic emulsification can tolerate a significant increase in viscosity of the droplet phase by changing the flow rates of the liquids (Nie *et al.*, 2008). On the other hand, the change in the interfacial energy of the mixture, in comparison with the host monomer, may have a dramatic effect on the formation of droplets and will be discussed below for the example of free-radical microfluidic polymerization of copolymer microbeads (Lewis *et al.*, 2005; Seo, 2005).

In the subsequent sections, we review the preparation of composite polymer particles, namely, copolymer particles, hybrid polymer particles loaded with low molecular weight inorganic additives, and hybrid polymer particles loaded with low molecular weight organic additives. Microfluidic syntheses of composite particles with a core-shell structure and a Janus morphology are discussed in Chapters 9 and 10, respectively.

7.7.1 Copolymer Particles

To obtain copolymer particles, typically, two or more co-monomers are pre-mixed in the desired ratio, the liquid mixture is emulsified, and the droplets are polymerized to generate solid polymer particles (Lewis *et al.*, 2005; Nie *et al.*, 2005; Nie *et al.*, 2006; Huang *et al.*, 2008). Figure 7.14 illustrates microfluidic synthesis of carboxylated microbeads from a mixture of TPGDA and acrylic acid (AA) (Lewis *et al.*, 2005). Acrylic acid is a hydrophilic monomer and mixing of TPDGA with AA has three

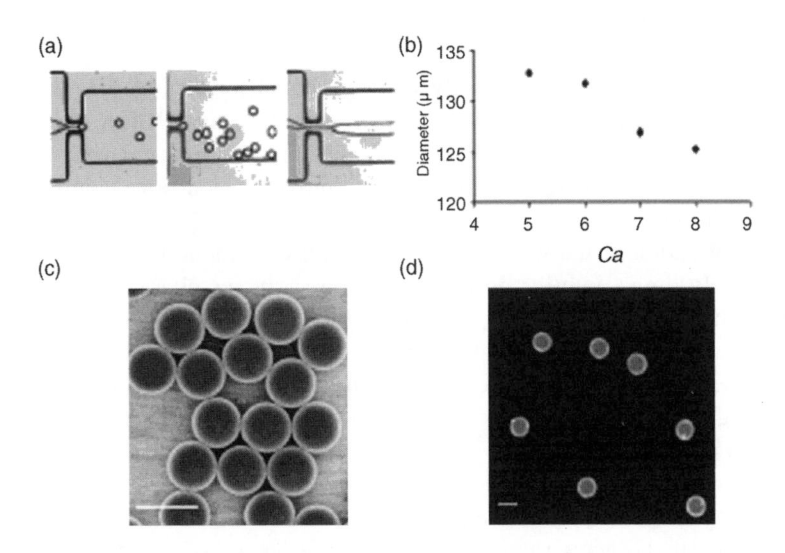

Figure 7.14 *(a) Emulsification of TPGDA mixed with different amounts of acrylic acid (AA) in a microfluidic flow-focusing device. Concentration of acrylic acid, C_{AA}, in the monomer mixture from left to right: 5, 8, and 15 wt%, respectively. (b) Variation in the diameter of droplets plotted as a function of C_{AA}. (c) Scanning electron microscopy image of poly (TPGDA-AA) particles obtained by photopolymerization of TPGDA-AA droplets containing 5 wt% of AA. (d) Fluorescence microscopy image of copolymer beads as in (c) conjugated with FITC-BSA. For the emulsification of the TPGDA-AA mixture, the flow rates of the droplet and continuous phases were 0.01 and from 0.5 to 2.0 mL h^{-1}. Scale bar is 100 μm. (Adapted, with permission, from Lewis et al., 2005). Reprinted with permission from Macromolecules, Continuous Synthesis of Copolymer Particles in Microfluidic Reactors by P.C. Lewis, R.R. Graham, Z. Nie et al., **38**, 10, 4536–4538 Copyright (2005) American Chemical Society*

important implications for the microfluidic emulsification of the mixture in a flow-focusing droplet generator. Two of these effects are shown in Figure 7.14a,b.

Addition of AA to TPDGA results in reduced viscosity of the mixture and a reduction in its interfacial tension with water thereby increasing the wettability of the walls of the microchannels fabricated in hydrophilic polyurethane. At low concentrations of AA (below 8 wt%) the stream of the droplet phase did not adhere to the walls of the orifice and broke up in the flow-focusing regime to release monodipersed droplets. When the concentration of AA exceeded 8 wt%, the size of droplets reduced, the distribution of sizes broadened, and the stream of the droplet phase showed a trend of adhering to the walls of the orifice. In principle, the last problem could be overcome by fabricating a droplet generator in a more hydrophobic polymer than polyurethane, such as in PDMS (Seo, 2005; Seo *et al.*, 2007; Li *et al.*, 2007). At a concentration of AA above 10 wt%, the hydrophilicity of the monomer mixture further increased and the monomer thread did not break into droplets in the range of flow rates of liquids studied. Therefore, in the microfluidic synthesis of copolymer particles from the TPGDA-AA mixture, the maximum concentration of AA was limited to approximately 5 wt% (Lewis *et al.*, 2005).

Owing to the decrease in viscosity and interfacial tension of the TPGDA-AA mixture with the continuous phase, in comparison with pure TPDGA monomer, the size of droplets produced from the co-monomer mixture reduced with increasing the concentration of AA (Figure 7.14b), thereby providing additional control over the emulsification process.

The complication of the use of polar hydrophilic co-monomers, such as AA, is their relatively high solubility in the continuous aqueous phase. Despite the fact that, typically, the interval of time between the emulsification and the polymerization stages is relatively short, a polar monomer can diffuse out from the droplets into the aqueous continuous phase. As a result, the fraction of the polar polymer in the resulting particles reduces and the overall composition of particles changes, in comparison with the precursor co-monomer mixture. This effect can be minimized by shortening the time between the emulsification and polymerization stages and/or by decreasing the solubility of the polar co-monomer in the continuous phase. For example, one of the ways to suppress the migration of AA from droplets to the continuous phase is to emulsify the co-monomer mixture under acidic conditions under which the solubility of AA in the aqueous phase is reduced [pK_a for carboxylic groups is 4.25 (Beyer, Walter, and Lloyd, 1997)].

Microfluidic polymerization of the precursor droplets of the TPGDA-AA mixture allowed for the production of polymer particles carrying carboxylic surface functionalities. Such particles have many important applications, including the detection, immobilization, and isolation of biologically active species, such as proteins and cells. Droplets with a predetermined size were generated from the TPGDA-AA co-monomer mixture at a concentration of acrylic acid of 5 wt%, and the monomer mixture was photopolymerized *on-chip* under UV irradiation (Lewis *et al.*, 2005). Figure 7.14c shows poly(TPGDA-AA) microspheres synthesized under optimized conditions. Particles had a narrow size distribution, with coefficient of variance below 2%, that is, similar to polydispersity of the corresponding precursor droplets. By using X-ray photoelectron spectroscopy (XPS) it was found that the surface of the microspheres contained 12.3 mol% of acrylic acid. This amount of carboxylic groups on the surface of the copolymer microbeads was sufficient for the surface immobilization of biomolecules, that is, bioconjugation. The authors demonstrated conjugation of poly(TPGDA-AA) particles with bovine serum albumin protein molecules covalently labeled with fluorescein isothiocyanate (FITC-BSA) (Desai and Stramiello, 1993). Figure 7.14d shows a fluorescent microscopy image of the poly (TPGDA-AA) microbeads synthesized in the microfluidic reactors and conjugated with FITC-BSA.

7.7.2 Polymer Particles Loaded with Low-Molecular Weight Organic Additives

Composite polymer particles with varying low-molecular weight organic additives were synthesized by polymerizing monomers pre-mixed with small molecules: catalysts, fluorescent dyes, porogen organic solvents, or liquid crystals, such as 4-cyano-4′-pentylbiphenyl (5CB). Porogen organic solvents and 5CB were miscible with the host monomer when added in a relatively high volume fraction, however, following polymerization, the mixture underwent phase separation and the resulting polymer particles had a two-phase structure.

Figure 7.15a shows polarization microscopy images of the composite polyTPGDA microbeads comprising 5CB (Xu *et al.*, 2005). The morphology of the particles was

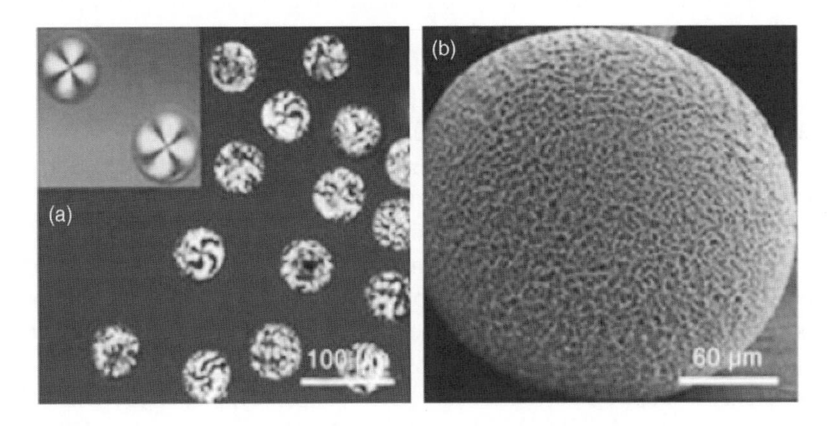

*Figure 7.15 (a) Polarization microscopy image of 4-cyano-4′-pentylbiphenyl (5CB)-polyTPGDA microspheres. Inset shows the morphology of particles generated at low polymerization rate (see text). (b) SEM image of a porous polyTPGDA microsphere. (Adapted, with permission from Xu et al., 2005). Reprinted from Angewandte Chemie Int. Edi., Generation of Monodisperse Particles by Using Microfluidics: Control over Size, Shape, and Composition by S. Hu, Z. Nie, E. Kumacheva et al., **44**, 724–728 (2005). Copyright Wiley-VCH Verlag GmbH & Co. KGaA. Reproduced with permission*

determined by the distribution of the liquid crystal and it was dependent on the relative rates of phase separation and polymerization. Fast polymerization led to the uniform distribution of the liquid crystal in the polymer particles, whereas when polymerization was slow the liquid crystals-segregated in the core of the particles (Figure 7.15a, inset).

Mixing of monomers with organic solvents acting as porogens provided the route to the synthesis of porous particles (Xu *et al.*, 2005; Dubinski *et al.*, 2008). Macroporous polymer particles have a broad range of applications, which include the utilization of porous microbeads as ion-exchange resins and sorbents, catalyst supports, and materials for biomedical applications. Control of the distribution of sizes of macroporous particles is vital in a number of applications. For example, polydisperse microbeads show reduced performance in chromatographic applications, due to the band broadening and increased column back-pressure (Li and Stover, 1998). Thus the intrinsic feature of microfluidic synthesis – the ability to generate polymer microbeads with a narrow distribution of sizes – may prove useful for chromatographic applications.

In the microfluidic production of macroporous copolymer particles, porogen solvents are pre-mixed with the monomers in a relatively high volume ratio. Following polymerization of the monomers and subsequent phase polymerization, the particles are collected at the outlet of the microfluidic reactor and washed to remove the solvent. Figure 7.15b shows the SEM image of the porous particle produced by microfluidic emulsification of TPGDA mixed with dioctyl phthalate (1 : 4 weight ratio), followed by UV-initiated free-radical polymerization of TPGDA and the subsequent removal of the porogen solvent with acetone. The mean size of the pores was 900 nm (Xu *et al.*, 2005).

The size of pores in the particles is another important factor in the synthesis of macroporous beads. A broad range of applications of macroporous copolymers requires a particular range of pore sizes. For instance, microbeads used in the size-exclusion chromatography of oligomers and small molecules requires the size of pores in the order of 100 Å, whereas protein separation and catalyst supports require materials with a size of pores of up to 4000 Å (Lloyd, 1991; Wernicke and Eisenbei, 1982).

The mean size of pores in macrobeads is determined by the relative rates of phase separation and polymerization, and is typically controlled by tuning the composition of the co-monomer and organic solvent mixture. For example, by selecting porogen solvents with a range of solubility parameters, it is possible to generate polymer particles with a range of pore sizes. This approach was undertaken in the microfluidic synthesis of macroporous particles from the mixture of glycidyl methacrylate (GMA) and ethylene glycol dimethacrylate (EGDMA). The size of pores was tuned by using various porogens, such as diethyl phthalate (DEP), diisobutyl phthalate (DBP), dioctyl phthalate (DOP), and diisodecyl phthalate (DDP) (Dubinsky *et al.*, 2008). These porogens had solubility parameters in a range of from 20.5 $(MPa)^{1/2}$ (DEP) to 14.7 $(MPa)^{1/2}$ (DDP) (Barton, 1991), whereas the polymer had a solubility parameter of $\sim$24 $(MPa)^{1/2}$.

Figure 7.16 shows the globular porous surface of particles obtained by free-radical polymerization of the GMA-EGDMA-porogen mixture. With increasing difference between the solubility parameter of the polymer and a porogen, the size of pores and

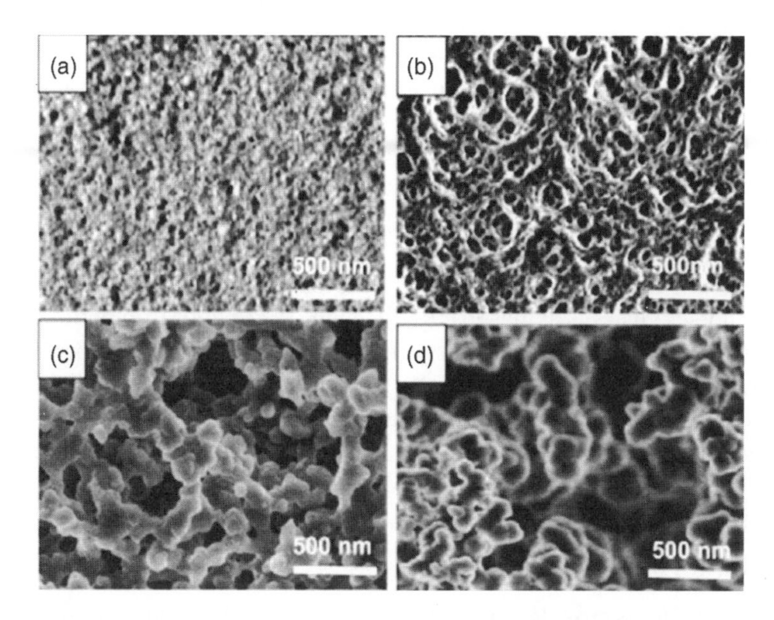

Figure 7.16 *SEM images of the surface of macroporous poly(GMA-EGDMA) particles after 60 s on-chip polymerization and 180 s off-chip polymerization of monomer mixtures comprising DEP (a), DBP (b), DOP (c), and DDP (d). (Adapted, with permission from Dubinsky et al., 2008). Reprinted with permission from Macromolecules, Microfluidic Synthesis of Macroporous Copolymer Particles by S. Dubinsky, H. Zhang, Z. Nie et al.,* ***41****, 10, 3555–3561 Copyright (2008) American Chemical Society*

the size of polymer globules (characterizing the extent of phase separation between the porogen solvent and the polymer) notably increased. The specific surface areas were 28.7, 13.9, 6.6, and 3.4 $m^2 g^{-1}$ for the microbeads synthesized by using DEP, DBP, DOP, and DDP, respectively.

Comparison of the structures of particles synthesized by the microfluidic synthesis and conventional UV-initiated suspension polymerization showed that the microbeads obtained by conventional synthesis had a significantly larger size of globules, especially, the microbeads synthesized in the presence of strongly non-polar porogens such as DOP or DDP (Dubinsky *et al.*, 2008). The difference in particle morphology was ascribed to the higher rate of polymerization achieved in the microfluidic polymerization, in comparison with conventional suspension polymerization: a uniform exposure of every monomer droplet to UV irradiation caused the formation of a greater number of free radicals and hence, a larger number of nuclei which aggregated in globules (Švec *et al.*, 1975; Švec and Frechet, 1995).

On the other hand, very similar specific surface areas and a mean pore radius were reported for the polymer microbeads prepared by microfluidic synthesis and conventional suspension polymerization of the mixture containing methacrylic acid and trimethylolpropane trimethacrylate or the mixture of methacrylic acid and EGDMA

(Zourob *et al.*, 2005). As the porous beads in two works (Zourob *et al.*, 2005; Dubinsky *et al.*, 2008) were prepared under different conditions, the effect of the relative rates of polymerization and phase separation in the microfluidic synthesis on the particle morphology needs further investigation.

Mixing of monomers with a small amount of a functional organic species was used for the synthesis of functionalized particles such as fluorescent dye-labeled microbeads (Xu *et al.*, 2005; Shepherd *et al.*, 2006; Nie *et al.*, 2005; Martin-Banderas *et al.*, 2006) or particles carrying biocatalysts such as glucose oxidase (GCO) or horseradish peroxidase (HRP) (Jeong *et al.*, 2005). Functional organic species, for example, fluorescent dyes, could also be selectively incorporated into one of the phases of the polymer particles (e.g., in the core of core–shell particles or in one of the halves of Janus particles) by generating multiphase precursor monomer droplets and compartmentalizing organic species in a particular droplet phase (see Chapter 10).

7.7.3 Hybrid Polymer–Inorganic Particles

Microfluidic synthesis provides a straightforward route to the continuous production of hybrid polymer–inorganic particles. The advantages of such synthesis include precise control of the concentration of the inorganic component in the particles, in comparison with dispersion or emulsion polymerizations, and a narrow polydispersity of hybrid particles in comparison with suspension polymerization. Examples of hybrid particles obtained by the continuous microfluidic synthesis include polymer microbeads carrying polymer or silica particles (Zhang *et al.*, 2006; Shepherd *et al.*, 2006), magnetic and semiconductor nanoparticles (Takeuchi *et al.*, 2005; Xu *et al.*, 2005), carbon black and titania dioxide pigments (Nisisako *et al.*, 2004).

Two more requirements have to be set for the emulsification of monomer mixtures with inorganic particles, in addition to those listed above. Firstly, when the monomer is mixed with inorganic particles, the concentration of the latter should not exceed a critical value at which the liquid mixture may clog the microchannels. Secondly, it is also important that the when droplets are formed, the particles at the tip of the stream of the liquid phase are not released into the continuous phase.

Inorganic particles can be dispersed uniformly throughout the entire polymer particle (Xu *et al.*, 2005) or localized in the specific compartment of the polymer particle, for example, in the shell of core–shell particles (Takeuchi *et al.*, 2005), or in one half of the Janus particles (Nisisako, Torii, and Higuchi, 2004; Shepherd *et al.*, 2006) (see Chapters 9 and 10). In the latter case, the surface chemistry of the nanoparticles favors their compatibility with a particular phase of the precursor droplets.

Figure 7.17 shows an optical fluorescence image of exemplary hybrid polymer-–inorganic particles generated by the microfluidic means: polyTPGDA microbeads comprising 4 nm-size CdSe quantum dots (Xu *et al.*, 2005). A monomer TPGDA pre-mixed with CdSe nanoparticles and a photoinitiator was emulsified in a microfluidic flow-focusing droplet generator to produce hybrid droplets, which were subsequently polymerized using photoinitiated polymerization. The hybrid particles maintained

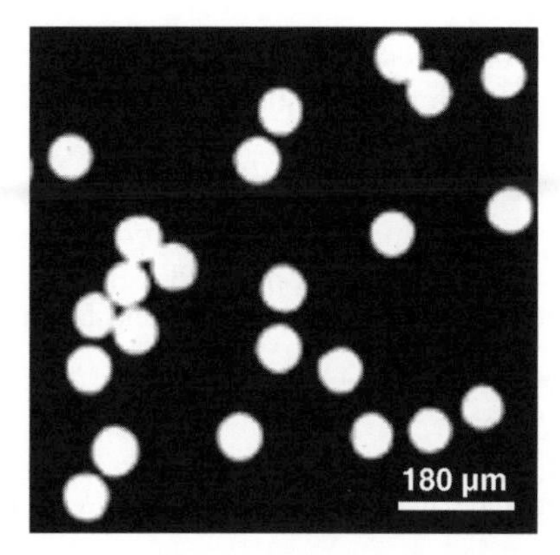

Figure 7.17 Fluorescence microscopy image of polyTPGDA microspheres loaded with 4 nm size CdSe quantum dots; $\lambda_{ex} = 502$ nm. (Adapted from Xu et al., 2005). Reprinted from Angewandte Chemie Int. Ed., Generation of Monodisperse Particles by Using Microfluidics: Control over Size, Shape, and Composition, **44**, 724–728 (2005). Copyright Wiley-VCH Verlag GmbH & Co. KGaA. Reproduced with permission

a narrow size distribution. The nanoparticles were uniformly dispersed in the hybrid microbeads and retained strong fluorescence characteristic for the CdSe quantum dots.

References

Anna, S.L., Bontoux, N., and Stone, H.A. (2003) *Appl. Phys. Lett.*, **82**, 364–366.

Anna, S.L. and Mayer, H.C. (2006) *Phys. Fluids*, **18**, 121512.

Anseth, K.S., Wang, C.M., and Bowman, C.N. (1994) *Macromolecules*, **27**, 650–655.

Barton, A.F.M. (1991) *CRC Handbook of Solubility Parameters and Other Cohesion Parameters*, 2nd edn, CRC Press, Boca Raton.

Basaran, O.A. (2002) *AIChE J.*, **48**, 1842–1848.

Beyer, H., Walter, W., and Lloyd, D. (1997) *Organic Chemistry: A Comprehensive Degree Text and Source Book*, Horwood Publishing Limited.

Bong, K.W., Pregibon, D.C., and Doyle, P.S. (2009) *Lab Chip*, **9**, 863.

Brandrup, J., Immergut, E.H., Grulke, E.A., Abe, A., and Bloch, D.R. (1999) *Polymer Handbook*, 4th edn, John Wiley & Sons, Inc., New York.

Decker, C. (1985) *Macromolecules*, **18**, 1241–1244.

Decker, C. (1998) *Polym. Int.*, **45**, 133–141.

Decker, C. (2002) *Macromol. Rapid Commun.*, **23**, 1067–1093.

Dendukuri, D., Tsoi, K., Hatton, T.A., and Doyle, P.S. (2005) *Langmuir*, **21**, 2113–2116.

Dendukuri, D., Pregibon, D.C., Collins, J., Hatton, T.A., and Doyle, P.S. (2006) *Nat. Mater.*, **5**, 365–369.

Dendukuri, D., Gu, S.S., Pregibon, D.C., Hatton, T.A., and Doyle, P.S. (2007) *Lab Chip*, **7**, 818–828.

Desai, M.C. and Stramiello, L.M.S. (1993) *Tetrahedron Lett.*, **34**, 7685–7688.

Dubinsky, S., Zhang, H., Nie, Z., Gourevich, I., Voicu, D., Deetz, M., and Kumacheva, E. (2008) *Macromolecules*, **41**, 3555–3561.

Günther, A. and Jensen, K.F. (2006) *Lab Chip*, **6**, 1487–1503.
Hettiarachchi, K., Talu, E., Longo, M.L., Dayton, P.A., and Lee, A.P. (2007) *Lab Chip*, **7**, 463–468.
Hines, M.A. and Guyot-Sionnest, P. (1996) *J. Phys. Chem., B*, **100**, 468–471.
Honda, T., Miyazaki, M., Nakamura, H., and Maeda, H. (2005) *Lab Chip*, **5**, 812–818.
Huang, S.H., Khoo, H.S., Chang, S.Y., and Tseng, F.G. (2008) *Microfluid. Nanofluid.*, **5**, 459.
Iwasaki, T. and Yoshida, J. (2005) *Macromolecules*, **38**, 1159–1163.
Jeong, W.J., Kim, J.Y., Kim, S., Lee, S., Mensing, G., and Beebe, D.J. (2004) *Lab Chip*, **4**, 576–580.
Jeong, W.J., Kim, J.Y., Choo, J.B., Lee, E.K., Han, C.S., Beebe, D.J., Seong, G.H., and Lee, S.H. (2005) *Langmuir*, **21**, 3738–3741.
Kaczmarek, H. and Decker, C. (1994) *J. Appl. Polym. Sci.*, **54**, 2147–2156.
Kim, J.-W., Utada, A.S., Fernandez-Nieves, A., Hu, Z., and Weitz, D.A. (2007) *Angew. Chem. Int. Ed.*, **46**, 1819–1822.
Lewis, P.C., Graham, R.R., Nie, Z., Xu, S., Seo, M., and Kumacheva, E. (2005) *Macromolecules*, **38**, 4536–4538.
Li, W.-H. and Stover, H.D.H. (1998) *J. Polym. Sci., Part A: Polym. Chem.*, **36**, 1543–1551.
Li, W., Nie, Z., Zhang, H., Paquet, C., Seo, M., Garstecki, P., and Kumacheva, E. (2007) *Langmuir*, **23**, 8010–8014.
Li, W., Pham, H.H., Nie, Z., MacDonald, B., Guenther, A., and Kumacheva, E. (2008) *J. Am. Chem. Soc.*, **130**, 9935–9941.
Lloyd, L.L. (1991) *J. Chromatogr.*, **544**, 201–217.
Loscertales, I.G., Barrero, A., Guerrero, I., Cortijo, R., Marquez, M., and Ganan-Calvo, A.M. (2002) *Science*, **295**, 1695–1698.
Martin-Banderas, L., Rodriguez-Gil, A., Cebolla, Á., Chávez, S., Berdún-Álvarez, T., Garcia, J.M.F., Flores-Mosquera, M., and Gañán-Calvo, A.M. (2006) *Adv. Mater.*, **18**, 559–564.
Miyazaki, K. and Horibe, T.J. (1988) *J. Biomed. Mater. Res.*, **22**, 1011–1022.
Moore, J.E. (1997) Photopolymerization of multifunctional acrylates and methacrylates. *In*: *Chemistry and Properties of Crosslinked Polymers* (ed. Labana, S.S.), Academic Press, New York, p. 535.
Nakashima, T., Shimizu, M., and Kukizaki, M. (1991) *Key Eng. Mater.*, **61/62**, 513–516.
Nie, Z.H., Xu, S.Q., Seo, M., Lewis, P.C., and Kumacheva, E. (2005) *J. Am. Chem. Soc.*, **127**, 8058–8063.
Nie, Z.H., Li, W., Seo, M., Xu, S.Q., and Kumacheva, E. (2006) *J. Am. Chem. Soc.*, **128**, 9408–9412.
Nie, Z., Seo, M., Xu, S., Lewis, P.C., Mok, M., Kumacheva, E., Garstecki, P., Whitesides, G.M., and Stone, H.A. (2008) *Microfluid. Nanofluid.*, **5**, 585–594.
Nisisako, T., Torii, T., and Higuchi, T. (2004) *Chem. Eng. J.*, **101**, 23–29.
Nisisako, T., Torii, T., Takahashi, T., and Takizawa, Y. (2006) *Adv. Mater.*, **18**, 1152–1156.
Ottino, J.M. and Wiggins, S. (2004) *Phil. Trans. R. Soc. London A*, **362**, 923–935.
Quevedo, E., Steinbacher, J., and McQuade, D.T. (2005) *J. Am. Chem. Soc.*, **127**, 10498–10499.
Seo, M., Nie, Z., Xu, S., Mok, M., Lewis, P.C., Graham, R., and Kumacheva, E. (2005) *Langmuir*, **21**, 11614–11622.
Seo, M., Paquet, C., Nie, Z., Xu, S., and Kumacheva, E. (2007) *Soft Matt.*, **3**, 986–992.
Seo, M. (2008) PhD thesis, University of Toronto.
Serra, C., Sary, N., Schlatter, G., Hadziioannou, G., and Hessel, V. (2005) *Lab Chip*, **5**, 966–973.
Shepherd, R.F., Conrad, J.C., Rhodes, S.K., Link, D.R., Marquez, M., Weitz, D.A., and Lewis, J.A. (2006) *Langmuir*, **22**, 8618–8622.
Sperling, L.H. (1981) *Interpenetrating Polymer Networks and Related Materials*, Plenum Press, New York, pp. 1–10.
Sperling, L.H. and Mishra, V. (1996) *Polym. Adv. Technol.*, **7**, 197–208.
Stone, H.A., Stroock, A.D., and Ajdari, A. (2004) *Rev. Fluid. Mech.*, **36**, 381–411.

Sugiura, S., Nakajima, M., Itou, H., and Seki, M. (2001) *Macromol. Rapid Commun.*, **22**, 773–778.
Sugiura, S., Nakajima, M., and Seki, M. (2002) *Ind. Eng. Chem. Res.*, **41**, 4043–4047.
Sultan, W. and Busnel, J.-P. (2006) *J. Therm. Anal. Calorim.*, **83**, 355–359.
Švec, F., Hradil, J., Čoupek, J., and Kálal, J. (1975) *Angew. Macromol. Chem.*, **48**, 135–143.
Švec, F. and Frechet, J.M.J. (1995) *Macromolecules*, **28**, 7580–7582.
Takeuchi, S., Garstecki, P., Weibel, D.B., and Whitesides, G.M. (2005) *Adv. Mater.*, **17**, 1067–1072.
Thorsen, T., Reberts, R.W., Arnold, F.H., and Quake, S.R. (2001) *Phys. Rev. Lett.*, **86**, 4163–4166.
Wernicke, R. and Eisenbei, F. (1982) *Chromatography*, **15**, 347–350.
Wu, T., Mei, Y., Cabral, J.T., Xu, C., and Beers, K.L. (2004) *J. Am. Chem. Soc.*, **126**, 9880–9881.
Wyss, H.M., Blair, D.L., Morris, J.F., Stone, J.F., and Weitz, D.A. (2006) *Phys. Rev. E*, **74**, 061402.
Xu, S.Q., Nie, Z.H., Seo, M.S., Lewis, P., Kumacheva, E., Stone, H.A., Garstecki, P., Weibel, D.B., Gitlin, I., and Whitesides, G.M. (2005) *Angew. Chem. Int. Ed.*, **44**, 724–728.
Zhang, H., Tumarkin, E., Peerani, R., Nie, Z., Sullan, R.M.A., Walker, G.C., and Kumacheva, E. (2006) *J. Am. Chem. Soc.*, **128**, 12205.
Zhao, B., Viernes, N.O.L., Moore, J.S., and Beebe, D.J. (2002) *J. Am. Chem. Soc.* **124**, 5284.
Zourob, M., Mohr, S., Mayes, A.G., Macaskill, A., Pérez-Moral, N., Fielden, P.R., and Goddard, N.J. (2006) *Lab Chip* **6**, 296–301.

8

Microfluidic Production of Hydrogel Particles

CHAPTER OVERVIEW

8.1 Introduction

Polymer gels are networks of macromolecules that are inflated with a solvent. Gels that are swollen with organic solvents or with water are called organogels and hydrogels, respectively. Gels can be formed from synthetic polymers, biopolymers, or their mixtures. In all these systems, network structures are formed by physical bonding or

Microfluidic Reactors for Polymer Particles. Eugenia Kumacheva and Piotr Garstecki.

chemical crosslinking of polymer molecules. Examples of polymer networks formed by strong physical bonding include microcrystals, glassy nodules, and double helixes (Rubinstein and Colby, 2003). Strong physical bonds are essentially permanent and can only be broken by changing the ambient conditions, for example, by changing the temperature, pH, or the ionic strength of the system. Weak physical gelation occurs by temporary association of polymer molecules, and the resulting bonds have a finite lifetime. Weak physical bonding occurs between block copolymers, ionomers, and polymers interactiong via hydrogen bonding (Rubinstein and Colby, 2003). Covalent bonding between polymer molecules leads to chemical gelation. Typically, gelation occurs via polycondensation, addition polymerization, or vulcanization reactions, which produce permanent polymer networks. In gels formed by both chemical and physical gelation, the density of crosslinking of networks affects the porosity and the mechanical properties of the gel. High crosslinking density reduces mesh size in polymer networks and increases their mechanical strength.

The intrinsic feature of gels – their open network structures – results in their ability to undergo large changes in volume. The reversible changes in volume of microgels are governed by the imbalance between repulsive and attractive forces acting in gels. Increase in volume (or swelling) occurs when electrostatic repulsion and osmotic forces exceed the attractive forces, which include hydrogen bonding, van der Waals interactions, and hydrophobic or specific interactions.

Microgels are gel particles with dimensions in the range from submicrometers to tens of micrometers. Owing to the small size, large volume transitions occur in microgels within a relatively short time, in comparison with macroscopic gels. This feature, in addition to the facile synthesis and functionalization of microgels (e.g., bioconjugation), and the ability to control their colloid stability and dimensions in a wide range of dimensions have made these particles useful in product-oriented research.

Large volume phase transitions in microgels are particularly useful in developing "smart gels", which under particular conditions undergo shrinkage or expansion, thereby acting as sensors. Alternatively, shrinkage of microgel particles loaded with cargo molecules leads to the release of the encapsulated component in the surrounding environment when the ambient conditions are changed. Microgels have rapidly gained importance in materials science owing to their potential applications in drug delivery (Murthy, 2002, 2003; Lopez and Snowden, 2003; Nayak *et al.*, 2004; Nolan *et al.*, 2005; Varma, Kaushal, and Garg 2005; Lopez, Hadgraft, and Snowden, 2005; Bromberg, 2003a; Das *et al.*, 2006; Morris, Vincent, and Snowden, 1997; LaVan and Langer, 2002); sensing (Morris, Vincent, and Snowden, 1997; Retama, Lopez-Ruiz, and Lopez-Cabarcos 2003; Guo, Sautereau, and Kranbuehl, 2005); the fabrication of photonic crystals (Xu, 2003; Lyon *et al.*, 2002, 2004; Jones and Lyon, 2003); template-based synthesis of inorganic nanoparticles (Zhang, 2004); and separation and purification technologies (Bromberg, 2003b). Traditional applications of aqueous microgels include their use in coatings, the food industry, and industrial processing. Examples of commonly used microgels include those derived from starch, gums, agarose, and alginate, crosslinked poly(sodium methacrylate), and poly-*N*-isopropyl amide.

Encapsulation of living cells is another important application of polymer hydrogel beads (Orive *et al.*, 2003). Encapsulation of cells in biopolymer hydrogel beads creates an environment that allows one to mimic the three-dimensional biological environment of tissues. When hydrogel particles are appropriately designed, the products of nutrition and cell secretion can continuously diffuse through the walls of a particle, without disturbing the encapsulated cell. Encapsulation of human cells helps to overcome problems associated with an immunological barrier between the host and the transplanted cells (Dove, 2002; Calafiore *et al.*, 2004). Accordingly, the necessity of immunosuppressive drugs is avoided. In the last decade, encapsulation of cells has led to promising therapeutic treatments for diabetes, hemophilia, and cancer (see, e.g., Sun *et al.*, 1996; Hortelano *et al.*, 1996).

Among the biological hydrogels, alginate is the most frequently used material for cell encapsulation. Other biopolymers used for cell entrapment include κ-carrageenan, chitosane, gelatin, agarose, and cellulose derivatives. The utilization of synthetic polymers for the encapsulation of cells has also been explored. For example, polyacrylamide or polyacrylate hydrogel microbeads with predetermined porosity, charge, hydrophobicity, and toughness have been used for the immobilization of cells. However, the challenge in using synthetic polymers for the encapsulation of cells is to keep cells viable (Freeman and Aharonowitz, 1981; Freeman, 1987; Taylor and Cerankowski, 1975; Hoffman, Afrassiabi, and Dong, 1986; Bae *et al.*, 1987; Galaev and Mattiasson, 1993; Stevenson and Sefton, 1993; Pathak, Sawhney, and Hubbell, 1992). Currently, the encapsulation of cells is dominated by the use of biological polymers, or of hybrids of biological polymers and synthetic polymers.

Microgel particles can also serve as model cells and accordingly, can be used in biomechanical experiments mimicking the behavior of real cells, such as blood cells, when they flow through microchannels (Fiddes *et al.*, 2007). The surfaces of the microgels can be conjugated with proteins, in order to examine the role of "cell" interactions with microchannel walls on their flow behavior. On the other hand, the mechanical properties of the "surrogate cells" can be controlled by varying the crosslinking density of the particles (Fiddes *et al.*, 2009). Soft and deformable microgels exhibit greater reduction in speed, when they passed microchannels under confinement, than more rigid particles, due to the formation of large-area adhesive contact with the microchannel walls.

8.2 Methods Used for the Production of Polymer Microgels

This section describes the methods that have been exploited for the preparation of hydrogel particles (microgels) from biopolymers and synthetic polymers. The strategies for the production of hydrogel particles can be classified into three groups:

(a) by the method that is used to "shape" hydrogels in particles, as opposed to macroscopic polymer gels;

(b) by the mechanism of formation of the polymer network. Microgels can be formed by chemical or physical bonding of polymer molecules or by the combination of these two methods;
(c) by the morphology of microgels. Microgels can have a uniform, chemically and structurally homogeneous morphology, or a core–shell (capsular) structure.

8.2.1 Shaping Gels into Microgel Particles

Regardless of the microgel structure and the method used for the formation of the polymer network, the methods used for the preparation of gel beads can be tentatively divided into two large groups.

In the *top–bottom approach*, the preparation of gel beads starts from the emulsification of a solution of a precursor monomer or polymer in the continuous liquid phase. The droplets are then gelled, yielding microgel particles with a typical size of from hundreds of nanometers to millimeters. The size and the distribution of the sizes of the droplets determines the size and polydispersity of the resulting microgels, respectively. Polydisperse droplets transform into microgels with a broad size distribution. Generally, the emulsification process and the distribution of sizes of the droplets is optimized by using surfactants, co-solvents or non-solvents, and various types of stabilizers, which may include high molecular weight polymers or small particles. Emulsification is typically accomplished by the simultaneous formation of a large number of droplets by applying energy to two largely immiscible macroscopic liquids (a continuous phase and a droplet phases) that are brought into contact. Alternatively, the formation of "one droplet at a time" is achieved by breaking up threads of aqueous polymer solutions of biopolymers in droplets via mechanical cutting (Prüsse *et al.*, 1997), dripping (Fundueanu *et al.*, 1999; Bugarski *et al.*, 1994), or jetting methods (Levee *et al.*, 1994; Brandenberger and Widmer, 1999). The resulting droplets are injected in a liquid that may be miscible or immiscible with the droplet phase. In the former case, droplets are formed in air and then introduced into the continuous phase where fast polymer gelation counteracts the dissolution of the drops in the continuous phase. This process is frequently used for the generation of alginate microgels.

In the *bottom–top* approach to microgels, the starting system is typically a macroscopic solution of a monomer, a crosslinking agent, an initiator, and sometimes, a stabilizing agent. The synthesis of microgels occurs via polymerization of the monomer through the process of homogeneous nucleation. The growing polymer molecules reach a critical chain length at which they loose solubility in water and phase separate (collapse) to form "precursor" particles. Larger polymer particles grow by either deposition of the seed particles onto existing polymer particles, or by the aggregation of seed particles. This process is frequently called "precipitation polymerization", and it is generally used for the generation of microgels from synthetic polymers (Pelton, 2000; Hoare and Pelton, 2004). Under optimized conditions, the synthesis yields microgel particles with a narrow size distribution.

The typical dimensions of microgels vary from approximately 100 nm to several micrometers.

8.2.2 Forming a Polymer Network

Among the methods used for chemical crosslinking, free-radical polymerization is the strategy that is most frequently used for the synthesis of microgel particles. This method applies to both the bottom–top and top–bottom approaches to microgels (Pelton, 2000; Park and Hoffman, 1990). Free-radical polymerization leads to the formation of linear polymer chains that are crosslinked by a multifunctional reagent. Initiation of the free-radical polymerization is generally performed by heating or irradiation. Both methods lead to the dissociation of an initiator of the polymerization reaction. Typical chemical initiators used in microgel synthesis are persulfates, V-50, β-dimethylaminopropionitrile and *N,N,N′,N′*-tetramethylenediamine (TEMED), whereas sodium hydrosulfite, riboflavin and TEMED are used as the initiators of photochemical polymerization. Other less frequently synthetic routes include redox polymerization, and polycondensation.

Physically-induced gelation relies on the association of polymer or oligomer molecules governed by the formation of non-covalent bonds. Originally, a polymer is dissolved in a suitable solvent. Gelation is triggered by changing the temperature, pH, or the ionic strength of the polymer solution, by adding a physical crosslinker, by evaporating the solvent, or by adding a non-solvent. Reduction in polymer solubility in the solution triggers favorable polymer–polymer interactions and hence, gelation.

8.2.3 Controlling the Structure of Polymer Microgels

Gel microbeads can have a uniform, a gradient, or a capsular structure. A microcapsule contains a well-defined core and a gel shell, in comparison with microgels that have a uniform structure and composition throughout the particle. Gradient microgels have a structure and/or a composition gradually changing throughout the particle. For example, the crosslinking density can increase from the microgel center to the microgel exterior. Microgel particles with a uniform, a gradient, or a capsular structure can be obtained by both bottom–top and top–bottom methods and by both chemical and physical crosslinking.

Capsules have the most interesting structure with respect to microgel applications in the encapsulation of biologically active species. Microgel capsules are typically produced by gelling a polymer layer around a liquid, a gas, or a solid core. Most frequently, capsules have a liquid core. The coacervation/precipitation method is one of the typical methods for the generation of hydrogel capsules. In this method, first, an emulsion of oil droplets is obtained in an aqueous polymer-rich solution of two oppositely charged polyelectrolytes. A change in the polymer solubility causes the polyelectrolytes to gel through the electrostatic attraction between the polymer molecules. The crosslinked polymer precipitates on the surface of the droplets. Alternatively, polymer phase separation and precipitation of the gel on the surface of the droplets can occur without the use of polyelectrolytes: the reduction in the solubility of a non-charged polymer in the continuous phase can be accomplished by changing the temperature of the solution or by adding a non-solvent.

Layer-by-layer deposition of water-soluble polyelectrolytes has recently attracted a great deal of interest. Generally, the process includes sequential deposition of multiple layers of oppositely charged polymer layers onto the particles, droplets, or bubbles (Peyratout and Dähne, 2004). Electrostatic attraction and entanglement of the polymer molecules, and mutual interpenetration of the polyelectrolyte layers make capsular shells mechanically strong. When polyelectrolytes are deposited on the surface of rigid particles, the rigid cores may be removed and replaced with water or a water solution. In addition, forces other than of electrostatic origin, such as hydrogen bonding or hydrophobic interactions between the polymers, can be exploited for the layer-by-layer deposition of polymer layers.

The use of double emulsions is another important method for the production of microgel capsules. To generate gel capsules, droplets engulfing smaller droplets are first formed. Gelation is typically achieved by changing the temperature or removing the solvent from the shell of the compound (in the case of physical gelation) or by the means of free-radical or redox polymerization (chemically mediated gelation) (Chu *et al.*, 2007, Shah *et al.*, 2008).

8.3 Microfluidic Synthesis and Assembly of Polymer Microgels

In principle, all features of polymer microgels discussed above are applicable to the microgel particles prepared by microfluidic methods. Nevertheless, microfluidic emulsification brings specific requirements to the preparation of polymer microgels. The conditions for microfluidic synthesis of polymer microgels, which are somewhat distinct from the synthesis of rigid polymer particles, are discussed below.

Microfluidic synthesis and assembly of polymer microgels are generally separated into two distinct stages: microfluidic emulsification and gelation of the precursor droplets. Direct microfluidic emulsification of solutions of rapidly gelling polymer solutions is a challenge, due to their high viscosity and extensional flow of the polymer.

Figure 8.1 illustrates a typical problem encountered in the attempt to breakup a thread of gelled polymer into droplets (Zhang *et al.*, 2006). The system under study was a gel obtained by ionically crosslinking sodium alginate with Ca^{2+} ions. Emulsification was conducted in a five-channel microfluidic planar flow-focusing device (Anna, Bontoux, and Stone, 2003). An aqueous solution of sodium alginate was supplied to the central channel, an aqueous solution of $CaCl_2$ was injected in the intermediate channels, and mineral oil was introduced in the two outermost channels. At the exit from the central and intermediate channels, the aqueous solutions of sodium alginate and $CaCl_2$ mixed and formed a liquid thread that was forced into the orifice. Emulsification of the aqueous solution occurred when the concentrations of the polymer and/or the crosslinking agent were low (Figure 8.1a); however, the resulting weak microgel particles were mechanically weak and dissolved when they were transferred into an aqueous medium. When the concentration of the polymer and/or $CaCl_2$ was increased, rapid gelation of alginate with

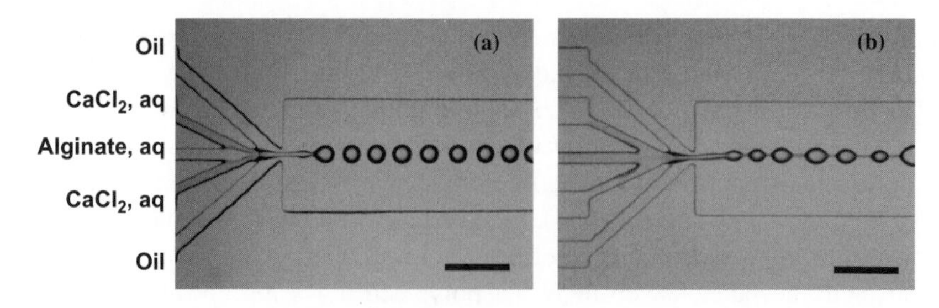

Figure 8.1 (a) Emulsification of 0.25 wt% aqueous solution of sodium alginate mixed with 0.05 wt% solution of $CaCl_2$. Scale bar is 200 μm. (b) Formation of nodes on the liquid thread of 0.50 wt% aqueous alginate solution mixed with 0.10 wt% of $CaCl_2$ in co-flowing mineral oil in a five-channel microfluidic device. Scale bar is 300 μm. In (a) and (b) the flow rates of the alginate solution, the solution of $CaCl_2$ and mineral oil are 0.2, 0.05, and 10 mL h^{-1}, respectively. Image (b) is adapted, with permission, from Zhang et al., 2006. Reprinted with permission from H. Zhang et al., Microfluidic Production of Biopolymer Microcapsules with Controlled Morphology, Journal of the American Chemical Society, **128**, 12205. Copyright (2006) American Chemical Society

Ca^{2+} ions led to an instantaneous increase in viscosity of the thread, which in a broad range of flow rates of the liquids did not breakup to release droplets (Figure 8.1b). Instead, the thread featured nodes connected by narrow necks. The nodes had a broad distribution of sizes, which implied that even if further increase in the flow rate of the non-polar continuous phase resulted in the breakup of the jet, it would not lead to the generation of particles with a uniform size distribution.

The emulsification behavior illustrated in Figure 8.1 implies that when there is a possibility of rapid polymer gelation, it is imperative to separate, both spatially and temporarily, the emulsification and gelation processes. More generally, it is advisable to conduct gelation "by demand", that is, by triggering chemical reactions or physical processes in the droplets at a particular moment of time.

Currently used or emerging methods for the preparation of hydrogel microbeads by microfluidic methods are summarized below. In principle, with an appropriate modification these methods are applicable to the generation of organogel particles, however this is beyond the subject area of this chapter.

8.3.1 Microgels Produced by Chemically Mediated Gelation

The method resembles the microfluidic synthesis of solid polymer particles, which was described in Chapter 7. When the synthesis is conducted in a two-phase flow, the preparation of polymer microgels starts from the microfluidic emulsification of the low-viscosity solution of the precursor monomer or reactive polymer. This step is accompanied by *on-chip* or *off-chip* polymerization and/or crosslinking of the monomer or polymer. Chemically mediated gelation was achieved by either photoinitiated or redox-initiated polymerization of the precursor polymers or oligomers. Microgels were

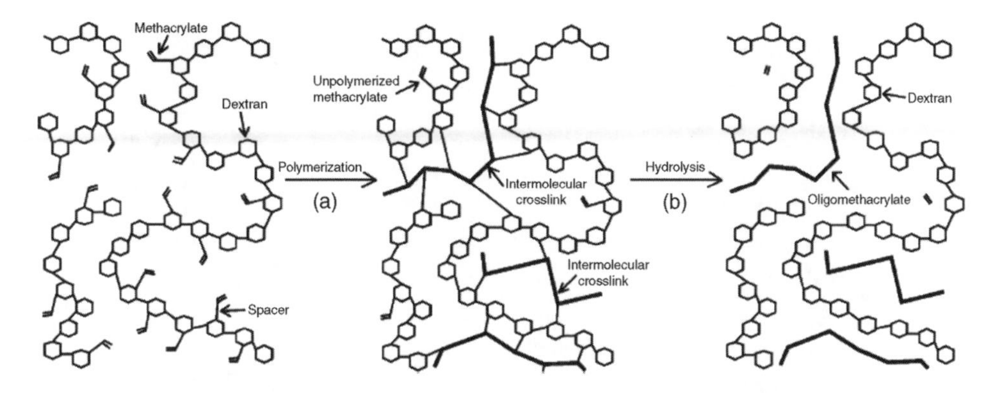

Figure 8.2 *Schematic representation of the polymerization of dex-HEMA (step A), leading to the formation of intra- and intermolecular crosslinks, which form the three-dimensional hydrogel network, and (step B) the hydrolysis of the dex-HEMA hydrogels leading to the formation of dextran chains and oligomethacrylates as degradation products. Adapted from De Geest et al., 2005. Reprinted with permission from B.G. De Geest, et al., Synthesis of Monodisperse Biodegradable Microgels in Microfluidic Devices, Langmuir,* ***21****, 10275. Copyright (2005) American Chemical Society*

synthesized from polyacrylamide (Shepherd *et al.*, 2006), poly-*N*-isopropylacrylamide) (Kim *et al.*, 2007; Shah *et al.*, 2008), and poly(dextran-hydroxyethyl methacrylate) [poly(dex-HEMA)] (De Geest *et al.*, 2005). Microfluidic synthesis was used to produce polymer microgels with dimensions in the range of from 10 to 100 μm. The method enabled the synthesis of particles with a uniform, Janus, and capsular structure.

One of the particularly useful examples is the generation of microgel particles from poly(dex-HEMA) (De Geest *et al.*, 2005). The dex-HEMA hydrogels are ideal candidates for the encapsulation and release of small molecules: when the polymer starts to degrade, the size of the pores between the dextran chains increases due to the cleavage of the crosslinks, and the diffusion of molecules entrapped in the hydrogels is enhanced. Figure 8.2 illustrates the reaction used for the synthesis of dex-HEMA hydrogels and their subsequent degradation by hydrolysis of the carbonate ester groups, which link methacrylate groups and the dextran fragments. The details of the synthesis and characterization of dex-HEMA have been reported elsewhere (van Dijk-Wolthuis *et al.*, 1997; van Dijk-Wolthuis, 1997b). The degradation rate of dex-HEMA polymers was controlled by varying the number of methacrylate groups per dextran chain and the initial content of water in the system.

Microgels were produced by photopolymerization of dex-HEMA compartmentalized in aqueous droplets. The emulsification of the aqueous solution of dex-HEMA is illustrated in Figure 8.3a. Under optimized conditions, monodisperse precursor droplets were generated in the mineral oil, with the addition of 4% (v/v) of a non-ionic (cetyl dimethicone copolymer) surfactant, ABIL EM-90 (Degussa), in the oil phase, in order to reduce the interfacial tension between the droplet and the continuous phases and to prevent subsequent coalescence of the droplets. A photoinitiator, Irgacure 2959 (Ciba

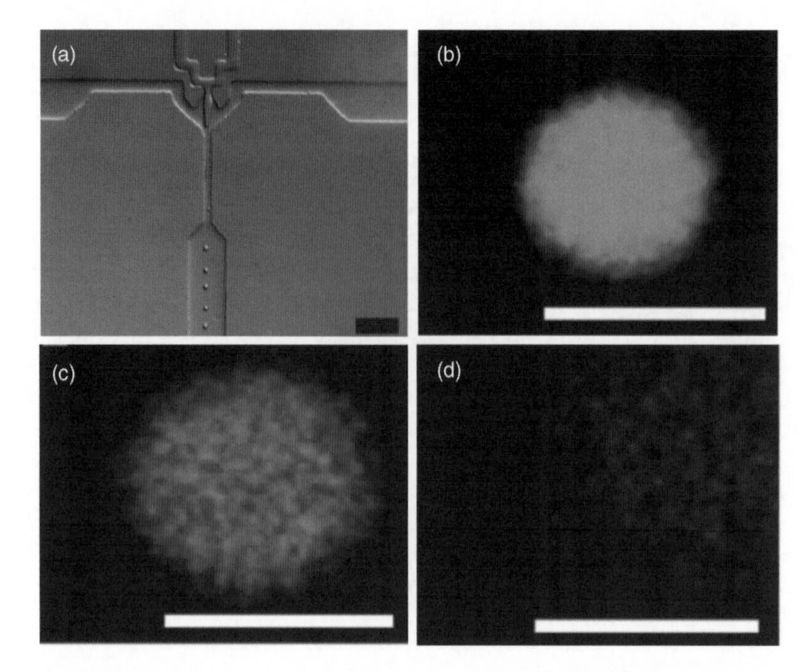

Figure 8.3 *(a) Optical microscopy image of the formation of aqueous dex-HEMA droplets in the continuous mineral oil phase. (b-d) Confocal microscopy images of dex-HEMA microgels containing FITC-BSA, respectively, before (b), during (c), and after (d) degradation. The time lapse between the images is 30 s. The scale bars represent 100 μm. Adapted, with permission from De Geest et al., 2005. Reprinted with permission from B.G. De Geest, et al., Synthesis of Monodisperse Biodegradable Microgels in Microfluidic Devices, Langmuir,* ***21****, 10275 Copyright (2005) American Chemical Society. See Plate 9*

Chemicals) was added to the aqueous dex-HEMA solution before its emulsification to enable photopolymerization of dex-HEMA. The droplets were gelled *off-chip* by collecting them at the exit of the droplet generator and immediately exposing them to UV-irradiation. Subsequently, the resulting dex-HEMA microgels were separated from the oil by centrifugation and washed with deionized water. The productivity of the method was 0.15 mg of monodisperse microgels per hour.

To explore potential applications of the microgels for the encapsulation of proteins, a fluorescein-labeled bovine serum albumin (FITC-BSA) was introduced into the dex-HEMA phase. The encapsulation efficiency of FITC-BSA was ensured by insolubility of the FITC-BSA in the oil phase. Sodium hydroxide was used to accelerate the degradation of the microgels. Figure 8.3b–d shows confocal microscopy images of dex-HEMA microgels containing FITC-BSA before, during, and after degradation, respectively.

Recently, chemically triggered gelation of droplets emulsified in microfluidic droplet generators was used to produce poly(*N*-isopropylacrylamide) (polyNIPAm) microgel capsules with controlled dimensions and morphology (Chu *et al.*, 2007; Shah *et al.*, 2008). These groups demonstrated the ability to encapsulate dyes and nanoparticles (e.g., semiconductor quantum dots and magnetic nanoparticles) in microgel particles, create a controlled number of voids in the microgel interior, and produce core–shell microgels (capsules) with aqueous or oil cores. The microfluidic approach to the generation of capsules is discussed in Chapter 9.

The polyNIPAm molecules forming the microgels contain both hydrophilic amide groups and hydrophobic isopropyl groups. In water, below approximately 32 °C, the interactions between the polymer chains and water are favored, due to the hydrogen bonding between the amide groups and water molecules, and the microgels are swollen with water. At higher temperatures, the hydrogen bonds between the water molecules and the amide groups are disrupted, water is expelled from the microgels, and the particles dramatically shrink (Saunders and Vincent, 1999). The temperature mediated shrinkage of polyNIPAm microgels was also demonstrated for the capsules produced by microfluidic synthesis (Chu *et al.*, 2007; Shah *et al.*, 2008). In particular, a complex core–shell structure of microgel particles has been exploited for potential applications in drug delivery. Figure 8.4a shows the capsules with a shell of thermosensitive polyNIPAm hydrogel that encapsulates an oil core containing several water droplets. With an increase in the temperature above the de-swelling temperature of the polymer, the hydrogel shell shrank. The incompressible oil core exerted a counter-stress on the inner wall of the hydrogel shell. The shell ruptured, and the encapsulated core of oil and water droplets were instantaneously released (Chu *et al.*, 2007).

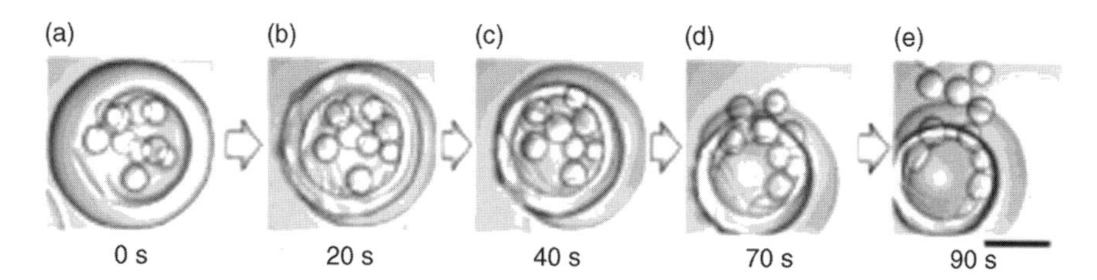

Figure 8.4 *Temperature-sensitive hydrogel microcapsule for pulsed release. (a) Optical micrograph of a microcapsule with a shell comprised of a thermosensitive hydrogel containing aqueous droplets dispersed in oil. Upon increasing the temperature, the hydrogel shell shrinks by expelling water. This capsule was generated from a triple emulsion, where the continuous liquid is oil, the hydrogel shell is aqueous, the inner middle fluid is also oil, and the innermost droplets are aqueous. (b–e) Optical micrograph time series showing the forced expulsion of the oil and water droplets contained within the microcapsule when the temperature is rapidly increased from 25 to 50 °C. The time series begins once the temperature reaches 50 °C. The extra layer surrounding the microcapsule in (b–e) is water that is squeezed out from the hydrogel shell as it shrinks. The coalescence of the expelled inner oil with the outer oil can not be resolved, as both liquids have the same index of refraction. The scale bar is 200 mm. Adapted, with permission, from Chu et al., 2007. Reprinted from Angewandte Chemie Int. Ed., Controllable Monodisperse Multiple Emulsions by L. Chu, A. S. Utada, R. K. Shah et al.,* ***46***, ***47***, *8970-8974(2007). Copyright Wiley-VCH Verlag GmbH & Co. KGaA. Reproduced with permission.*

In comparison with other microfluidic methods for the preparation of microgels, chemical crosslinking or polymerization has the advantage of producing mechanically strong microgels. However, without an appropriate optimization of the microgel composition, the resulting particles cannot be easily cleaved and metabolized. Furthermore, photo-irradiation has to be carefully optimized to avoid potential harm to the biological species (e.g., cells) to be encapsulated in the microgels. The usage of potentially cytotoxic monomers, crosslinking agents, surfactants, and initiators may also reduce cell viability. Therefore, this method has to be used with caution when microgel particles have biorelated applications.

8.3.2 Microgels Produced by Physically Mediated Gelation

This type of gelation is typically realized for biological polymers such as agarose, alginate, κ-carrageenan, carboxymethylcellulose, chitosan, gelatin, or pectin. The exact mechanism of physical gelation depends on the type of polymer, however, in general, gelation is caused by hydrophobic interactions (Forster and Antonietti, 1998; Nomura *et al.*, 2003), ionic interactions (Zhang *et al.*, 2006; Kuo, 2004) or hydrogen bonding (Sugiura *et al.*, 2001; Xu *et al.*, 2005).

8.3.2.1 Gelation Induced by a Change in Temperature

In thermally mediated microfluidic preparation of microgels, a low viscous precursor polymer solution is emulsified at a temperature that is greater than the gelation temperature of the polymer. Subsequent *in situ* (*on-chip*) or *off-chip* gelation of the polymer achieved by cooling of the droplets yields microgel particles. *Off-chip* gelation broadens the distribution of sizes of microgel beads due to the coalescence of the precursor droplets. Therefore, even if complete *in situ* gelation is not realized, "pre-gelation" conducted *on-chip* can be useful in the generation of microgels with a narrow size distribution.

In 2002 gelatin particles were generated by the microfluidic method (Iwamoto *et al.*, 2002). Gelatin is a fibrous protein that is produced from collagen. Gelatin gels are used extensively in the pharmaceutical, cosmetics, and food industry, owing to their remarkable mechanical properties and to the biological origin of the gel. In the work of Iwamoto *et al.* microfluidic emulsification of a low-viscous 5 wt% gelatin solution occurred at 40 °C (where gelatin molecules have a random coil configuration). Emulsification occurred in the terrace-like microfluidic device. Isooctane containing 5 wt% of the surfactant tetraglycerin condensed ricinoleic acid ester was used as the continuous phase. The emulsion droplets had an average particle diameter of about 40 μm and a polydispersity of 5.1%. The droplets were collected and gelled *off-chip* to 25 °C (where a transition from coil to helix occurs). Following overnight incubation, the average diameter of the particles reduced to about 32 μm, and their polydispersity increased 7.3%. The temperature was then lowered to 5 °C by rapid air and the gelatin beads were dried. Redispersion of the particles in iso-octane yielded microgels with an average diameter of 15.6 μm and polydispersityof 5.9% (Figure 8.5).

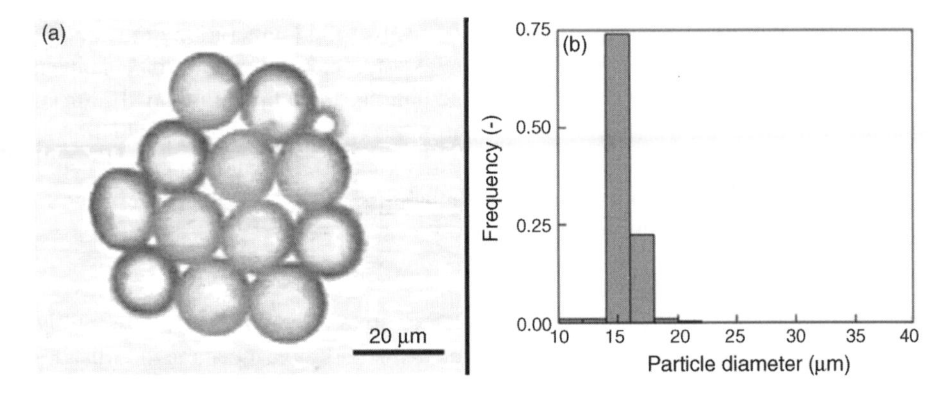

Figure 8.5 *(a) Photomicrograph of the dried gelatin microbeads resuspended in isooctane. (b) Particle diameter distribution (Iwamoto et al., 2002). Reprinted with permission from S. Iwamoto, et al., Preparation of Gelatin Microbeads With a Narrow Size Distribution Using Microchannel Emulsification, AAPS PharmSciTech **3**, article 25. Copyright (2002) American Association of Pharmaceutical Scientists*

The generation of biological microgels by thermally induced gelation was also demonstrated for agarose microgels (Xu *et al.*, 2005). Emulsification was carried out at a temperature greater than 37 °C (the gelling temperature of agarose). In contrast to the work of Iwamoto *et al.* (2002), the transformation of droplets into microgels was achieved *on-chip* by applying a temperature gradient to the microfluidic device. The droplets flowing through the downstream channel were cooled to the temperature below the gelation temperature of agarose. Gelation of agarose occurred through the formation and subsequent aggregation of double helices containing two galactose residues bonded through a rigid anhydro bridge. This method yielded monodisperse microgel particles with diameters in the size range of 50–250 μm.

Gelation induced by the change in temperature of the precursor droplets was also used for the generation of microgels of κ-carrageenan, in order to study the deformation of gelling droplets in a continuous flow process (Walther *et al.*, 2005). Temperature gradients in the device were produced by maintaining the aqueous κ-carrageenan solution at 100 °C and varying the temperature of the continuous phase (sunflower oil) from 54 °C down to room temperature, thereby increasing its viscosity. The rate of gelation of the droplets was controlled by the temperature of the continuous phase. It was found that upon cooling the droplets towards gelation of k-carrageenan, their deformation increased with increasing viscosity of the sunflower oil. However, after the droplets were completely gelled, the trend reversed and the deformation decreased.

When the generation of microgel particles using the thermosetting method is conducted *on-chip*, the necessity to maintain a large temperature gradient across a microfluidic device presents a significant limitation of this technique. The temperature in the emulsification compartment has to be sufficiently high to generate monodisperse droplets from the precursor polymer solution, however the temperature in the gelation compartment has to be sufficiently low to gel the polymer before the droplets leave the microfluidic reactor. Furthermore, small fluctuations in temperature in the droplet generator can affect the viscosities of the continuous and droplet phases, which would immediately affect the ability of the system to produce monodisperse precursor droplets. These technical challenges limit the application of this method for the fast continuous production of microgels of biopolymers.

8.3.2.2 Gelation Induced by Ionic Crosslinking

Ionic gelation occurs by crosslinking of polyelectrolyte polymer chains with multivalent ions or small ionized molecules. Typically, this gelation mechanism is used for biological polymers. Exemplary biopolymers include alginate, κ-carrageenan, carboxymethylcellulose, and chitosan. Among these polymers, anionic alginate polyelectrolyte crosslinked by Ca^{2+} cations binding to the guluronic acid blocks is the most frequently used system. Alginate gels crosslinked with Ca^{2+} ions have applications as drug carriers, as immobilization matrixes for bioreactor systems, and scaffolds in tissue engineering.

As ionic gelation may occur very rapidly, it is strongly advisable that the crosslinking process takes place *after* microfluidic emulsification, so that the increase in viscosity of the polymer solution does not interfere with the formation of droplets, as shown in Figure 8.1. The separation of the emulsification and gelation stages is achieved by introducing crosslinking agents into the precursor droplets (and not in the droplet phase prior to its emulsification). Alternatively, an ionic crosslinking agent can be added to the droplet phase in an inactive state but activated after the formation of the droplets. Below we describe the preparation of ionically gelled microgels using various methods of addition or activation of a crosslinking agent in the droplets.

Coalescence-Induced Gelation This method relies on the coalescence of two types of aqueous droplets, one containing a precursor polymer solution, and the other droplet containing a solution of the crosslinking agent (Sugiura *et al.*, 2005; Liu *et al.*, 2006; Um *et al.*, 2008). Formation of droplets occurs in two parallel microfluidic droplet generators. Coalescence of the two types of droplets leads to the rapid gelation of the polymer and the formation of hydrogel microbeads. The efficiency of this method [first proposed by Sugiura *et al.* (2005)] relies on the collisions of two different types of droplets: when their coalescence is not synchronized, the method produces a fraction of non-gelled particles or polydisperse beads with varying crosslinking densities, which are generated by the coalescence of more than two droplets.

Figure 8.6 illustrates two experimental setups used for the generation of aqueous microgels. In Figure 8.6a the extrusion of aqueous solutions of sodium alginate and $CaCl_2$ from two micro-nozzles leads to the formation of two types of droplets dispersed in the soybean oil (Sugiura *et al.*, 2005). Coalescence of the droplets containing alginate with the aqueous droplets of $CaCl_2$ results in the formation of ionically crosslinked alginate particles. In this work, the excess amount of $CaCl_2$ solution was expected to favor coalescence between the droplets of alginate and $CaCl_2$, in comparison with

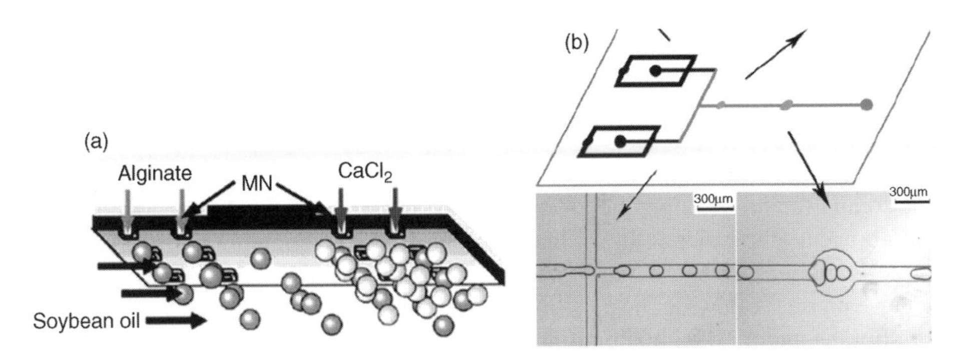

Figure 8.6 *(a) Schematic of a microfluidic setup for the formation of microgels by coalescence of droplets containing a solution of sodium alginate and droplets containing $CaCl_2$. The precursor droplets of the crosslinking solution and the polymer solution are formed in separate sections of the microfluidic device. Gelation occurs upon the collisions of the two types of droplets. Adapted, with permission, from Sugiura et al., 2005. Reprinted from S. Sugiura, et al., Size control of calcium alginate beads containing living cells using micro-nozzle array. Biomaterials* ***26****, 3327. Copyright (2005) with permission from Elsevier. (b) (Top) Schematic diagram of the microfluidic device. comprising one flow-focusing device for generating the sodium alginate droplets and another flow-focusing device for generating $CaCl_2$ droplets. The downstream channels contain two circular expansion chambers in which alginate and $CaCl_2$ droplets coalesce. (Bottom) Optical microscopy images of the droplets moving in the downstream channels (left) and coalescing in the circular expansion chamber (right). Adapted, with permission, from Liu and Chang, 2006. Reprinted from K. Liu, H.-J. Ding, J. Liu, Y. Chen, and X.-Z. Zhao, Shape-Controlled Production of Biodegradable Calcium Alginate Gel Microparticles Using a Novel Microfluidic Device, Langmuir,* ***22****, 9453. Copyright (2006) American Chemical Society*

coalescence between the alginate droplets. Therefore, the solution of $CaCl_2$ was supplied at the flow rate that was 20 times greater than that of the solution of sodium alginate solution. The drawback of this method was uncontrolled coalescence occurring in the downstream channel: the distribution of sizes of alginate microgels was broader than that of the precursor droplets.

Microfluidic generation of alginate microgels with enhanced control of coalescence of aqueous alginate and $CaCl_2$ droplets was demonstrated by Liu *et al.* (2006). Figure 8.6b shows the schematic drawing of the modified method for the coalescence-driven microfluidic preparation of ionically crosslinked microgels. As in the work of Sugiura *et al.* (2005) aqueous droplets of sodium alginate and $CaCl_2$ were generated in two microfluidic flow-focusing devices sharing a common downstream channel. The downstream channel contained two circular expansion compartments in which the droplets slowed down. Coalescence of droplets occurred in the expansion compartment. This method enabled better control of coalescence of the two distinct types of droplets, thereby producing particles with low polydispersity. In addition, this group exploited the ability to gel droplets in constrained conditions and produced alginate particles with a range of shapes such as plugs, disks, microspheres, rods, and threads.

Internal Gelation In this method, microfluidic emulsification produces droplets that compartmentalize an aqueous solution of the precursor polymer *and* a crosslinking agent (the latter is in an inactive state). The continuous non-polar phase, e.g., oil, contains an agent that is miscible with both the continuous phase and the disperse phase. Following the partition of this agent in the droplets, it "activates" the crosslinking agent. The ionic binding of the crosslinking agent to the polymer leads to polymer gelation (Zhang *et al.*, 2007; Tan and Takeuchi, 2007; Workman *et al.*, 2007).

One of the examples of the application of the internal gelation method is the formation of alginate microgels ionically crosslinked with Ca^{2+} ions. Figure 8.7a illustrates microfluidic generation of droplets containing an aqueous solution of sodium alginate and small $CaCO_3$ particles. The value of pH of the solution in the droplets is high and $CaCO_3$ does not dissolve in alginate solution. The droplets are dispersed in the continuous oil phase comprising acetic acid. Partition of the acetic acid in the droplets causes a local decrease in pH, which triggers the dissolution and dissociation of $CaCO_3$. The liberated Ca^{2+} ions bind to the deprotonated carboxylic groups of alginate molecules, causing polymer gelation.

Figure 8.7b shows an optical microscopy image of Ca^{2+}-crosslinked alginate microgels obtained by the method of internal gelation. The internal gelation method preserved the narrow distribution of sizes of microgel particles, which was close to that of the precursor droplets. The average diameter and polydispesity of the microgels were 139 μm and 3.6%, respectively (Figure 8.7c).

Internal gelation produces weakly crosslinked microgels, which slowly dissolve when transferred from the continuous phase into an aqueous phase (Zhang *et al.*, 2007). Furthermore, because $CaCO_3$ particles introduced into the feed-tubing or the microfluidic channels may aggregate or precipitate, this method has to be used with caution. A longer residence time of microgels in a microfluidic reactor, the utilization of small $CaCO_3$ particles, and/or efficient diffusion of acid into the aqueous droplets can enhance the gelation process and generate strong gels. Additional *off-chip* post-gelation of microgels by transferring the weakly gelled particles into an aqueous $CaCl_2$ bath can be used to "strengthen" the microgels (Chan *et al.*, 2006).

External Gelation In this method, droplets of the aqueous polymer solution are emulsified in a non-polar liquid containing a crosslinking agent in a dissolved or dispersed state. As the droplets move through the downstream channel, the crosslinking agent partitions in the droplets and initiates the gelling process (Zhang *et al.*, 2006, 2007). The distinct feature of this method is that both the emulsification and gelation occur *on-chip,* in comparison with other external gelation methods in which following emulsification, the droplets are gelled *off-chip* by dripping the water-in-oil emulsion in the aqueous solution of a crosslinking agent (see e.g., Capretto *et al.*, 2008).

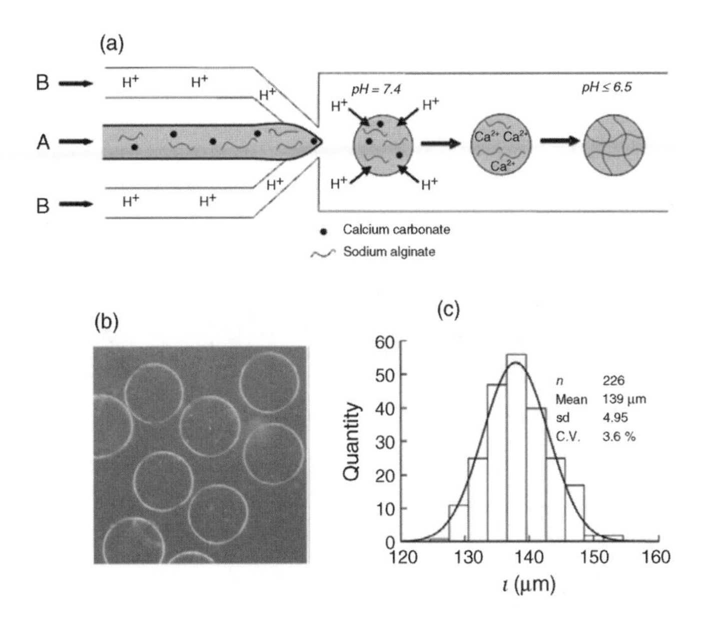

Figure 8.7 *Schematic of microfluidic generation of alginate microgels by an internal gelation. Alginate polymer solution mixed with $CaCO_3$ particles is injected into channel A. The continuous oil phase comprises acetic acid injected into channel B. Acetic acid diffuses into the emulsified alginate droplet, thereby locally reducing pH. $CaCO_3$ dissolves and dissociates releasing Ca^{2+} ions, which crosslink alginate molecules. Adapted, with permission, from Zhang et al., 2007. Reprinted from H. Zhang, et al., Exploring Microfluidic Routes to Microgels of Biological Polymers, Rapid Comm.* ***28****, 527. Copyright (2007) with permission from John Wiley and Sons. (b) Optical microscopy images of alginate microgels prepared by internal gelation. (Tan and Takeuchi, 2007). Reprinted from W.-H. T. Tan, and S. Takeuchi, Monodisperse Alginate Hydrogel Microbeads for Cell Encapsulation, Adv. Mater.* ***19****, 18, 2696 Copyright (2007) John Wiley & Sons, Inc. (c) Distribution of diameters of microgels shown in (b)*

Figure 8.8a shows a schematic of the application of the external gelation method for the continuous microfluidic production of alginate microgels (Zhang *et al.*, 2006). An aqueous solution of sodium alginate (the droplet phase) was supplied to the central channel of the microfluidic droplet generator as liquid A; an undecanol solution of CaI_2 (the continuous phase) was introduced in the side channels of the device as liquid B (Figure 8.8a). Following emulsification, the aqueous droplets of sodium alginate moved in the downstream channel where CaI_2 diffused from the continuous phase into the droplets, causing polymer gelation. The first indication of gelation was observed as the appearance of small islands of a "skin" on the surface of droplets (Figure 8.8b). With increasing time of residence of the droplets in the microfluidic device, the entire surface of droplets acquired a characteristic grain structure. The time interval between the formation of droplets and the first sign of "skinning" (typically, below 1 min) depended

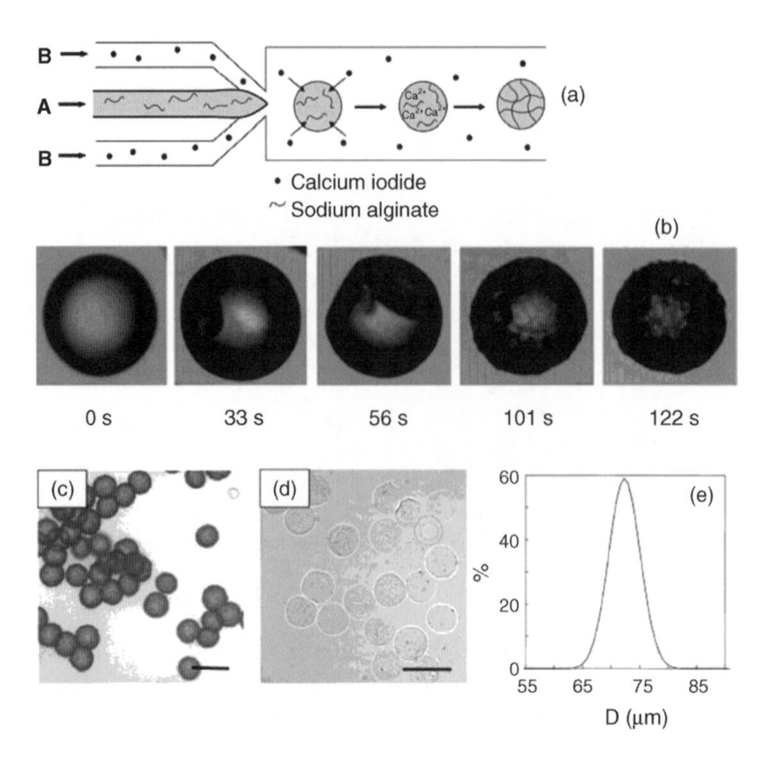

Figure 8.8 *(a) Schematic of a microfluidic setup for the formation of microgels by an external gelation method. An aqueous solution of sodium alginate is supplied to the central channel of the microfluidic droplet generator as liquid A; an undecanol solution of CaI_2 is introduced in the side channels as liquid B. When the aqueous droplets of sodium alginate move in the downstream channel, CaI_2 diffuses from the continuous phase to the droplets, where liberated Ca^{2+} ions bind to alginate molecules causing polymer gelation. (b) Optical microscopy images (top view) of gelling alginate droplets in the downstream channels following their exposure for different time intervals to a 0.15 wt% undecanol solution of CaI_2. (c) Micrograph of alginate microgels obtained by emulsifying 1.0 wt% alginate solution in 0.20 wt% undecanol solution of CaI_2. (d) Micrograph of alginate microgels transferred to a PBS solution at pH 7.4. (e) Size distribution of alginate microgels. Polydispersity of microgels is 3.8%. Adapted, with permission, from Zhang et al., 2006. Reprinted with permission from H. Zhang, et al., Microfluidic Production of Biopolymer Microcapsules with Controlled Morphology, Journal of the American Chemical Society,* ***128****, 12205. Copyright (2006) American Chemical Society*

on the concentration of CaI_2 in the undecanol. The microgels were collected in a large volume of the crosslinker-free solution, thereby quenching gelation of alginate.

Under conditions of laminar flow the diffusion of Ca^{2+} ions from undecanol to the aqueous alginate droplets is diffusion-controlled, thus gelation of alginate could be approximated with by Equation 8.1, which describes diffusion accompanied by an instantaneous chemical reaction (Crank, 1975), that is, binding of Ca^{2+} ions to the carboxylic groups on alginate, as follows:

$$\frac{\partial C}{\partial t} = D\frac{\partial^2 C}{\partial t^2} - \frac{\partial S}{\partial t} \tag{8.1}$$

where D is the diffusion coefficient of Ca^{2+} ions, C is the concentration of Ca^{2+} ions in undecanol, and S is the concentration of crosslinked Ca^{2+} ions.

Alginate microgels produced by the method of external gelation did not disintegrate in undecanol and upon their transfer into an aqueous phase using the centrifugation process (Figure 8.8c,d). The method yielded particles with dimensions from about 20 to 180 μm and narrow size distribution, as shown in Figure 8.8e. The method of external gelation was successfully used for the generation of κ-carrageenan and carboxymethylcellulose microgels (the latter polymer was crosslinked by Fe^{3+} ions) (Zhang *et al.*, 2006).

In subsequent work, undecanol and CaI_2 were replaced with less cytotoxic chemicals (Zhang *et al.*, 2007). Aqueous alginate droplets were dispersed in soybean oil comprising small $Ca(CH_3COO^-)_2$ particles. After the generation of alginate emulsions, as shown in Figure 8.8a, the particles of calcium acetate dissolved in aqueous droplets, thereby releasing Ca^{2+} ions and initiating ionotropic gelation. Owing to the use of $Ca(CH_3COO^-)_2$ particles, the process of gelation occurred significantly slower in the case of CaI_2 solution in undecanol.

In comparison with internal gelation, the method of external gelation allowed precise control of the microgel morphology by controlling the depth of penetration of Ca^{2+} ions in the droplets and hence, the thickness of the gelled shell (Zhang *et al.*, 2006). Because the diffusion of the crosslinking agent in the microgels was quenched off-chip, by varying the time of residence of the droplets of sodium alginate in the microfluidic reactor and the concentration of the crosslinking agent in the continuous phase, microgels with a capsular or gradient structure could be produced (see Chapter 9).

8.4 Microfluidic Encapsulation of Bioactive Species in a Microgel Interior

In principle, various bioactive species such as proteins, peptides, or enzymes, can be readily incorporated in a microgel interior by dissolving them in the droplet phase prior to microfluidic emulsification. This section focuses on the encapsulation of cells in microfluidically generated microgels, which turned out to be a more challenging process.

In the early 1960s, the encapsulation of cells in polymer microcapsules was suggested as a method to reduce the effects of immune rejection (Chang, 1964), thereby forming the basis of "cell therapy" for the treatment of various diseases. The encapsulation of cells in the interior of microgels with a uniform or capsular structure has led to the development of systems for the study and treatment of hormone or protein deficiency (Ross *et al.*, 2000), hepatic failure (Liu and Chang, 2006), cancer (Cirone, 2003), and diabetes (Lim and Sun, 1980). Microgel particles with a capsular structure were prepared by injecting droplets of an aqueous solution of alginate into an aqueous $CaCl_2$ solution, coating gelled particles with poly-L-lysine, and subsequently liquifying alginate by using, for example, sodium citrate.

Microfluidic encapsulation of cells in hydrogel particles offers advantages over traditional encapsulation methods such as a vibrating nozzle method (Koch *et al.*, 2003) and electrostatic droplet generation (Bugarski *et al.*, 1994), due to the relatively small microgel dimensions (typically, varying from tens to several hundred micrometers). Compared with larger gel particles, due to a larger surface to volume ratio, the utilization of smaller particles allows for an easier implantation of the microgels into the target site and better exchange of nutrients and oxygen between the encapsulated species and the external environment (Chicheportiche and Reach, 1988). Furthermore, microfluidic emulsification offers enhanced control over the number of encapsulated cells in the microgel interior (Edd *et al.*, 2008). Finally, microfluidic methods allow for the generation of microgel particles with a capsular structure (Kim *et al.*, 2007; Zhang *et al.*, 2006). The external membrane of the microgel capsules enables protection of the encapsulated species from harmful external stimuli (Dove, 2002) while a liquid core provides a suitable microenvironment (Breguet *et al.*, 2005).

Regardless of the method used for the preparation of the microgels, the microfluidic encapsulation of bioactive species, such as cells, enzymes, or proteins, in a particle interior has to meet the following requirements:

(i) The microgels have to be mechanically strong to withstand their transfer from the non-polar continuous phase to an aqueous media, such as in a buffer or cell culture media. Such transfer is often achieved by several rounds of centrifugation.
(ii) All reagents used in microfluidic encapsulation of bioactive species have to be biocompatible and not cytotoxic. This includes monomers, oligomers, or polymers used in the generation of microgels, surfactants, if any, the non-polar liquids forming the continuous phase, and the crosslinking agents. The effect of cytotoxicity of the reagents can be minimized by the short time contact with the bioactive species on chip, followed by the immediate transfer of the microgels to the different environment.
(iii) In order to preclude contamination of the system, the microfluidic device and the supplying tubing have to be sterilized before and after encapsulation.
(iv) The surface properties of the microfluidic device have to be optimized to minimize the influence of cell secretion products on droplet formation. The wetting of the device by the droplet-forming liquid may begin in the course of the encapsulation experiments, a possible result of the deposition of proteins, carbohydrates, or cell debris on the walls of microchannels (Tan and Takeuchi, 2007).
(v) In the encapsulation of cells in microgel particles, a high encapsulation efficiency of cells is highly desirable. For low and moderate cell concentrations in the droplet phase it is determined by the Poison distribution and a large fraction of cell-free particles counteracts the efficiency of microfluidic encapsulation cells. The ability

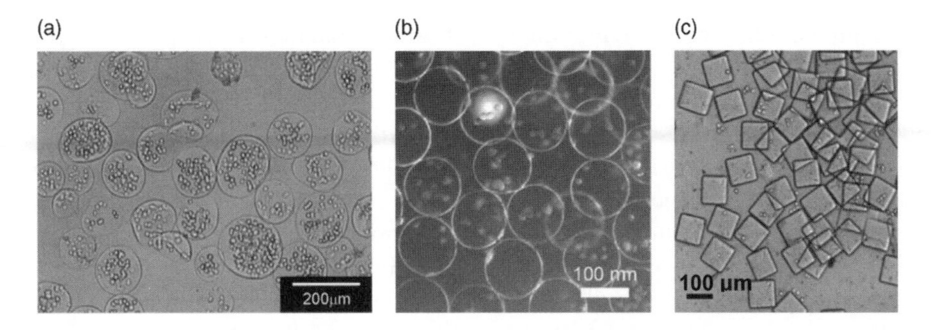

Figure 8.9 *Optical microscopy images of microgels generated in a microfluidic device by different gelation mechanisms. (a) Human kidney cells encapsulated in calcium alginate microgels generated by coalescence-driven gelation (Sugiura et al., 2005). Reprinted from S. Sugiura, et al., Size control of calcium alginate beads containing living cells using micro-nozzle array. Biomaterials* ***26****, 3327. Copyright (2005) with permission from Elsevier. (b) Calcium-crosslinked alginate microgel loaded with mammalian (Jurkat) cells generated via internal gelation method (Tan and Takeuchi, 2007). Reprinted from W.-H. T. Tan, and S. Takeuchi, Monodisperse Alginate Hydrogel Microbeads for Cell Encapsulation, Adv. Mater.* ***19****, 18, 2696 (2007). Copyright (2007) John Wiley and Sons (c) Images of mouse fibroblast cell-laden chemically crosslinked microgels generated through UV-polymerization of poly(ethylene glycol) diacrylate Panda et al., 2008. Reprinted from P. Panda, et al., Stop-flow lithography to generate cell-laden microgel particles. Lab on a Chip.* ***8****, 1056. Copyright (2008) with permission from Royal Society of Chemistry (http://dx.doi.org/10.1039/B804234A)*

of streamlines in microfluidic channels to separate cells with well-defined distances proved useful in efficient and controllable encapsulation of cells in droplets (Edd *et al.*, 2008).

For the first time, microfluidic encapsulation of cells in microgels was demonstrated for Ca^{2+} crosslinked alginate particles produced by the coalescence-driven gelation (Sugiura *et al.*, 2005). Figure 8.9a shows alginate microgels loaded with human kidney cells. The authors reported 70% viability for the loading in the microgels. Owing to the insufficient control of coalescence-driven gelation, this process generated alginate microgels with large polydispersity.

Figure 8.9b illustrates the encapsulation of mammalian (Jurkat) cells using emulsification in a T-junction microfluidic device and an internal gelation method to produce alginate microgels (Tan and Takeuchi, 2007). Polydispersity of the microgels did not exceed 3.6%. Based on the explanation given by the authors, with increasing concentration of $CaCO_3$ cell viability increased for two reasons. Firstly, with increasing content of Ca^{2+} ions the strength of the microgels increased, which helped to protect the encapsulated cells from mechanical stresses during the formation of the gel network. Secondly, the presence of CO_3^{2-} ions helped to regulate the value of pH in the droplets from becoming too acidic, thereby maintaining mild conditions for the encapsulated cells.

Until recently, no successful encapsulation of cells has been reported in chemically crosslinked microgels produced by microfluidic methods: because of the use of the photoinitiator and/or prepolymer the system was cytotoxic (Williams *et al.*, 2005). Photo-irradiation of the cells and exothermic polymerization reactions also counteracted the utilization of this method. Recently, the above mentioned limitations were overcome by using the stop-flow lithography method to generate microgels loaded with mouse fibroblasts (Figure 9.9c) (Panda *et al.*, 2008). The cell-laden aqueous solution of poly(ethylene glycol) diacrylate was introduced into the microchannel, stopped, and exposed through a mask to the UV-radiation source for a time interval varying from 800 ms to several seconds. Short-time irradiation led to the polymerization of poly(ethylene glycol) diacrylate and the formation of microgel particles encapsulating cells. Once polymerization was complete, the microgels moved from the device. Following the encapsulation, the microgels were transferred into cell culture media to reduce the exposure of the encapsulated cells to the solution of poly (ethylene glycol) diacrylate. In order to maximize cell viability, a wide range of monomer and photoinitator concentrations was examined and the protocol of the encapsulation procedure was optimized. Under optimized conditions, cell viability was reported to be 68%.

To summarize, over the past several years, microfluidic generation of polymer particles has expanded to the production of polymer microgels. Highly monodisperse microgels with varying dimensions, compositions, and morphologies have been successfully produced using microfluidic techniques. Currently, methods that rely on both chemical and physical polymer crosslinking have been utilized for the microfluidic synthesis and assembly of microgels, each with characteristic advantages and drawbacks. In particular, the use of biopolymers such as alginate, carageenan, or agarose for the preparation of biomicrogels is gaining in momentum, due to their biocompatibility, low toxicity, and ability to degrade under external stimuli.

One of the most promising applications of polymer microgels produced by microfluidic methods is the encapsulation of cells. Although most methods discussed in this chapter can be used for the encapsulation of bioactive species, some methods offer better control of the microgel properties (e.g., size distribution or morphology) and better cell viability compared with other methods. Microfluidic methods will be particularly useful in exploring cell fate in different biochemical environments by encapsulating cells with various biologically active species, such as enzymes, cytokines, or drugs. Future work in this field will no doubt yield greater cell viability and better control over the number of encapsulated cells per microgel as well as the rate of encapsulation.

With rising interest in the use of microfluidic techniques for high-frequency screening of encapsulated cells, it is important to develop methods further to reproducibly control the number of encapsulated species per particle. By reliably controlling the number of encapsulated species, future work in this area will focus on studies of interactions between the multiple cells encapsulated in the microgel interior. Furthermore, the generation of microgels comprising two chemically or physically distinct hemispheres will allow for the study of the influence of distinct environments on the behavior of cells entrapped on differing sides of the same microgel.

References

Anna, S.L., Bontoux, N., and Stone, H.A. (2003) *Appl. Phys. Lett.*, **82**, 364.
Bae, Y.H., Okano, T., Hsu, R., and Kim, S.W. (1987) *Rapid Commun.*, **8**, 481.
Brandenberger, H.R. and Widmer, F. (1999) *Biotechnol Progr.*, **15**, 366.
Breguet, V., Gugerli, R., Pernetti, M., von Stockar, U., and Marison, I.W. (2005) *Langmuir*, **21**, 9764.
Bromberg, L. and Alakhov, V. (2003a) *J. Control. Release*, **88**, 11.
Bromberg, L., Temchenko, M., and Hatton, T.A. (2003b) *Langmuir*, **19**, 8675.
Bugarski, B., Li, Q.L., Goosen, M.F.A., Pincelet, D., Neufeld, R.J., and Vunjak, G. (1994) *AIChE J.*, **40**, 1026.
Calafiore, R., Basta, G., Luca, G., Calvitti, M., Calabrese, G., Racanicchi, L., Macchiarulo, G., Mancuso, F., Guido, L., and Brunetti, P. (2004) *Biotech. Appl. Biochem.*, **39**, 159.
Capretto, L., Mazzitelli, S., Balestra, C., Tosi, A., and Nastruzzi, C. (2008) *Lab Chip*, **8**, 617.
Chan, A.W.J., Mazeaud, I., Becker, T., and Neufeld, R.J. (2006) *Enzy. Microb. Technol.* **38**, 265.
Chang, T.M.S. (1994) *Science*, **146**, 524.
Chicheportiche, D. and Reach, G. (1988) *Diabetologia*, **31**, 54.
Chu, L.Y., Utada, A.S., Shah, R.K., Kim, J.W., and Weitz, D.A. (2007) *Angew. Chem. Int. Ed.* **46**, 8970.
Cirone, P., Bourgeois, J.M., and Chang, P.L. (2003) *Hum. Gene. Ther.* **14**, 1065.
Crank, J. (1975) *The Mathematics of Diffusion*, Clarendon Press, Oxford, p. 414.
Das, M., Marydani, S.C., Chan, W.C.W., and Kumacheva, E. (2006) *Adv. Mater.*, **18**, 80.
De Geest, B.G., Urbanski, J.P., Thorsen, T., Demeester, J., and De Smedt, S.C. (2005) *Langmuir*, **21**, 10275.
Dove, A. (2002) *Nat. Biotechnol.*, **20**, 339.
Edd, J.F., Di Carlo, D., Humphry, K.J., Koster, S., Irimia, D., Weitz, D.A., and Toner, M. (2008) *Lab Chip*, **8**, 1262.
Fiddes, L., Young, L., Wheeler, A., and Kumacheva, E. (2007) *Lab Chip*, **122**, 863.
Fiddes, L., Chan, H.K., Wyss, K., Simmons, C.A., Wheeler, A., and Kumacheva, E. (2009) *Lab Chip*, **9**, 286.
Forster, S. and Antonietti, M. (1998) *Adv. Mat.*, **10**, 195.
Freeman, A. (1987) *Methods Enzymol.*, **135**, 216.
Freeman, A. and Aharonowitz, Y. (1981) *Biotechnol. Bioeng.*, **23**, 2747.
Fundueanu, G., Nastruzzi, C., Carpov, A., Desbrieres, J., and Rinaudo, M. (1999) *Biomaterials*, **20**, 1427.
Galaev, I.Y. and Mattiasson, B. (1993) *Enzyme Microb. Technol.*, **15**, 354.
Guo, Z., Sautereau, H., and Kranbuehl, D.E. (2005) *Macromolecules*, **38**, 7992.
Hoare, T. and Pelton, R. (2004) *Macromolecules*, **37**, 2544.
Hoffman, A.S., Afrassiabi, A., and Dong, L.C. (1986) *J. Control. Release*, **4**, 213.
Hortelano, G., Al-Hendy, A., Ofosu, F.A., and Chang, P.L. (1996) *Blood*, **87**, 5095.
Iwamoto, S., Nakagawa, K., Sugiura, S., and Nakajima, M. (2002) *AAPS PharmSciTech*, **3**, article 25.
Jones, C.D. and Lyon, L.A. (2003) *J. Am. Chem. Soc.*, **125**, 460.
Kim, J.-W., Utada, A.S., Fernandez-Nieves, A., Hu, Z., and Weitz, D.A. (2007) *Angew. Chem. Int. E*d.*, **46**, 1819.
Koch, S., Schwinger, C., Kressler, J., Heinzen, C., and Rainov, N.G. (2003) *J. Microencapsulation*, **20**, 303.
Kuo, S.M., Niu, G.C.-C., Chang, S.J., Kuo, C.H., and Bair, M.S. (2004) *J. Appl. Polym. Sci.*, **94**, 2150–57.
LaVan, D.A.L., Lynn, D.M., and Langer, R. (2002) *Nat. Rev. Drug Discovery*, **1**, 77.
Levee, M.G., Lee, G.M., Paek, S.H., and Palsson, B.O. (1994) *Biotechnol. Bioeng.*, **43**, 734.

Lim, F. and Sun, A.M. (1980) *Science*, **210**, 908.
Liu, K., Ding, H.-J., Liu, J., Chen, Y., and Zhao, X.-Z. (2006) *Langmuir*, **22**, 9453.
Liu, Z.C. and Chang, T.M.S. (2006) *Liver Transpl.*, **12**, 566.
Lopez, V.C., Hadgraft, J., and Snowden, M.J. (2005) *Int. J. Pharm.*, **292**, 137.
Lopez, V.C. and Snowden, M.J. (2003) *Drug Deliv. Syst. Sci.*, **3**, 19.
Lyon, L.A., Debord, J.D., Debord, S.B., Jones, C.D., McGrath, J.G., and Serpe, M.J. (2004) *J. Phys. Chem., B*. **108**, 19099.
Lyon, L.A., Kong, S.B., Eustis, S., and Debord, J.D. (2002) *Polym. Prepr.*, **43**, 24.
Morris, G.E., Vincent, B., and Snowden, M.J. (1997) *Progr. Colloid Polym. Sci.*, **105**, 16.
Murthy, N., Thng, Y.X., Schuck, S., Xu, M.C., and Frechet, J.M.J. (2002) *J. Am. Chem. Soc.*, **124**.
Murthy, N., Xu, M., Schuck, S., Kunisawa, J., and Frechet, N. (2003) *Proc. Natl. Acad. Sci.*, **100**, 4995.
Nayak, S., Lee, H., Chmielewski, J., and Lyon, L.A. (2004) *J. Am. Chem. Soc.*, **126**, 10258.
Nolan, C.M., Reyes, C.D., Debord, J.D., Garcia, A.J., and Lyon, L.A. (2005) *Biomacromolecules*, **6**, 2032.
Nomura, S.M., Tsumoto, K., Hamada, T., Akiyoshi, K., Nakatani, Y., and Yoshikawa, K. (2003) *Chembiochem*, **4**, 172.
Orive, G., Hernández, R.M., Gascón, A.R., Calafiore, R., Chang, T.M., De Vos, P., Hortelano, G., Hunkeler, D., Lacík, I., Shapiro, A.M., and Pedraz, J.L. (2003) *Nat. Med.*, **9**, 104.
Panda, P., Ali, S., Lo, E., Chung, B.G., Hatton, T.A., Khademhosseini, A., and Doyle, P.S. (2008) *Lab Chip*, **8**, 1056.
Park, T.G. and Hoffman, A.S. (1990) *Biotechol. Bioeng.* **35**, 152.
Pathak, C.P., Sawhney, A.S., and Hubbell, J.A. (1992) *J. Am. Chem. Soc.*, **114**, 8311.
Pelton, R. (2000) *Adv. Colloid Interface Sci.*, **85**, 1.
Peyratout, C.S. and Dähne, L. (2004) *Angew. Chem. Int. Ed.*, **43**, 3762.
Prüsse, U., Fox, B., Kirchhoff, M., Bruske, F., Breford, J., and Vorlop, K.D. (1998) *Chem. Eng. Technol.*, **21**, 29.
Retama, J.R., Lopez-Ruiz, B., and Lopez-Cabarcos, E. (2003) *Biomaterials*, **24**, 2965.
Ross, C.J.D., Bastedo, L., Maier, S.A., Sands, M.S., and Chang, P.L. (2000) *Hum. Gene Ther.*, **11**, 2117.
Rubinstein, M. and Colby, R.H. (2003) *Polymer Physics*, Oxford University Press, Oxford, 456 pp.
Saunders, B.R. and Vincent, B. (1999) *Adv. Colloid Interface Sci.*, **80**, 1.
Shah, R.K., Kim, J.-W., Agresti, J.J., Weitz, D.A., and Chu, L.-Y. (2008) *Soft Mat.*, **46**, 2303.
Shepherd, R.F., Conrad, J.C., Rhodes, S.K., Link, D.R., Marquez, M., Weitz, D.A., and Lewis, J.A. (2006) *Langmuir*, **22**, 8618.
Stevenson, W.T.K. and Sefton, M.F. (1993) *Fundamentals of Animal Cell Encapsulation and Immobilization*, CRC Press, Boca Raton, FL, pp. 143–182.
Sugiura, S., Nakajima, M., Ushijima, H., Yamamoto, K., and Seki, M. (2001) *J. Chem. Eng. Jpn.*, **34**, 757.
Sugiura, S., Odad, T., Izumidaa, Y., Aoyagid, Y., Satakef, M., Ochiaie, A., Ohkohchid, N., and Nakajimaa, M. (2005) *Biomaterials*, **26**, 3327.
Sun, Y.L., Ma, X.J., Zhou, D.B., Vacek, I., and Sun, A.M. (1996) *J. Clin. Invest.*, **98**, 1417.
Tan, W.-H.T. and Takeuchi, S. (2007) *Adv. Mater.*, **19**, 2696.
Taylor, L.D. and Cerankowski, L.D.J. (1975) *Polym. Sci. Polym. Chem.*, **13**, 2551.
Um, E., Lee, D.-S., Pyo, H.-B., and Park, J.-K. (2008) *Microfl. Nanofluid.*, **5**, 541.
van Dijk-Wolthuis, W.N.E., Hoogeboom, J.A.M., van Steenbergen, M.J., Tsang, S.K.Y., and Hennink, W.E. (1997a) *Macromolecules*, **30**, 4639.
van Dijk-Wolthuis, W.N.E., Tsang, S.K.Y., Kettenes-vanden Bosch, J.J., and Hennink, W.E. (1997b) *Polymer*, **38**, 6235.
Varma, M.V.S., Kaushal, A.M., and Garg, S. (2005) *J. Control. Release*, **103**, 499.

Walther, B., Cramer, C., Tiemeyer, A., Hamberg, L., Fischer, P., Windhab, E.J., and Hermansson, A.M. (2005) *J. Colloid Interface Sci.*, **286**, 378.
Williams, C.G., Malik, A.N., Kim, T.K., Manson, P.N., and Elisseeff, J.H. (2005) *Biomaterials*, **26**, 1211.
Workman, V.L., Dunnett, S.B., Kille, P., and Palmer, D.D. (2007) *Biomicrofluidics*, **1**, 014105.
Xu, S., Zhang, J., Paquet, C., and Kumacheva, E. (2003) *Adv. Funct. Mater.*, **13**, 468.
Zhang, J., Xu, S., and Kumacheva, E. (2004) *J. Am. Chem. Soc.*, **26**, 7908.
Xu, S., Nie, Z., Seo, M., Lewis, P.C., Kumacheva, E., Garstecki, P., Weibel, D., Gitlin, I., Whitesides, G.M., and Stone, H.A. (2005) *Angew. Chemie Int. Ed.*, **44**, 724.
Zhang, H., Tumarkin, E., Peerani, R., Nie, Z., Sullan, R.M.A., Walker, G.C., and Kumacheva, E. (2006) *J. Am. Chem. Soc.*, **128**, 12205.
Zhang, H., Tumarkin, E., Sullan, R.M.A., Walker, G.C., and Kumacheva, E. (2007) *Macromol. Rapid Commun.*, **28**, 527.

9

Polymer Capsules

CHAPTER OVERVIEW

9.1 Polymer Capsules with Dimensions in Micrometer Size Range

A polymer microcapsule is a particle that contains a well-defined core and a shell. Classification of polymer capsules can be based on the method of their preparation, on the nature of the core, which can be liquid, solid, or gaseous, and on the porosity of the shell. With respect to the last feature, macroporous membranes have pores of approximately 0.1–1 μm, whereas microporous membranes are characterized by the size of pores of about 100–500 Å. Gel shells or membranes have the size of "pores" of the order of the molecular dimensions. The "pores" in gels are formed by entanglement and/or crosslinking of the polymer molecules and the mesh size is determined by the space between these chains.

A broad range of methods that are currently used for the preparation of polymer capsules includes interfacial polymerization, complex coacervation, precipitation,

Microfluidic Reactors for Polymer Particles. Eugenia Kumacheva and Piotr Garstecki.

gelation, salting out, and solvent removal by evaporation or extraction (Mathiowitz, 1999). A narrow distribution of sizes of polymer capsules is an important factor in their utilization, especially, in biomedical applications. For example, in drug delivery the targeting efficiency and the release profile of a drug through a non-erodible (intact) shell of a microcapsule are greatly influenced by the uniformity is capsular sizes. Colloidal stability of the capsules is also dependent on the ability to control the size of the particles. Currently, most of the methods used for producing polymer capsules lack control over particle size distribution: the typical polydispersity of polymer capsules produced by conventional methods is between 10 and 50%. Furthermore, control of the structure of capsules, such as the relationship between the diameter of cores and the thickness of shells is beneficial in a number of applications of polymer particles.

Recently, several simple and straightforward approaches to polymer microcapsules have been developed using microfluidic synthesis and assembly of polymer particles. Control of the diameters of capsules was achieved by microfluidic emulsification of monomers or polymer solutions. The precursor droplets with a very narrow distribution of sizes and precisely controlled morphologies were subsequently solidified or gelled. The transformation of droplets into polymer capsules occurs via different methods, which include polymerization of shells in double emulsions, interfacial condensation reactions, external gelation, and self-assembly. In a different approach, capsules were produced by the deposition of molecules of particles on the surface of droplets or bubbles generated by the microfluidic means. Microfluidic strategies for the the synthesis and assembly of capsules are described in the present chapter.

9.2 Microfluidic Methods for the Generation of Polymer Capsules

9.2.1 From Double Emulsions to Polymer Capsules

Structured fluids that contain droplets comprising smaller droplets of another liquid are called double emulsions. Droplets with a core–shell structure offer control over compartmentalization of the inner liquid: the fluid "shell" acts as an extra barrier that separates the innermost fluid from the continuous phase. Double emulsions have potential applications for the controlled release of pharmaceutical, cosmetic, nutrition, and food ingredients. Furthermore, double emulsions can be used as "precursors" for the preparation of polymer particles: the interior droplets and/or shell can be polymerized or gelled, thereby transforming a droplet into a particle.

In conventional (non-microfluidic) methods double emulsions are produced in two consecutive emulsification steps, which generate polydisperse droplets with poorly defined structures. In contrast, microfluidic emulsification allows precise control of the dimensions of primary and core–shell droplets, the thickness of the shells, and the number of inner droplets per large droplet (Okushima *et al.*, 2004; Nisisako, Okushima, and Torii, 2005; Utada *et al.*, 2005; Nie *et al.*, 2005; Seo *et al.*, 2007). Control of these features is achieved by independently varying the flow rates and the flow-rate ratios of three liquids (the two liquids forming a droplet and the continuous phase) and the macroscopic properties of the liquids.

Microfluidic production of double emulsions is described in Chapter 5. This section is focused only on the experimental work and several designs of microfluidic devices that have been used for the generation of polymer capsules with rigid and gelled shells.

Figure 9.1 shows a microcapillary geometry that was used for the generation of precursor double emulsions by Utada *et al.* (2005). The device consists of two cylindrical capillary tubes placed in a square glass tube. The innermost liquid is supplied through a tapered cylindrical capillary tube, and the middle liquid is injected through the outer coaxial region, thereby forming a coaxial flow of two droplet phases. These two liquids are largely immiscible, at least, on the time scale of the emulsification. A third fluid (the continuous phase) is introduced through the outer coaxial region from the opposite direction. All three liquids are forced through the orifice formed by the remaining inner tube, which results in flow-focusing and the formation of droplets with different structures.

The device illustrated in Figure 9.1 (top) was used to generate core–shell droplets, which served as precursors to polymer particles with a capsular structure. In the first approach, Utada *et al.* (2005) used water as the inner liquid (forming the cores of the droplets) and a 30 wt% solution of a polymer, Norland Optical Adhesive (NOA), in acetone was used as the intermediate liquid. The continuous phase was water. After generating the double emulsions, NOA was photopolymerized by UV-irradiating the droplets moving through the collection tube. Bright field images of the precursor double emulsion droplet and the resulting particle with a solid shell are shown in Figure 9.1 (bottom, a and b, respectively). Polymerization of NOA was confirmed by crushing the resulting microspheres between two microscope cover slides and imaging the particles using scanning electron microscopy (SEM) (Figure 9.1c). The microfluidic device shown in Figure 9.1, top, was also used to make polymer vesicles, or polymerosomes from double emulsions with a single internal drop and dissolved diblock copolymers in the intermediate fluid. A volatile fluid used as the intermediate phase subsequently evaporated. Water-in-oil-in-water double emulsions were generated, in which a middle fluid contained a diblock copolymer, poly(butyl acrylate)-*b*-poly(acrylic acid) in a mixture of 70% toluene and 30% tetrahydrofuran (Figure 9.1d). When the solvent evaporated, the amphiphilic polymer self-assembled into layers on both interfaces, thereby forming a polymerosome (Figure 9.1e). The polymerosome structure was dilated with osmotic stress by adding a sucrose solution (0.1 M) to the continuous fluid (Figure 9.1f). More discussion on polymersomes is given in Section 9.2.2.

Double emulsions have also been used for the generation of hydrogel capsules (Kim *et al.*, 2007). Figure 9.2a shows a fluorescence microscope image of the individual capsule. The precursor droplets were generated in the continuous-phase liquid – a silicon oil with viscosity of 125 mPa s. Each droplet contained a single oil droplet and a shell comprising an aqueous solution of *N*-isopropylacrylamide (polyNIPAm), a crosslinker (*N,N′*-methylenebisacrylamide), the initiator ammonium persulfate, and two co-monomers allylamine and [2-(methacryloyloxy) ethyl] trimethyl ammonium

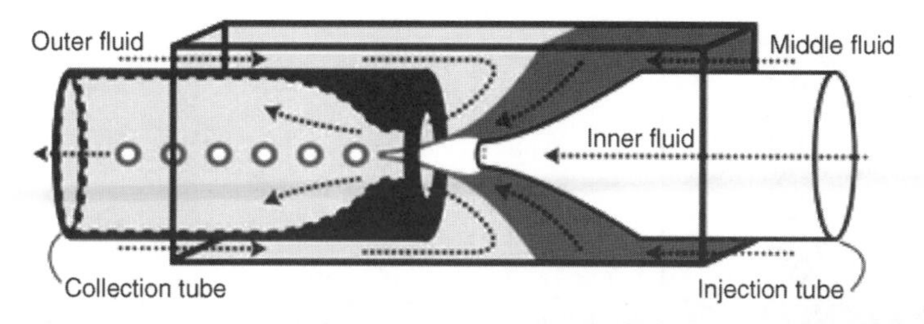

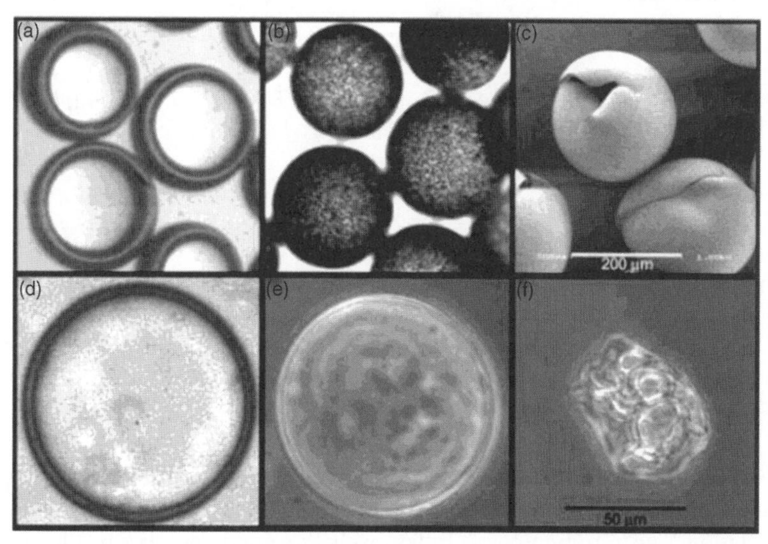

Figure 9.1 *(Top) Schematic of the coaxial capillary microfluidic device. The typical inner dimension of the square tube is 1 mm; this matches the outer diameter of the untapered regions of the collection tube and the injection tube. Typical inner diameters of the tapered end of the injection tube range from 10 to 50 μm. Typical diameters of the orifice in the collection tube vary from 50 to 500 μm. Adapted, with permission from Utada et al., 2005. Reproduced from Angewandte Chemie Int. Ed., Fabrication of Monodisperse Gel Shells and Functional Microgels in Microfluidic Devices by J. W. Kim, A. S. Utada, A. Fernandez-Nieves et al.,* ***46****, 11, 1819–1822. (2007). Copyright Wiley-VCH Verlag GmbH & Co. KGaA. Reproduced with permission. (Bottom) Core–shell structures fabricated from double emulsions. (a) Optical photomicrograph of the water-in-oil-in-water double emulsion precursor to solid spheres. The oil consists of 70% NOA and 30% acetone, with a viscosity of approximately 50 mPa s. (b) Optical photomicrograph of capsular particles produced by photopolymerizing the NOA. (c) SEM of the shells shown in (b) after they have been mechanically crushed. The scale bar in (c) also applies to (a) and (b). (d) Bright-field photomicrograph of the water-in-oil-in-water double emulsion precursor to a polymer vesicle. The oil phase consists of a mixture of toluene and tetrahydrofuran at 70/30 v/v with dissolved diblock copolymer poly(butyl acrylate)-b-poly (acrylic acid) at 2% w/v. The viscosity of this mixture was approximately 1 mPa s. (e) Phase-contrast image of the diblock copolymer vesicle after evaporation of the organic solvents. (f) Phase-contrast image of the deflated vesicle after osmotic stress was applied through the addition of 0.1 M sucrose to the outer fluid. The scale bar in (f) also applies to (d) and (e). Reproduced with permission from Science, Monodisperse Double Emulsions Generated from a Microcapillary Device by A. S. Utada, E. Lorenceau, D. R. Link, 537 Copyright (2005) American Association for the Advancement of Science*

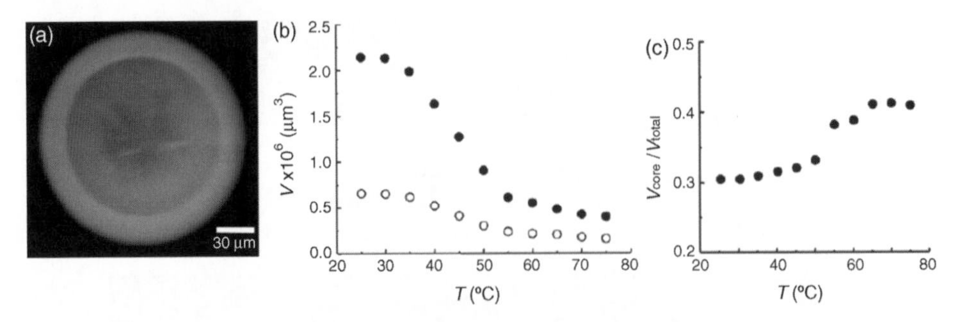

Figure 9.2 *(a) A fluorescence microscope image of an FITC-labeled microgel capsule that is prepared from a pre-microgel double emulsion which contained a single silicon oil droplet. (b) Volume changes of the overall core–shell polyNIPAm microgel (●) and its internal void (○). (c) Volume of the internal void scaled by the volume of the whole microgel as a function of temperature. The core–shell microgels were prepared under the flow rate of the internal phase of 100 μL h^{-1}, the flow rate of the middle phase of 300 μL h^{-1}, and the flow rate of the outer (continuous) phase of 2000 mL h^{-1}. All swelling measurements were carried out in water. Adapted, with permission, from Kim et al., 2007. Reproduced with permission from Angewandte Chemie Int. Ed., Fabrication of Monodisperse Gel Shells and Functional Microgels in Microfluidic Devices by J.W. Kim, A.S. Utada, A. Fernandez-Nieves et al.,* ***46****, 11, 1819–1822 Copyright (2007) Wiley-VCH*

chloride. The latter co-monomer favored increase in the coil-to-globule transition temperature of water-soluble polyNIPAm and thereby assisted in room temperature polymerization. The introduction of allylamine in the monomer mixture allowed further functionalization of the polymer shells with amino groups. The accelerator *N,N,N′,N′*-tetramethylethylenediamine was dissolved in the inner oil phase. After the formation of compound droplets, the accelerator diffused from the oil core into the surrounding aqueous monomer solution layer. This step allowed emulsification without complications, that is, rapid increase in viscosity caused by the rapid gelation of the droplet phase. After collection of the microgel capsules, the oil phase was extracted by washing the particles with an excess of isopropyl alcohol, and the capsules were transferred into deionized water. The overall diameter of the microgel capsules and the thickness of shells were tuned by changing the relative flow rates of three liquid phases introduced into the microfluidic device.

The response of the capsules to the changes in temperature was probed by measuring the variations in the dimensions of the entire capsule and the capsular core as a function of temperature (Figure 9.2b). As both the core and the shell contained thermoresponsive polyNIPAm, their dimensions decreased with increasing temperature, following polymer deswelling. No hysteresis was observed in the repeating cooling and heating cycles. The cores of the capsular particles reached a minimum volume at approximately 50 °C, however the overall dimensions of the capsules continued to decrease at higher temperatures. Figure 9.2c shows the variation in the volume of the core, which was normalized by the total capsular volume. The increase in this ratio at a temperature above about 50 °C suggested that in this temperature range the shell of the capsule shrank while the core volume remained intact. The authors explained this effect by the increasing hydrophobicity of the capsular shell which reduced its permeability to water and limited the degree to which the volume of the core could change (Kim *et al.*, 2007).

The method of Kim *et al.* was utilized to produce microgel capsules by using triple water-in-oil-in-water-in-oil emulsions (Chu *et al.*, 2007). In the compound droplets, the outer aqueous layer contained a monomer NIPAm, a crosslinker, and an initiator. An accelerator was added to the inner oil phase, from which it diffused into the outer aqueous shell and speeded up polymerization of the monomer. In the resulting microcapsule, a shell of the thermosensitive hydrogel polyNIPAm encapsulated several oil droplets.

A two-step emulsification process was also used to generate gas–liquid–liquid emulsions, in which the droplets encapsulating microbubbles were transformed into polymer capsules (Wan *et al.*, 2008). In the first step, a microfluidic flow-focusing device was used to generate microbubbles in an aqueous phase comprising acrylamide. The dispersion of bubbles was then passed through either a second flow-focusing device, or a T-junction device with oil as the continuous phase, thereby producing microbubble-in-water-in oil emulsion droplets. Acrylamide in the aqueous phase was subsequently photopolymerized to yield polymer particles encapsulating many bubbles. This method offers the ability to control the elasticity of microparticles by changing the internal porosity: the elastic modulus, E, of the porous polyacrylamide particles was $E = (3.6 \pm 1.2) \times 10^7$ Pa) compared with $E = (8.6 \pm 1.7) \times 10^8$ Pa for identical particles without bubbles.

Polymer capsules have also been produced by generating double emulsions in a single-step process (Nie *et al.*, 2005). In this approach, three immiscible liquids (a silicone oil, a photopolymerizable monomer, and an aqueous phase) were passed through the narrow orifice of the planar microfluidic flow-focusing device, so that the oil and the monomer phases formed a coaxial jet surrounded by the aqueous continuous phase. The coaxial jet became unstable to perturbations with wavelengths larger than its circumference and it reduced its surface area by breaking up into segments, due to Rayleigh–Plateau hydrodynamic instability (Probstein, 1989). Under the action of interfacial tension, the liquid segments acquired a spherical shape (Figure 9.3a). Excellent control was achieved over the size of the inner droplets and the dimensions of core–shell droplets. The authors found the regime of the relative flow rates of three liquids, in which droplets with a precisely controlled number of inner droplets were formed (Figure 9.3b).

Continuous on-chip photopolymerization of the monomer shells in the droplets was carried out in the extension wavy channel (polymerization compartment) of the microfluidic reactor. The monomer phase was either ethyleneglycol diacrylate, or tripropyleneglycol diacrylate mixed with a photoinitiator, 1-hydroxycyclohexyl phenyl ketone. Free-radical polymerization of the monomers transformed compound droplets into polymer capsules with oil cores (Figure 9.3c,d).

Conceptually similar single-step emulsification was employed to produce precursor droplets for the poly(D,L-lactic acid-*co*-glycolic acid) (PLGA) microcapsules (Martin-Banderas *et al.*, 2005). In a flow-focusing device, two concentric needles placed in the

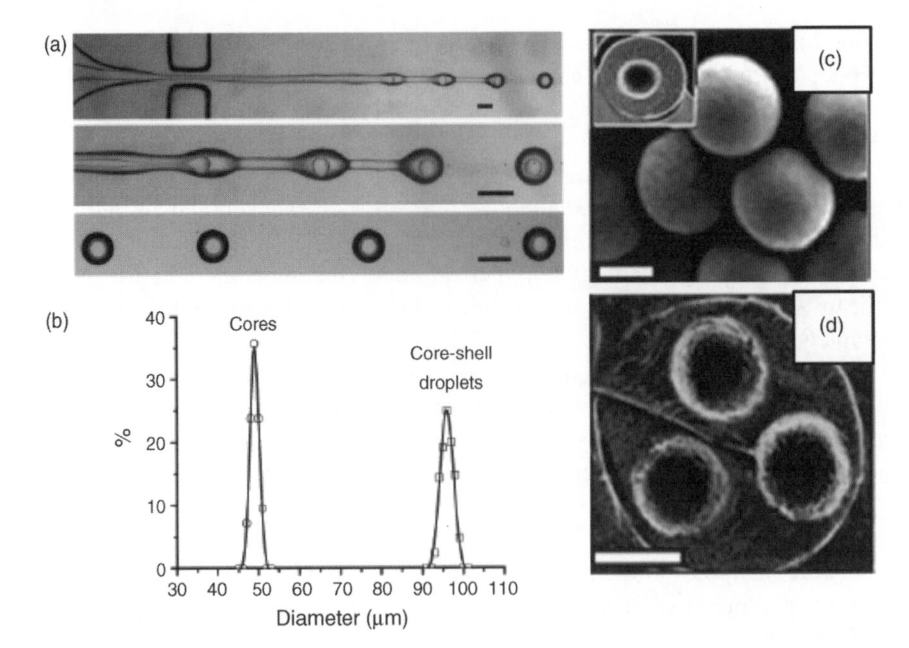

__Figure 9.3__ (a) Typical optical microscopy images of the breakup of a liquid jet formed by a coaxial thread of silicone oil (viscosity 10 cst) and EGDMA, and the resulting core–shell droplets flowing in the downstream channel. (b) Size distributions of the SO (silicone oil) cores and SO–EGDMA core–shell droplets obtained at $Q_w = 26\,mL\,h^{-1}$, $Q_m = 0.30\,mL\,h^{-1}$, and $Q_{SO} = 0.045\,mL\,h^{-1}$. Scale bar is 100 μm. (c) Scanning electron microscopy image of polymer capsules produced by photopolymerizing TPGDA shell in core–shell droplets with a single silicone oil core. Inset shown a cross-section of the capsule with a single silicone oil core. (d) Scanning electron microscopy image of the cross-section of a polyTPGDA particle with three silicone oil cores. The particle is embedded in epoxy glue. Scale bar is 40 μm. Adapted, with permission, from ref. Nie et al., 2005. Reproduced with permission from Journal of the American Chemical Society, Polymer Particles with Various Shapes and Morphologies Produced in Continuous Microfluidic Reactors by Z. Nie, S. Xu, M. Seo, __127__, 22, 8058–8063 Copyright (2005) American Chemical Society

surrounding continuous phase focused a stream of two coaxially flowing liquids, which broke up to release core–shell droplets. The continuous fluid phase was either a liquid, or a gas. PLGA microparticles encapsulating an aqueous antibiotic gentamycin were produced by using a gas as the focusing phase. The outside layer of the coaxial jet was formed by the solution of PLGA dissolved in an organic solvent; the inner stream of the coaxial jet was composed of an aqueous solution of gentamycin. After breakup, the organic solvent evaporated, leaving a PLGA shell surrounding the gentamycin solution. In addition to exploiting the evaporation techniques, Martin-Banderas *et al.* (2005) generated polymer capsules with multiple cores by forcing three immiscible fluids through a small orifice and polymerizing the photopolymer SK9 localized in droplet shells (Figure 9.4).

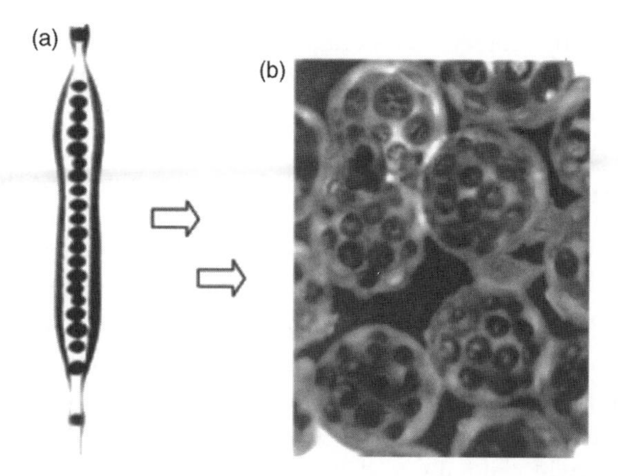

Figure 9.4 *(a) In-flight photograph of the natural breakup of a coaxial jet of ink surrounded by a photopolymer (SK9). (b) Optical microscopy image of the resulting multi-core microcapsules produced after UV-curing of SK9. Adapted, with permission, from Martin-Banderas et al., 2005. Reproduced with permission from Small, Flow-Focusing: a Versatile Technology to Produce Size-Controlled and Specific-Morphology Microparticles by L. Martin-Banderas, M. Flores-Mosquera, P. Riesco-Chueca et al.,* ***1****, 7, 688–692 Copyright (2005) Wiley-VCH. See Plate 10*

9.2.2 Polymersomes

Diblock copolymers self-assemble in selective solvents and produce structures that are similar to vesicles formed from bilayers of phospholipids (Discher *et al.*, 1999; Ahmed and Discher, 2005). Polymer vesicles or so-called polymerosomes are core–shell polymer particles with a core formed by a solvent and a shell composed of a block copolymer amphiphile. Polymerosomes may find applications in the encapsulation of nano- to picoliter volumes of fluids and delivery and release of molecules encapsulated in the capsular interior. The assembly of polymerosomes offers an attractive strategy in creating useful structures for encapsulation by controlling the structure and properties of diblock copolymers. Currently, polymerosomes are prepared via phase segregation of the block copolymers in solution, which is triggered by adding a poor solvent for one of the polymer blocks (Shen and Eisenberg, 1999) or by rehydrating a dried film of the copolymers (Antonietti and Förster, 2003).

Microfluidics offers an alternative approach to the production of polymerosomes with a very narrow size distribution by generating precursor droplets of the solutions of block copolymers. Recently, two microfluidic strategies were utilized for the assembly of polymerosomes. In the first method, polymerosomes were prepared using water-in-oil-in-water precursor droplets. The double emulsions were formed by breaking up concentric fluid streams of immiscible water and oil fluids and releasing compound droplets. The intermediate oil phase in these droplets was a solution of the diblock copolymer in a volatile organic solvent. The removal of the solvent from the shell of the droplets, followed by solvent evaporation (Alex and Bodmeier, 1990; Cohen *et al.*, 1991) resulted in the formation of capsules with rigid block copolymer walls.

This method was used to form polymerosomes from the copolymer poly(n-butyl acrylate)–poly(acrylic acid) (PBA–PAA) (Lorenceau *et al.*, 2005). The molecular weights of the hydrophobic PBA block and the hydrophilic PAA block were 4000 and 1500, respectively. The solution of PBA–PAA in a mixture of tetrahydrofuran and toluene formed an intermediate layer in the compound droplets. Dissolution of the co-solvent mixture from the middle layer of the droplets and ultimate evaporation of the solvent led to the formation of polymerosomes. With the removal of the solvent from the droplet, the shell in the droplets became thinner and the concentric interfaces tended to disappear. Index matching of the inner and outer liquids caused the drop to fade in bright field microscopy, as shown in Figure 9.5a–c. As the thin wall of the vesicle could not be visualized using bright field microscopy, phase contrast microscopy was used to enhance the index-of-refraction mismatch between the block copolymer and the surrounding fluid (Figure 9.5d). Visualization of the block copolymer membrane was further improved by adding CdSe quantum dots to the inner fluid before the formation of the double emulsion. Following the formation of polymerosomes, the quantum dots were trapped on the inner side of the walls of the polymerosome and showed the location of the wall of the block copolymer membrane.

Assuming that all the solvent evaporates, it was estimated that the final thickness of the shells was approximately 1.5 μm (Lorenceau *et al.*, 2005). As the expected thickness of a bilayer of PBA–PAA block copolymer is about 40 nm, it was concluded that the membrane was not unilamellar and it was approximately 100 layers thick. The authors suggested that the thickness of the wall of polymerosomes could be further controlled by adjusting the initial concentration of the diblock copolymers in the cosolvent mixture.

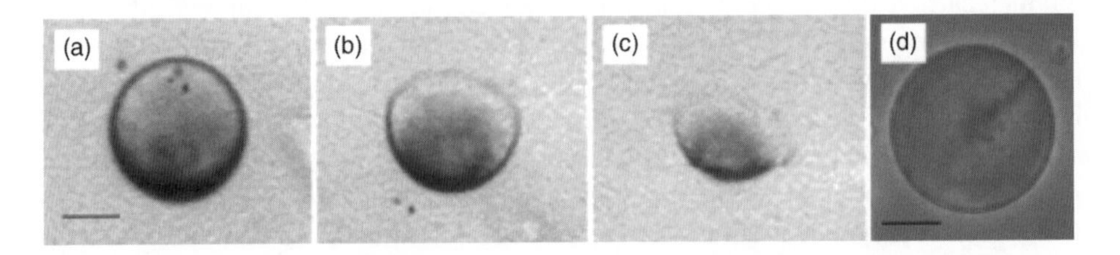

Figure 9.5 *Evaporation and dissolution of a THF–toluene mixture from the oil phase of a double emulsion. Evaporation of the organic solvents allows the PBA–PAA diblock to organize into a vesicle. (a) The middle fluid layer is clearly visible. (b and c) Index matching of the inner and outer fluids causes the drop to fade in bright field microscopy as the solvent leaves the middle layer. Panels (a–c) are separated by 3 min. The scale bar in (a) represents 40 μm. (d) Polymersome imaged with phase contrast microscopy. The scale bar in (d) represents 30 μm. Adapted, with permission, from Lorenceau et al., 2005. Reproduced with permission from Journal of the American Chemical Society, Interfacial Polymerization within a Simplified Microfluidic Device: Capturing Capsules by E. Quevedo, J. Steinbacher, and D. T. McQuade, 127,* ***30****, 10498 Copyright (2005) American Chemical Society*

Another microfluidic method for the generation of polymerosomes relied on the self-assembly of the block copolymer poly(styrene-*b*-methylmethacrylate) at the interface between an aqueous phase and a block copolymer solution in an organic solvent (Abraham *et al.*, 2006). Capsules of partly hydrolysed poly(styrene-*b*-methylmethacrylate) were prepared using primary droplets of the solution of the block copolymer in dichloromethane. The droplets were dispersed in an aqueous solution of water using a T-junction microfluidic device. Subsequently, the droplets were drained, dispersed on a silicon wafer, and dried by slow evaporation to remove the solvent from their interior. The resulting particles had an inner cavity and a porous outer surface. The efficiency of these polymer microcapsules as containers for the storage and controlled release applications was evaluated by encapsulating the microcapsules with Congo Red dye and investigating the release of the dye by UV–VIS spectroscopy.

9.2.3 Interfacial Polymerization

A different approach to the microfluidic production of polymer capsules exploits interfacial polymerization reactions. Formation of Nylon-6,6 capsules generated via interfacial condensation of 1,6-diaminohexane and adipoyl chloride was described in Chapter 7 (Takeuchi *et al.*, 2005). Briefly, an aqueous solution of 1,6-diaminohexane was emulsified in hexadecane. Then, a solution of adipoyl chloride in dichloroethane and hexadecane was introduced into the microchannel. Polymerization between adipoyl chloride and 1,6-diaminohexane occurred at the surface of the droplets and resulted in the formation of a rigid Nylon shell around the droplets. The polymerization reaction was stopped by quenching unreacted adipoyl chloride by collecting particles in a solution of dodecane-1-ol in hexadecane. Clogging of the microfluidic reactor was avoided by preventing contact between the particles and the walls of microchannels with a coaxial sheath of fluid. The authors demonstrated the ability to incorporate 50 nm-size magnetic nanoparticles in the interior of the Nylon capsules and in this manner to manipulate the magnetic microbeads using a magnetic field. After polymerization the polymer membrane was mechanically strong and did not break under the action of the magnetic field. In addition, semi-permeability of Nylon membranes to water molecules was demonstrated in swelling–deswelling experiments by dehydrating the capsules or by tuning the osmotic pressure across the polymer shell.

Shortly after the report by Takeuchi *et al.* (2005), a simple microfluidic T-junction device for the continuous synthesis of polyamide capsules was proposed, which consisted of a tubing and a small-gauge needle (Quevedo *et al.*, 2005). Figure 9.6a shows a schematic of this device. A continuous phase is pumped through the poly(vinyl chloride) (PVC) tubing, and an orthogonally placed needle introduces a disperse phase into the continuous flow.

Polyamide capsules were formed by introducing poly (ethylene imine) into the continuous phase and adding sebacoyl chloride and 1,3,5-benzene tricarboxylic acid chloride in a chloroform–cyclohexane disperse phase. The interfacial polycondensation reaction started when the two phases met and yielded well-defined polyamide capsules with robust shells and polydispersity varying between 3.3 and 8.6% (Figure 9.6b).

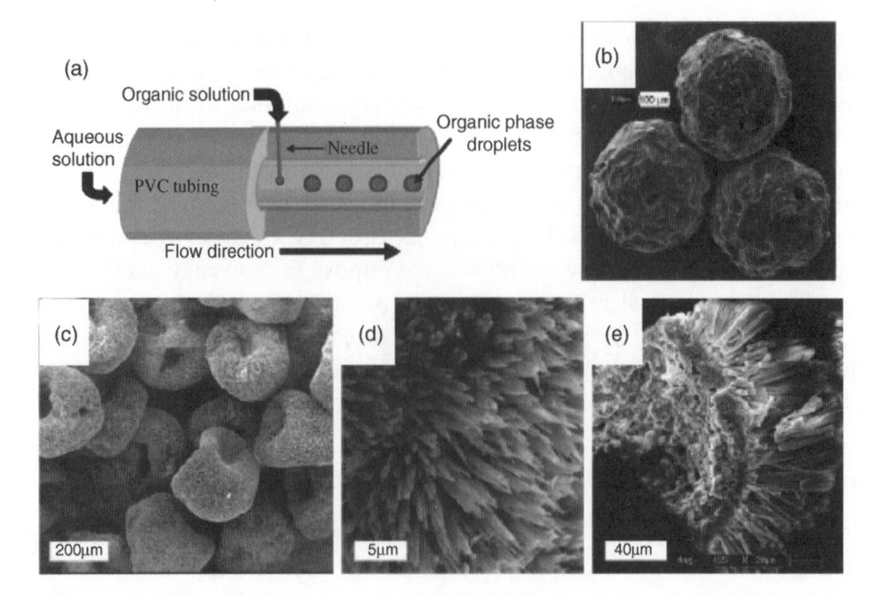

Figure 9.6 *(a) Schematic of the microfluidic device. The continuous phase flows along the tubing, and a disperse phase is introduced through small-gauge needles. Both phases are driven by syringe pumps. The aqueous phase is contained within a 50 mL syringe and flows through poly(vinyl chloride) (tubing (1/16 in. i.d.); the organic solution is dispensed from a 1 or 5 mL syringe and introduced via a 30-gauge needle inserted through the wall of the tubing and situated in the middle of the channel. Reprinted, with permission, from Quevedo, 2005. Reproduced with permission from Journal of the American Chemical Society, Interfacial Polymerization within a Simplified Microfluidic Device: Capturing Capsules by E. Quevedo, J. Steinbacher, and D.T. McQuade,* ***127****, 30, 10498 Copyright (2005) American Chemical Society (b) SEM image of microcapsules prepared using an aqueous flow rate of 13.0 mL min^{-1}, organic flow rate of 0.141 mL min^{-1}. Magnification is 133×, scale bar is 100 μm. (c–e) SEM images of organosilane microcapsules produced in a microfluidic device. (c) A population of typical, spheroidal microcapsules; (d) a higher magnification image of the spiny surface, and (e) a microcapsule fragment showing the inner, amorphous shell and the outer, spiny layer. Adapted, with permission, from Steinbacher et al., 2006. Reproduced with permission from Journal of the American Chemical Society, Rapid Self-Assembly of Core-Shell Organosilicon Microcapsules within a Microfluidic Device by J.L. Steinbacher, R.W.Y. Moy, K.E. Price et al.,* ***128****, 29, 9442 Copyright (2006) American Chemical Society*

The dimensions of capsules were tuned from 313 to 865 μm by carrying the flow rates of the liquids. The capsular shells had an unusual fibrous structure.

The simplified microfluidic device proposed by Quevedo *et al.* (2005) showed promising applications in the utilization of interfacial polycondensation in the production of polymer particles with a capsular structure, with dimensions in the order of hundreds of micrometers. The authors pointed out that despite a very simple design of the tube-and-needle microfluidic device, it had two important advantages (Quevedo,

2005). Firstly, similarly to other microfluidic setups which exploited the sheath flow, in the device of Quevedo *et al.* the disperse phase was entirely surrounded by the continuous phase and the droplets did not wet the walls of the tubing. Secondly, if the device was clogged with polymer particles, the tubing could be easily replaced, and the system became functional, again, within seconds.

In the next step, the tube-and-needle microfluidic device was utilized to generate hierarchical organosilicon capsules composed of oligomeric diphenylsilanediol (Steinbacher *et al.*, 2006). To prepare the spinulose microparticles, neat dichlorodiphenylsilane (Cl_2SiPh_2) was used as a droplet phase and an aqueous glycerol solution was used as the continuous phases. At the interface between the droplet and continuous phases, the chlorosilane hydrolyzed to diphenylsilanediol. The latter compound oligomerized and crystallized, thereby yielding a solid capsular shell. The SEM images of the organosilane capsules are shown in Figure 9.6c–e. The walls of the capsules consisted of an inner amorphous layer surrounded by an outer layer of approximately 10 μm long spinules. Examination of the shell under high magnification revealed that the spinules were oriented perpendicular to the surface of the capsule. Solid state and solution nuclear magnetic resonance (NMR) spectroscopy on the microcapsules revealed that they contained diphenylsilanediol [$Ph_2Si(OH)_2$] with decreasing amounts of the dimeric 1,1,3,3-tetraphenyldisiloxane, the associated trimer, and higher order oligomers. A significant amount of glycerol was physically incorporated into the capsular walls. Steinbacher *et al.* envisaged potential applications of the organosilicone microcapsules as templates for the formation of metal particles with a high surface area (Steinbacher *et al.*, 2006).

9.2.4 External Gelation

The method of external ionic gelation conducted in microfluidic reactors for polymer particles was described in Chapter 8. Briefly, when droplets of an aqueous polymer solution are dispersed in a non-polar liquid containing a crosslinking agent, the latter diffuses into the droplets and initiates the gelling process that propagates from the droplet interface with the continuous phase to the droplet interior (Zhang, 2006, 2007). Gelation is then quenched by collecting gel particles in a large volume of crosslinking agent-free organic phase, thereby generating gel capsules.

The external gelation method was used in the microfluidic production of alginate, κ-carageenan, and carboxymethyl cellulose particles with a capsular structure. For example, alginate capsules were generated by emulsifying an aqueous solution of sodium alginate in a solution of CaI_2 in undecanol. Following emulsification, gelation was caused by the diffusion of CaI_2 in the droplets moving through the downstream microchannel, subsequent liberation of Ca^{2+} ions in an aqueous phase, and their binding to the alginate molecules. This process could be approximated with Equation 8.1 (see Chapter 8) (Crank, 1975).

The depth of penetration of Ca^{2+} ions into the droplets and thus the thickness of a gelled layer in the capsules was determined by the concentration of the crosslinking agent in the continuous phase and the time of diffusion of the crosslinker into the droplets. The latter factor was controlled by the time of residence of the droplets in the microfluidic device, that is, by their velocity and the length of the microchannels.

Figure 9.7 shows the structure and properties of alginate capsules prepared by the method of external gelation (Zhang *et al.*, 2006). The thickness of gelled shells in the capsules was determined by the depth of penetration of Ca^{2+} ions in the droplets. Fluo-3 (a marker for Ca^{2+} ions) was introduced into the aqueous sodium alginate solution and the resulting particles were imaged using confocal fluorescence microscopy. Figure 9.7a–f shows that with increasing concentration of CaI_2 in the continuous phase and with increasing time of gelation, the thickness of Ca^{2+}-rich zones in the

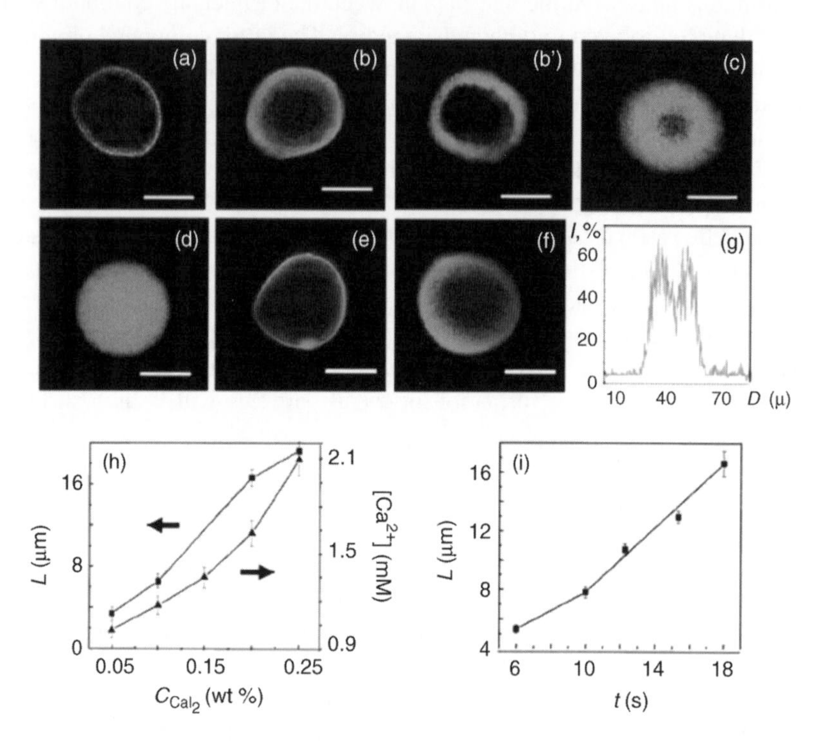

Figure 9.7 *Confocal fluorescence microscopy images of alginate microgels obtained by emulsification of 1.0 wt% alginate solution in a solution of CaI_2 in undecanol with concentrations, wt%: (a) 0.05, (b, b′) 0.10, (c, e, f) 0.20, and (d) 0.25. In (b′) microgels were stored for 5 days in CaI_2-free undecanol. Gelation time was 18 s for (a–d) and 6 and 10 s for (e) and (f), respectively. (g) Representative profile of fluorescent density for the alginate microcapsule shown in (c). (h) Variation in depth of penetration of Ca^{2+} ions in microgels (■) and concentration of Ca^{2+} ions (▲) in microgels. (i) Depth of penetration of Ca^{2+} ions in microgels versus gelation time. Scale bar is 20 μm. In (h, i) the lines are added for visual guidance. Adapted, with permission, from Zhang et al., 2006. Reproduced with permission from Journal of the American Chemical Society, Microfluidic Production of Biopolymer Microcapsules with Controlled Morphology by H. Zhang, E. Tumarkin, R. Peerani et al.,* ***128****, 37, 12205–12210 Copyright (2006) American Chemical Society. See Plate 11*

microgels increased until a uniform distribution of Ca^{2+} ions in the microgel particles was reached.

The capsular structure of the particles was preserved after microgel storage for five days in pure undecanol or in an aqueous environment. The thickness, L, of the capsular shells was determined by plotting fluorescence line profiles, as in Figure 9.7g. The thickness of the gel shells increased with increasing concentration of CaI_2 in the continuous phase and with diffusion time (Figure 9.7h and i), in agreement with Equation 8.1. For Fluo-3 the detection sensitivity of Ca^{2+} ions is as low as 0.325 μM, thus it was assumed that Ca^{2+} ions were mostly localized in the bright zones of the microgels and the concentration of Ca^{2+} ions in the capsular shells was estimated to be from 1.64 to 2.38 mM. Such a concentration of Ca^{2+} ions was sufficient to induce gelation of alginate. Figure 9.7 explicitly shows that control of the microgel structure – from capsules to particles with a uniform structure – could be achieved by varying either the concentration of the crosslinking agent in the continuous organic phase, or the time of gelation, or by changing two variables simultaneously.

A study of the mechanical properties of the capsules using atomic force microscopy showed that the elastic modulus of the microgels with a uniform structure was about 17% higher than that of the capsules; however the values of elastic moduli of 167–190 kPa were close to the values reported for the alginate beads prepared by dropping a solution of sodium alginate into the solution containing Ca^{2+} ions (Ouwerx *et al.*, 1998).

The potential application of the method of external gelation for the encapsulation of a controlled number of cells per capsule was explored by encapsulating polystyrene beads in the interior of microcapsules (Zhang *et al.*, 2006). The average number of particles per capsule was controlled by varying the diameter of precursor droplets (by tuning the flow rates of the droplet and the continuous phases). The average number of beads per capsule was 1–7, although a large population of small 30 μm-size capsules produced at high flow rates of the continuous phase was particle-free.

9.2.5 Capsules with Colloidal Shells

Capsules formed by encapsulating droplets or bubbles with a shell of colloidal particles (a colloidal "armor") can be readily generated in microfluidic devices. Such structures (generally known as Pickering emulsions) show remarkable stability to coalescence and the ability to acquire anisotropic shapes (Binks and Horozov, 2005; Subramaniam *et al.*, 2005; Studart *et al.*, 2007). These properties make Pickering systems useful for the fabrication of thermal and acoustic insulators, and materials with high structural stability (Alargova *et al.*, 2004; Binks and Murakami, 2006; Scheffler and Colombo, 2005).

Bubbles coated with a colloidal shell were first generated via microfluidic means by Subramaniam *et al.* (2005). Figure 9.8a–c illustrates the formation of armored bubbles in a three-channel microfluidic flow-focusing device. The continuous phase is supplied to the outer channels, and the inner channel carries the dispersed phase (a gas or a liquid) (Figure 9.8a,b). The interface between the continuous and dispersed phases serves as the

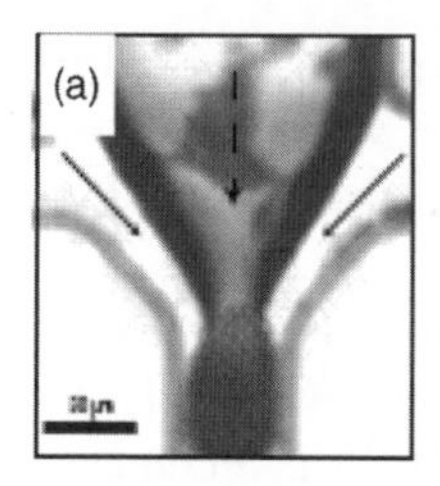

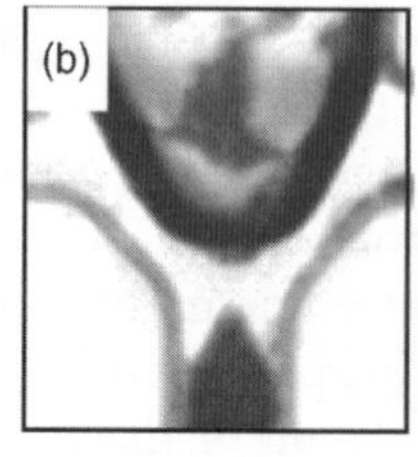

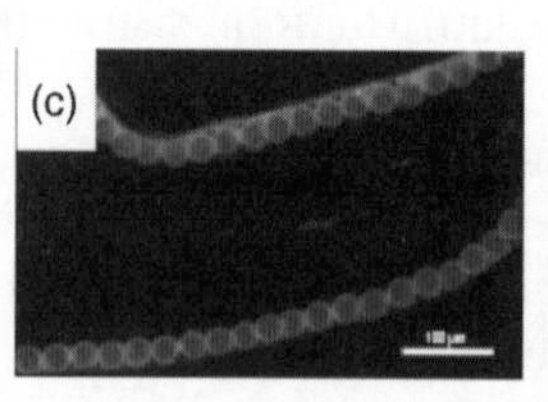

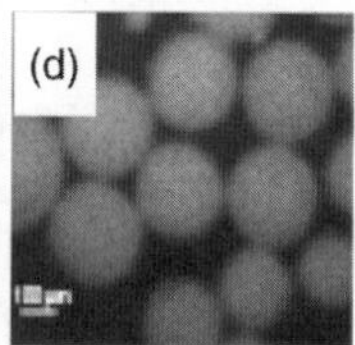

Figure 9.8 *(a, b) Hydrodynamic focusing for the continuous formation of droplets with colloidal armor. The direction of dispersed-phase flow is indicated by the dashed arrow, and the direction of the colloidal–suspension flow is indicated by the solid arrows. Flow of the continuous-phase liquid focuses the dispersed-phase fluid into a narrow thread and targets the particles onto the interface. The speed of the particles and the frequency of shedding are controlled by tuning the difference in driving pressures of the inner and outer channels (typical values are: suspension, 1.92 psi.; gas, 1.69 psi; 1 psi = 6894.76 Pa). (c) The interfacial crystal consisting of 4 μm diameter charge-stabilized fluorescent polystyrene beads grows and subsequently experiences greater shear, which results in the ejection of a jammed shell. (d) The armored bubbles coated with a conductive armor of gold particles. In (c, d) the scale bar is 100 μm. Adapted, with permission, from Subramaniam et al., 2005. Reproduced with permission from Langmuir, Generation of Polymerosomes from Double-Emulsions by E. Lorenceau, A. S. Utada, D. R. Link et al.,* ***21****, 20, 9183–9186 Copyright (2005) American Chemical Society*

substrate for the deposition of colloid particles dispersed in the continuous phase. Figure 9.8c shows the attachment of polystyrene beads as a monolayer on a gas thread, which becomes unstable and breaks to form a bubble coated with a colloidal shell.

Particle-encapsulated bubbles formed reproducibly at high linear particle velocities of $10\,cm\,s^{-1}$, which suggested that shear-assisted particle adsorption at the interface occurred at a timescale of tens of microseconds. The jammed colloidal shell underwent shear-induced liquid-to-solid transitions. As a result, particle-coated droplets and bubbles were extremely stable to coalescence. Importantly, the topologically induced jamming of the beads was independent of the nature of the particle–particle interactions. Therefore, a variety of colloidal materials could be used to form particle-encapsulated droplets and bubbles. As an example, Figure 9.8d shows gas bubbles encapsulated with a conductive armor of gold nanoparticles.

The approach developed by Subramaniam *et al.* (2005) paved the way for the generation of hierarchical structures with precisely controlled combinations and relative positions of particles on the fluid (liquid–liquid or gas–liquid) interface. For example, armored particles with two distinct hemi-shells, or Janus armors, were produced by introducing two types of microspheres into the two outer microchannels of the microfluidic device.

The original microfluidic approach to producing capsules with a colloidal shell was further modified by introducing poly(divinylbenzene–methacrylic acid) [poly (DVB–MAA)] particles in the droplet phase, rather than in the carrier liquid (Nie *et al.*, 2008). The ratio between the concentration of particles and the dimensions of droplets (determining the surface area of the droplets) was tuned to provide the deposition of a monolayer of particles at the liquid–liquid interface and to avoid excess of particles in the droplet interior. Assuming that the diameter of the colloidal particles is much smaller than the diameter of the droplets, the surface coverage, δ, of droplets was estimated as $\delta = A_p/A_d = C_p \cdot \rho_d \cdot a_d/(4\rho_p \cdot a_p)$ ($\delta < 1.0$), where A_d and A_p are the surface area of the droplets and the area of the droplets coated with particles, respectively; ρ_d and a_d are the density of the droplet phase and the radius of droplets, respectively; and C_p, ρ_p, and a_p are the concentration, the density, and the radius of particles, respectively. Thus the variation in the surface coverage of the droplets with particles was achieved by increasing the radius, a_d, of droplets, or the concentration, C_p, of the particles in the droplet phase. In Figure 9.9a the broken line shows the estimated variation in the concentration of polymer particles, C_p, with the radius of droplets, that is required to achieve a complete coverage of the droplets of 0.906 (assuming hexagonal packing of the particles on the droplet surface). Below the line, the amount of particles was not sufficient to form a close-packed layer on the liquid–liquid interface, whereas above the line the microbeads completely covered the surface of the droplets and the excess of particles was localized in the droplet interior.

Both oil-in-water and water-in-oil emulsions were exploited in the inside-out approach to particle-stabilized capsules (Nie *et al.*, 2008). In the former case, the dispersion of poly(DVB–MAA) microbeads in an ethanol–water mixture was emulsified in hexadecane. In the latter case, the droplets were formed in an aqueous phase from TPGDA mixed with 14 wt% poly(divinylbenzene–methacrylic acid) microbeads and 4 wt% of a photoinitiator. Following the formation of particle-coated droplets, TPGDA was photopolymerized, yielding particle-stabilized polymer particles (Nie *et al.*, 2008).

Figure 9.9b shows optical microscopy and confocal fluorescence microscopy images of droplets coated with poly(DVB–MAA) particles, which were obtained in three regimes: at non-complete coverage, full coverage, and at an excess of particles in the droplet phase, as predicted in Figure 9.9a. The experimental results were in agreement with the estimation. A zoomed-in inset in Figure 9.9b shows that on the surface of the droplets the polymer beads were packed in a hexagonal lattice, the feature that was not evident in the droplets with non-complete coverage.

9.3 Emerging Applications of Polymer Capsules Produced by Microfluidic Methods

One of the most straightforward applications of polymer capsules generated by microfluidic methods includes the encapsulation and delivery of drugs, nutrients, food or cosmetics ingredients. Prior to microfluidic emulsification, a drug is dissolved in the droplet phase. Following the formation of precursor droplets and subsequently, polymer capsules, the drug is localized in the interior of the particles. The release of the drug through a non-erodible (intact) capsular shell occurs by diffusion. The rate of release depends on the internal and external concentrations of the drug. Currently, the

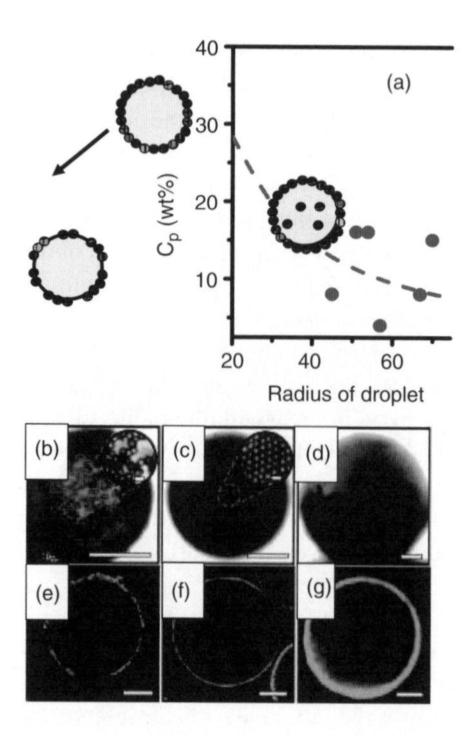

Figure 9.9 *(a) A diagram for the control over the surface coverage of droplets with a colloidal shell. The broken line characterizes the condition for the production of Pickering emulsions with complete coverage of droplets with particles used in the present work. Filled circles show the experimental data points. (b–d) Optical and (e–g) oil-immersion confocal microscopy images of the water–ethanol droplets armored with a shell of poly (DVB–MAA) particles at the not-complete (b, e) and complete (c, f) surface coverage, and at the excess of particles in the droplet interior (d, g). In panel (d) an excess of particles appears as the large dark region on the background of the droplet coated with a monolayer of particles. In (e–g) poly(DVB–MAA) microbeads were copolymerized with anthryl methacrylate. Confocal microscopy images show the plane located in the center of the droplets.* $C_p = 8$ *wt%.* $\lambda_{ex} = 380$ *nm. Scale bars in panels (b–f) and (g) are 50 and 100* μ*m, respectively. Scale bars in insets are 5* μ*m. Reprinted with permission from Nature Materials, Controlled assembly of jammed colloidal shells on fluid droplets by A. B. Subramaniam, M. Abkarian, and H. A. Stone,* ***4****, 553–556 Copyright (2005) Macmillan Publishers Ltd*

utilization of microfluidically generated polymer capsules as drug delivery vehicles is limited to the proof of concept, in which a drug such as the antibiotic gentamycin, is localized in the capsular core (Martin-Banderas *et al.*, 2005).

Alternatively, instead of molecules, small droplets can be encapsulated and released from the interior of microcapsules by breaking the capsular shell. This delivery mechanism was demonstrated for microgel capsules derived from triple W/O/W/O emulsions (Chu *et al.*, 2007). The microcapsules had a thermosensitive polyNIPAm hydrogel shell and an oil core containing several water droplets (see Figure 8.4a). Figure 8.4b–e illustrates the release of the inner water droplets from the capsule. Upon heating the system from 25 to 50 °C, the polyNIPAm moves through the lower critical solution temperature, expels water, and shrinks. Because of the incompressibility of the inner oil core, the hydrogel shell breaks and the innermost water droplets are released into the continuous oil phase, as shown in Figure 8.4a–e. The advantage of this microfluidic method included the ability to generate in a controlled manner capsules with multiple internal cores (potentially, containing biologically active substances) and releasing them using external stimuli.

Another important application of capsules produced by microfluidic methods involves the encapsulation of cells in the interior of polymer capsules and/or the encapsulation of cells with enzymes, cytokines, and an extracellular matrix. The feasibility of microfluidic encapsulation of cells in hydrogel particles with a uniform, non-capsular structure has been demonstrated by a number of research groups (see Chapter 8). Currently, the encapsulation of cells in gel particles with a capsular structure has not been demonstrated, although hydrogel capsules with a liquid core provide a more natural biological environment to the cells, which may prove important during cell growth, renewal, and differentiation.

Because during microfluidic emulsification, the incorporation of cells into precursor droplets is a random process, the encapsulation of n cells in a microcapsule can be evaluated by using Poisson statistics (the same considerations apply to the encapsulation of cells in particles with a uniform composition and structure). It has been established that experimental results on cell encapsulation are well described as $f(\lambda, n) = \lambda^n e^{-\lambda}/n!$, where n is the expected number of cells per capsule and λ is the average number of cells per droplet (Köster *et al.*, 2008).

To date, the encapsulation of cells in polymer capsules has not been demonstrated, however, the possibility of such a process has been shown by encapsulating polystyrene (PS) particles in alginate capsules (Zhang *et al.*, 2006). Figure 9.10 shows typical micrographs and histograms for the distribution of PS particles in alginate capsules that were produced by the external gelation method. For the constant concentration of PS particles in the droplet phase, the average number of microspheres per precursor droplet was controlled by varying the diameters of the alginate capsules, which was achieved by tuning the ratio of the flow rates of the continuous and droplet phases. As shown in Figure 9.10a′,b′, the average number of PS microbeads per capsule was 7 and 3 for the 150 and 100 μm-size capsules, respectively. A large population of 30 μm-size microgels contained a single PS particle, and approximately 36% of these capsules were particle-free (Figure 9.10c′).

Finally, polymer capsules generated by the method of external gelation were used to model the behavior of cells in topographically and chemically patterned microfluidic devices (Fiddes *et al.*, 2007, 2009), paving the way for fundamental studies of the flow of suspensions of cells in biological microenvironments. Figure 9.11a,b shows the "surrogate cells" and the constriction in the microchannel in which three interesting phenomena were observed and examined: (i) the effect of confinement, (ii) the role of interactions between the hydrogel capsules and the channel surface, and (iii) the effect

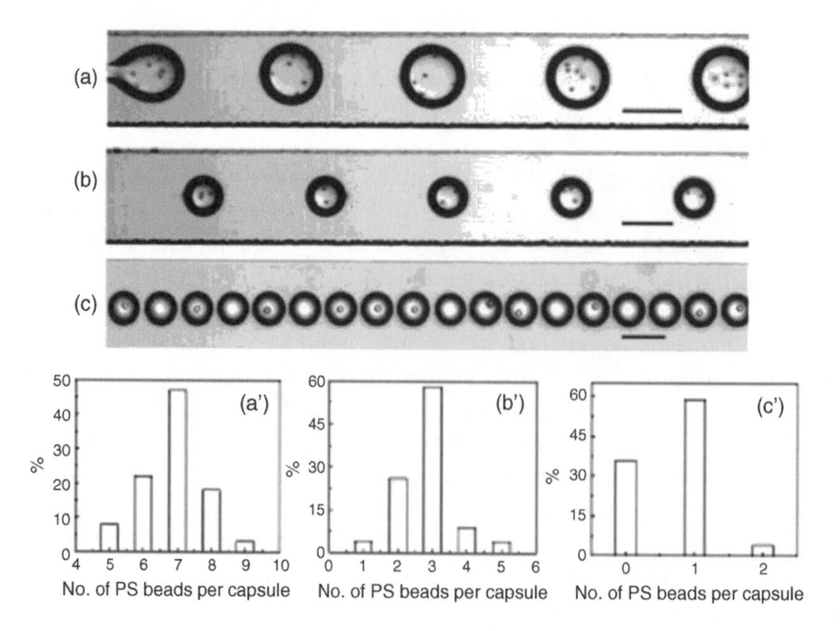

Figure 9.10 *(a–c) Optical microscopy images of polystyrene (PS) beads encapsulated in alginate capsules with the average diameter 150 (a), 100 (b), and 30 μm (c). Scale bars are 150 μm (a, b) and 50 μm (c). (a′–c′): Histograms of distribution of the number of PS beads per individual alginate capsule, as shown in (a–c), respectively. Concentration of alginate was 1.0 wt%. Concentration of PS beads in aqueous alginate solution was 2×10^6 particles per mL. The flow rate of the continuous aqueous phase, Q_w, was 0.05, 0.08, and 0.10 mL h^{-1} and the flow rate of the non-polar droplet phase, Q_o, was 0.5, 0.8, and 1.2 mL h^{-1} for (a–c), respectively. Adapted, with permission, from Zhang et al., 2006. Reproduced with permission from Journal of the American Chemical Society, Microfluidic Production of Biopolymer Microcapsules with Controlled Morphology by H. Zhang, E. Tumarkin, R. Peerani et al.,* ***128****, 37, 12205–12210 Copyright (2006) American Chemical Society*

of the velocities of the capsules prior to their passage through an orifice on their velocity in the constriction (Fiddes *et al.*, 2007, 2009). The normalized velocity, ν_{max}, of the capsules in the orifice was determined by normalizing their velocity in the orifice with respect to the velocity in the channel-at-large. It should be noted that in the constriction all capsules experienced an increase in the linear velocity because of the smaller cross-sectional area of the constriction than that of the channel-at-large; however, interactions with channel walls influenced these effects to a different extent. For example, in one series of experiments cationic and anionic hydrogel capsules were driven through microchannels carrying a negative surface charge. It was found that anionic capsules showed a larger (up to 26-fold) increase in velocity than cationic capsules, due to electrostatic repulsion between the capsules and the microchannel walls. This effect was exaggerated at lower initial velocities of the capsules. On the other hand, attraction forces between the cationic capsules and negatively charged microchannel walls did not result in the notable change in the velocity of capsules in the constriction.

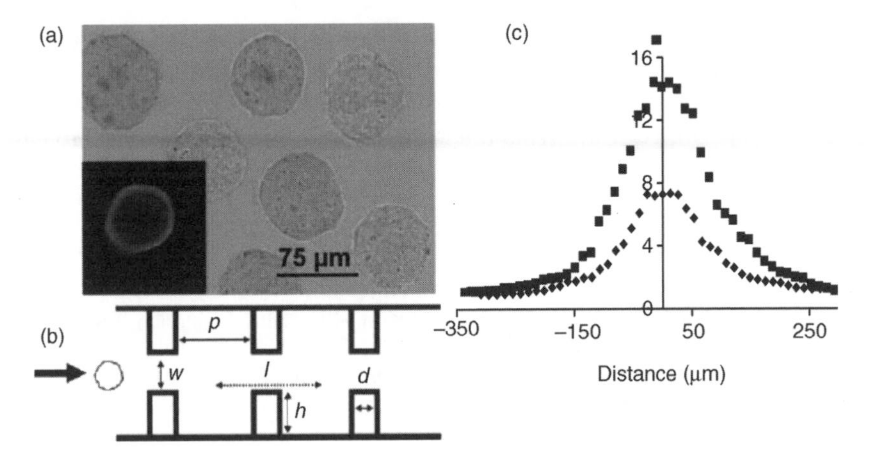

__Figure 9.11__ (a) Optical microscopy image of alginate capsules in PBS solution. Confocal fluorescence microscopy image of alginate capsule labelled with Ca^{2+} indicator, Fluo-3 (inset, adapted, with permission, from Zhang et al., 2006). Reproduced with permission from Journal of the American Chemical Society, Microfluidic Production of Biopolymer Microcapsules with Controlled Morphology by H. Zhang, E. Tumarkin, R. Peerani et al., __128__, 37, 12205–12210 Copyright (2006) American Chemical Society (b) Schematic of the ten-orifice microchannel, with dimensions: d = 100 μm, h = 500 μm, and p = 2 mm. The depth of the microchannel is 150 μm. The width, w, of the orifice varied from 76 to 200 μm. The mean diameter of the microgel capsule is 75 μm. The path, l, over which microgel velocity was measured, is 1 mm. Adapted, with permission, from Fiddes, 2007. Reprinted with permission from Lab on a Chip, Flow of Microgel Capsules through Topographically Patterned Microchannels by L.K. Fiddes, E.W. Young, E. Kumacheva and A.R. Wheeler, 7, 863–867 Copyright (2007) Royal Society of Chemistry (c) Experimentally measured profiles of normalized velocity of 40 μm diameter rigid (■) and soft (◆) biotinylated microgels passing through a 40 μm wide orifice modified with avidin. Adapted, with permission, from Fiddes, 2009. Reprinted with permission from Lab on a Chip, Flow of microgel capsules through topographically patterned microchannels by L. K. Fiddes, E. W. Young, E. Kumacheva and A. R. Wheeler, 7, 863-867 Copyright (2007) Royal Society of Chemistry

In the next step, Fiddes *et al.* used a microfluidic device with a similar geometry to explore the effect of attractive receptor–ligand interactions on the flow of biotinylated microgel capsules in avidin-functionalized microchannels (Fiddes *et al.*, 2009). With increasing biotin concentration, the velocity of the microgels in the constrictions was reduced relative to several control systems. This effect was observed below a critical velocity of the microgels in the channel-at-large. Furthermore, the mechanical properties of the microgel capsules influenced their flow through chemically and topographically patterned microchannels. Soft compliant capsules showed a lower velocity in the constrictions, relative to rigid capsules carrying the same amount of biotin molecules on the surface. The normalized velocity profiles of soft and rigid microgels are shown in Figure 9.11c. The soft microgels (shear moduli of 1.63 kPa) had ν_{max} of

7.3 ± 1.4, while the rigid microgels (with shear moduli 2.99 kPa), had a ν_{max} of 14.4 ± 1.3. The difference in the flow behavior of the capsules was ascribed to the ability of soft particles to deform in the constriction and maximize the area of adhesive contact with the orifice walls. This effect suppressed microgel motion through the constriction.

References

Abraham, S., Jeong, E.H., Arakawa, T., Shoji, S., Kim, K.C., Kim, I., and Go, J.S. (2006) *Lab Chip*, **6**, 752–756.

Ahmed, F. and Discher, D.E. (2005) *Annu. Rev. Biomed. Eng.*, **8**, 323–341.

Alargova, R.G., Warhadpande, D.S., Paunov, V.N., and Velev, O.D. (2004) *Langmuir*, **20**, 10371–10374.

Alex, R. and Bodmeier, R. (1990) *J. Microencapsul.*, **7**, 347–355.

Antonietti, M. and Förster, S. (2003) *Adv. Mater.*, **15**, 1323–1333.

Binks, B.P. and Horozov, T.S. (2005) *Angew. Chem. Int Ed.*, **117**, 3788–3791.

Binks, B.P. and Murakami, R. (2006) *Nat. Mater.*, **5**, 865–869.

Chu, L., Utada, A.S., Shah, R.K., Kim, J., and Weitz, D.A. (2007) *Angew. Chem. Int. Ed.*, **46**, 1–6.

Cohen, S., Yoshioka, T., Lucarelli, M., Hwang, L.H., and Langer, R. (1991) *Pharmaceutical Res.*, **8**, 713.

Crank, J. (1975) *The Mathematics of Diffusion*, 2nd edn, Clarendon Press, Oxford, p. 479.

Discher, B.M., Won, Y.Y., Ege, D.S., Lee, J.C-M., Bates, F.S., Discher, D.E., and Hammer, D.A. (1999) *Science* **284**, 1143.

Fiddes, L.K., Young, E.W., Kumacheva, E., and Wheeler, A.R. (2007) *Lab Chip*, **7**, 863–867.

Fiddes, L., Chan, H.K., Wyss, K., Simmons, C.A., Wheeler, A., and Kumacheva, E. (2009) *Lab Chip*, **9**, 286–290.

Kim, J.W., Utada, A.S., Fernandez-Nieves, A., Hu, Z., and Weitz, D.A. (2007) *Angew. Chem. Int. Ed.*, **46**, 1819–1822.

Köster, S., Angilè, F.E., Duan, H., Agresti, J.J., Wintner, A., Schmitz, C., Rowat, A.C., Merten, C. A., Pisignano, D., Griffith, A.D., and Weitz, D.A. (2008) *Lab Chip*, **8**, 1110–1115.

Lorenceau, E., Utada, A.S., Link, D.R., Cristobal, G., Joanicot, M., and Weitz, D. (2005) *Langmuir*, **21**, 9183–9186.

Martin-Banderas, L., Flores-Mosquera, M., Riesco-Chueca, P., Rodriguez-Gil, A., Cebolla, A., Chavez, S., and Ganan-Calvo, A.M. (2005) *Small*, **1**, 688–692.

Mathiowitz, E. (ed.) (1999) *Encyclopedia of Controlled Drug Delivery*, vol. 2, John Wiley and Sons, Inc., New York, p. 565.

Nie, Z., Xu, S., Seo, M., Lewis, P.C., and Kumacheva, E. (2005) *J. Am. Chem. Soc.*, **127**, 8058–8063.

Nie, Z., Park, J.I., Li, W., Bon, S., and Kumacheva, E. (2008) *J. Am. Chem. Soc.*, **130**, 16508–16509.

Nisisako, T., Okushima, S., and Torii, T. (2005) *Soft Matt.*, **1**, 23–27.

Okushima, S., Nisisako, T., Torii, T., and Higuchi, T. (2004) *Langmuir*, **20**, 9905–9908.

Ouwerx, C., Velings, N., Mestdagh, M.M., and Axelos, M.A.V. (1998) *Polym. Gels Networks*, **6**, 393–408.

Probstein, R.F. (1989) *Physicochemical Hydrodynamics: An Introduction*, 2nd edn, Butterworths Series in Chemical Engineering, Butterworth-Heinemann Ltd, p. 368.

Quevedo, E., Steinbacher, J., and McQuade, D.T. (2005) *J. Am. Chem. Soc.*, **127**, 10498–10499.

Scheffler, M. and Colombo, P. (eds) (2005) *Cellular Ceramics: Structure, Manufacturing, Properties and Applications*, Wiley-VCH, Weinheim, p. 645.

Seo, M., Paquet, C., Nie, Z., Xu, S., and Kumacheva, E. (2007) *Soft Matt.*, **3**, 986–992.
Shen, H. and Eisenberg, A. (1999) *J. Phys. Chem. B*, **103**, 9473–9487.
Steinbacher, J.L., Moy, R.W.Y., Price, K.E., Cummings, M.A., Roychowdhury, C., Buffy, J.J., Olbricht, W.L., Haaf, M., and McQuade, D.T. (2006) *J. Am. Chem. Soc.*, **128**, 9442.
Studart, A.R., Gonzenbach, U.T., Akartuna, I., Tervoort, E., and Gauckler, L.J. (2007) *J. Mater. Chem.*, **17**, 3283–3289.
Subramaniam, A.B., Abkarian, M., Mahadevan, L., and Stone, H.A. (2005) *Nature*, **438**, 930.
Subramaniam, A.B., Abkarian, M., and Stone, H.A. (2005) *Nature Mater.*, **4**, 553–556.
Takeuchi, S., Garstecki, P., Weibel, D.B., and Whitesides, G.M. (2005) *Adv. Mater.*, **17**, 1067–1072.
Utada, A.S., Lorenceau, E., Link, D.R., Kaplan, P.D., Stone, H.A., and Weitz, D.A. (2005) *Science*, **308**, 537–541.
Wan, J., Bick, A., Sullivan, M., and Stone, H.A. (2008) *Adv. Mater.*, **20**, 3314–3318.
Zhang, H., Tumarkin, E., Peerani, R., Nie, Z., Sullan, R.M.A., Walker, G.C., and Kumacheva, E. (2006) *J. Am. Chem. Soc.*, **36**, 117–142.
Zhang, H., Tumarkin, E., Sullan, R.M.A., Walker, G.C., and Kumacheva, E. (2007) *Macromol. Rapid Commun.*, **28**, 527–538.

10

Microfluidic Synthesis of Polymer Particles with Non-Conventional Shapes

CHAPTER OVERVIEW

10.1 Generation of Particles with Non-Spherical Shapes

Polymer particles frequently have interesting applications that are largely governed by their non-spherical shapes. For example, microbeads with non-spherical shapes undergo field-induced orientation and assemble in topologically complex lattices (Alargova *et al.*, 2004; Lu, Yin, and Xia, 2001; Breen *et al.*, 1999; Bowden *et al.*, 1999). Colloid crystals formed by asymmetric colloidal particles may show interesting photonic properties (Hosein *et al.*, 2009).

Generally, conventional suspension, emulsion, and dispersion polymerizations generate spherical polymer particles. Synthesis of polymer particles with non-spherical shapes is rather unusual, although it has been achieved by applying shear (Alargova *et al.*, 2006), by using colloidal jamming (Bon *et al.*, 2007), or combining thermodynamic and kinetic factors (Skjeltorp, Ugelstad, and Ellingsen, 1986; Cho and Lee, 1985; Okubo, Fujibayashi, and Terada 2005). Furthermore, controllable and

Microfluidic Reactors for Polymer Particles. Eugenia Kumacheva and Piotr Garstecki.

highly reproducible synthesis of polymer particles with complex morphologies remains a synthetic challenge.

Recently, in addition to other remarkable features, microfluidic synthesis provided the ability to synthesize non-spherical particles with a variety of shapes and morphologies. Such synthesis has been realized in single-phase and multiphase flow. Currently used microfluidic approaches used for the syntehsis and assembly of polymer non-spherical particles are described in the present chapter.

10.1.1 Two-Phase Synthesis of Non-Spherical Polymer Particles

The first microfluidic approach to non-spherical polymer particles relied on the formation of droplets of polymerizable liquids, which were confined to microchannels with at least one dimension smaller than an unperturbed diameter of the droplets. Under such conditions, because of confinement, breakup of the thread of the polymerizable liquid was not followed by shape relaxation of droplets into spheres. *In situ* (on-chip) polymerization of these droplets produced non-spherical particles with shapes determined by the extent of confinement.

Figure 10.1 illustrates schematically the approach to the microfluidic synthesis of non-spherical particles from the monomer tripropyleneglycol diacrylate (TPGDA) (Xu *et al.*, 2005; Seo *et al.*, 2005). The shape of the precursor droplets was determined by the relationship between the diameter of an undeformed droplet and the dimensions of the microchannel in which polymerization took place. The diameter of an undeformed spherical droplet was given by $d_u = (6V_d/\pi)^{1/3}$ where V_d is the volume of the droplet. Droplets with spherical shapes formed when the value of d_u was smaller than the width, w, and/or the height, h, of the microchannel. For $w > d$ and $h > d$, the droplets minimized their surface energy by acquiring a spherical shape (Figure 10.1a) and when TPGDA was polymerized, yielded spherical polymer particles (Figure 10.1a'). In wide shallow microchannels with $w > d_u$, $h < d_u$, the droplets assumed a discoid shape (Figure 10.1b) and formed two circular interfaces with the top and bottom walls of the microchannel.

Polymerization of TPGDA produced the discoid polymer particles shown in Figure 10.1b′. For $w < d_u$, $h < d_u$, the droplets assumed a rod-like shape with the length l_R given by $l_R = V/wh$ (Figure 10.1c). Polymerization of the TPGDA rod-like plugs resulted in the formation of polymer rods. The aspect ratios of the non-spherical particles could be controlled independently by changing the volume of the droplets (by varying the flow rates of the droplet and the continuous phases) and the dimensions of the polymerization compartment. In addition, ellipsoidal particles were produced by generating precursor droplets of TPGDA at a high flow rate of the continuous aqueous phase and polymerizing them "on the fly" (Figure 10.1d) (Xu *et al.*, 2005). In this case, rather than confinement, the strong shear stress imposed on the droplets determined their non-spherical shape.

A similar approach was used to synthesize disk-shaped particles from the UV-sensitive pre-polymer NOA 60 (Dendukuri *et al.*, 2005). Droplets of NOA 60 were confined to microchannels with appropriate geometries and subsequently transformed into solid particles by photopolymerization.

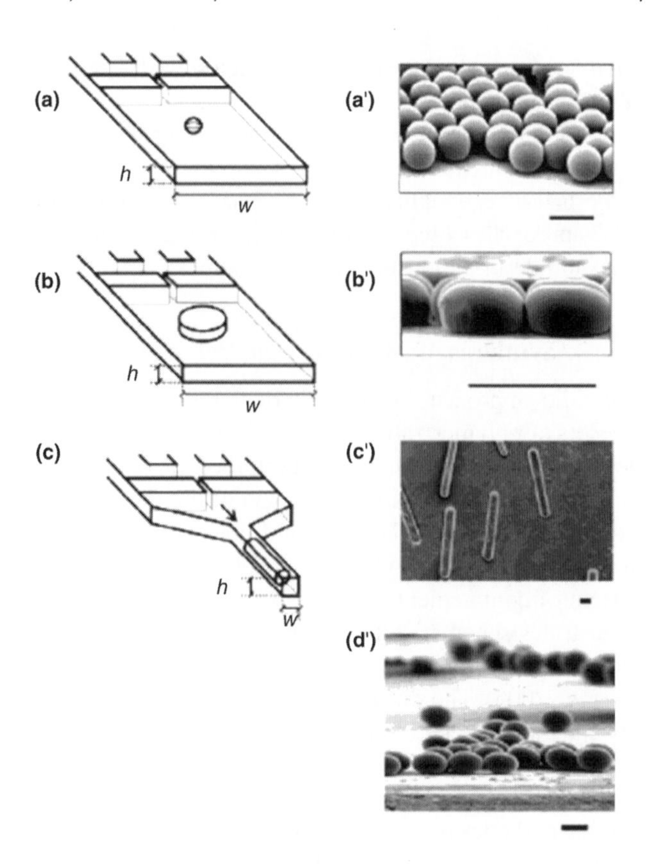

Figure 10.1 *Schematic (a–c) and optical microscopy (a′–c′, d) images of polyTPGDA particles with different shapes: microspheres (a, a′), disks (b, b′), (c, c′) rods, and (d) ellipsoidal polyTPGDA particles obtained via photopolymerization of droplets produced at $Q_w = 8\,mL\,h^{-1}$, $Q_m = 0.1\,mL\,h^{-1}$, and $c_{in} = 4\,wt\%$. Scale bar is 50 μm. Adapted, with permission, from Seo, 2005. Reprinted with permission from Langmuir, Microfluidics: from Dynamic Lattices to Periodic Arrays of Polymer Disks by M. Seo, S. Xu, Z. Nie et al.,* ***21****, 11, 4773-4775 Copyright (2005) American Chemical Society*

In the confinement-based approach to the microfluidic synthesis of particles with non-spherical shapes two factors prevented clogging of microchannels with polymerized microbeads. Typically, following polymerization, the dimensions of the "precursor" droplets reduced by approximately 5–10%. The shrinkage of droplets led to the formation of a thin lubricating aqueous layer between the moving particles and the channel walls, thereby reducing friction between the particles and the walls of the microchannels. The second factor relied on the inhibition of free-radical polymerization reactions at the interface of plugs with the walls of a microfluidic device fabricated in poly(dimethyl siloxane) (PDMS). The inhibition was caused by oxygen diffusing into the microchannel through the PDMS (Decker and Jenkins, 1985). This factor, in addition to the first one, resulted in the existence of the lubricating layer between the solidifying droplets and the channel walls. Recently, the approach exploiting confinement in microchannels was also employed for the generation of gel fibers (Sugiura *et al.*, 2008).

In the second approach, non-spherical polymer particles were synthesized by generating precursor droplets comprising two immiscible organic phases: a liquid monomer and a non-polymerizable liquid (Nie *et al.*, 2005). Following polymerization of the monomer, the non-polymerizable liquid was removed, thereby leading to the controllable and reproducible production of non-spherical polymer microparticles. In this method, microfluidic synthesis offered the ability to control the morphologies of precursor droplets (and hence the corresponding polymer particles) by the hydrodynamic means, that is, by varying the flow rates of the two immiscible liquids and thus the volume fractions of each liquid in the droplets.

The schematic drawing in Figure 10.2a illustrates the approach to the formation of biphasic precursor droplets. Emulsification occurred in a three-channel microfluidic flow-focusing device (Nie *et al.*, 2005). The oil and the monomer phase were supplied to the central channels of the droplet generator as liquids A and B, respectively, whereas the continuous aqueous phase (liquid C) was introduced in two side channels of the microfluidic device. The coaxial oil–monomer jet extended into the downstream channel and broke up into segments, which under the action of interfacial tension acquired a spherical shape.

Formation of precursor biphasic droplets with varying relative fractions of the monomer and oil phases was achieved by tuning the relative frequency of breakup of the oil and monomer threads. These frequencies determined the formation of an inner oil droplet and a compound monomer–oil droplet, respectively, and they were controlled by varying the flow rates of liquids A, B, and C.

The formation of droplets with different morphologies was conveniently represented in the phase-like diagram plotted in Figure 10.2b. In order to meet the requirement of ternary diagrams (the sum of three variables is constant and equal to 1) each axis shows the ratio of flow rate of a particular liquid (water, oil, or monomer phase) to the total flow rate of the three liquids. The central part of the diagram (filled symbols) shows the formation of core–shell droplets with a monomer shell and a different number of oil cores. These droplets served as the precursors to polymer capsules (see Chapter 9). Open symbols show the outer region of the diagram, in which the droplets had an asymmetric morphology. For example, in an early stage of evolution of a monomer droplet (and after close-to-complete emergence of an oil droplet), droplets with a small monomer inclusion adjacent to the surface of oil droplet were formed (region A). In the later stages of the formation of a monomer droplet, the fraction of the monomer inclusion increased (region B). After passing through the region of the core–shell droplets (region D), in an early stage of the evolution of breakup of an oil droplet, the coaxial jet generated droplets with a small oil inclusion and a large fraction of the monomer (region C).

Polymer particles with asymmetric shapes were produced by photopolymerizing a monomer in the compound droplets and removing the silicone oil phase with acetone. Figure 10.2c–e shows SEM images of truncated microspheres and hemispheres

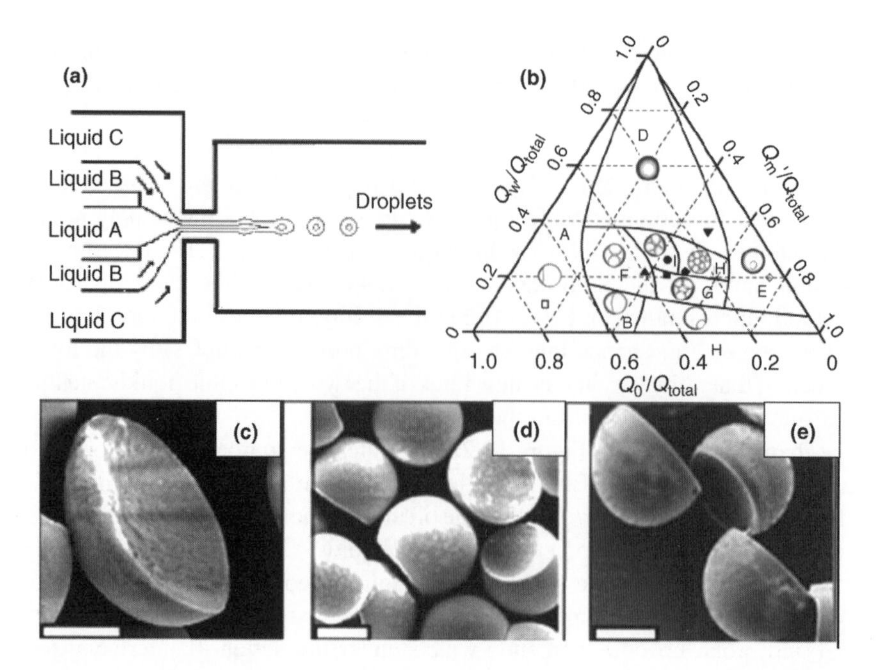

Figure 10.2 (a) Schematic of the production of droplets in a microfluidic flow-focusing device by laminar co-flow of silicone oil (A), monomer (B), and aqueous (C) phases. (b) Ternary phase-like diagram of hydrodynamic conditions used in the production of droplets with various morphologies. To accommodate all flow-rate ratios on the same diagram we used $Q'_o = 240Q_o$, $Q'_m = 120Q_m$, $Q_{total} = Q'_o + Q'_m + Q_w$. Filled symbols correspond to hydrodynamic conditions used for the production of core–shell droplets. Droplets with asymmetrically aligned cores (△) were obtained at $Q_w = 18\,mL\,h^{-1}$, $Q_m = 0.50\,mL\,h^{-1}$, $Q_o = 0.012\,mL\,h^{-1}$. Droplets with a small monomer inclusion adjacent to the oil surface (□) were produced at $Q_w = 6.0\,mL\,h^{-1}$, $Q_m = 0.025\,mL\,h^{-1}$, $Q_o = 0.20\,mL\,h^{-1}$. (c–e) Scanning electron microscopy images of polymer microbeads obtained by polymerizing TPGDA in droplets obtained in regimes A, B, and C [as in (c)], after removing a silicone oil core. Adapted, with permission, from Nie et al., 2005. Reprinted with permission from Journal of the American Chemical Society, Polymer Particles with Various Shapes and Morphologies Produced in Continuous Microfluidic Reactors by Z. Nie, S. Xu, S. Meo et al., **127**, 22, 8058–8063 Copyright (2005) American Chemical Society

generated by polymerizing TPGDA in the droplets formed in regions A and B of the phase-like diagram.

In the another method, non-spherical precursor droplets were generated by combining hydrodynamic and thermodynamic means. Droplets with a multiphase morphology were generated by breaking-up co-flowing streams of a photopolymerizable acrylate monomer 1,6-hexanediol diacrylate (HDDA) and a silicone oil, as shown in Figure 10.3a (Nisisako and Torii, 2007). The two liquids introduced into the two arms of the Y-shaped microchannel generated two parallel adjacent streams in the continuous aqueous phase. The binary stream was broken up via a shear-rupturing mechanism, and

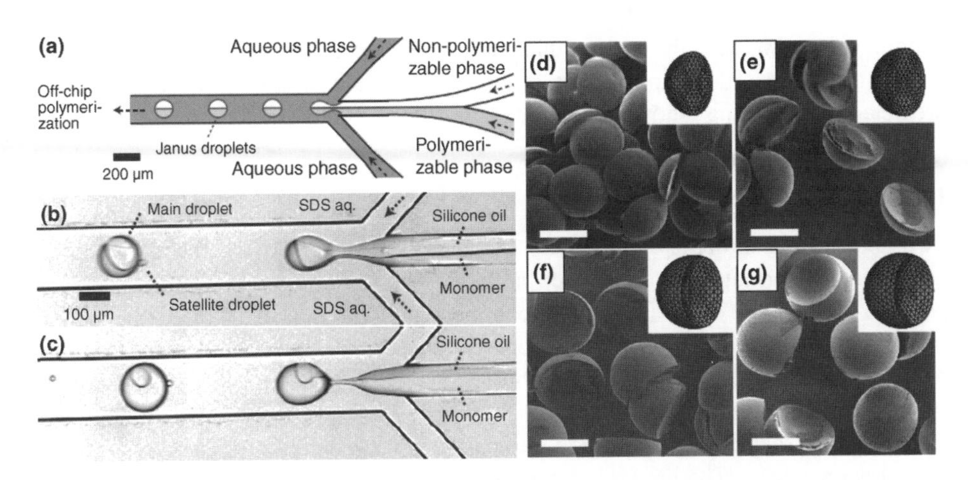

Figure 10.3 (a) Schematic of the sheath-flowing geometry used to fabricate biphasic droplets. (b) Formation of biphasic droplets when the flow rates of the monomer (Q_m) and silicone oil (Q_o) are equal; $Q_m = Q_o = 0.5\,\text{mL h}^{-1}$. The breakup rate is about 260 drops s^{-1}. Particles of various shapes are engineered by varying the ratio between the flow rates Q_m and Q_o. (c) Formation of biphasic droplets containing different volume ratios of silicone oil and the monomer; $Q_m = 0.9\,\text{mL h}^{-1}$, $Q_o = 0.1\,\text{mL h}^{-1}$. (d) Thin, truncated shape, $Q_m/Q_o = 0.2/0.8$. (e) $Q_m/Q_o = 0.3/0.7$. (f) $Q_m/Q_o = 0.5/0.5$. (g) $Q_m/Q_o = 0.6/0.4$. $Q_m + Q_o = 1.0\,\text{mL h}^{-1}$. The insets show the estimated shapes based on calculations of the three interfacial energies. The scale bars represent 100 μm. Adapted, with permission, from Nisisako and Torii, 2007. Reprinted from Advanced Materials, Formation of Biphasic Janus Droplets in a Microfabricated Channel for the Synthesis of Shape-Controlled Polymer Microparticles by T. Nisisako and T. Torii, **19**, 1489–1493 (2007). Copyright Wiley-VCH Verlag GmbH & Co. KGaA. Reproduced with permission

produced the biphasic droplets (Figure 10.3a–c). Because of the difference in interfacial tensions between the organic liquids and the aqueous phase (HDDA has a lower interfacial energy than silicone oil) the droplets acquired an asymmetric morphology, which was governed by the balance of the interfacial tension at the monomer/water interface (γ_{mw}), the interfacial tension at the silicone–oil/water interface (γ_{sw}), and the interfacial tension at the monomer/silicone–oil interface (γ_{ms}). By varying the ratio of the flow rates of the two organic liquids, the authors were able to tune their relative volume fractions in the precursor biphasic droplets (Figure 10.3b and c).

Nisisako and Torii (2007) calculated the spreading parameter $S = \gamma_{sw} - (\gamma_{mw} + \gamma_{ms}) < 0$ and suggested that for $S > 0$ a thin HDDA film formed between the silicone oil and the aqueous phase, which reduced the interfacial energy at this interface. The results of computer simulations of the 3D-shapes and the convex/concave geometries in the biphasic droplets simulations were in agreement with the experimental results. Photopolymerization and the removal of the oil phase were performed outside the microfluidic device. Figure 10.3d–g shows scanning electron microscopy images of polymer non-spherical particles derived from the biphasic droplets. The variation in the ratio of flow rates of the two organic phases led to the generation of biphasic droplets with varying volume ratios of the two organic phases, and hence particles with various shapes (Figure 10.3d–g). The shapes of the engineered particles were similar to the results predicted by modeling (shown in the insets in Figure 10.3d–g).

10.1.2 One-Phase Synthesis

Two-phase synthesis of polymer particles includes the formation of the precursor droplets and their transformation into polymer particles. The resulting non-spherical shapes of polymer particles are determined by the confinement of droplets in the microchannels, or by the multiphase morphology of the droplets. Recently, the range of possible shapes of polymer particles produced by microfluidic synthesis was extended by one-phase polymerization (Dendukuri *et al.*, 2006). The synthesis exploited lithographic techniques, so that the shapes of polymer particles were determined by the features of transparency masks. The method of continuous flow lithography (CFL) is described in Chapter 7. Briefly, the synthesis was implemented on an inverted microscope using projection photolithography. The ability to synthesize an array of particles was based on the inhibition of free-radical polymerization reactions at the interface with the PDMS device, which resulted in the formation of the non-polymerized "lubrication layer" adjacent to the walls of the microchannels. Arrays of polymeric particles were patterned into a stream of pre-polymer mixed with a photoinitiator, which flowed through the microchannels. The particles dispersed in a stream of non-polymerized oligomer were collected at the outlet of the device.

The polymerization process led to the synthesis of particles with a variety of shapes. Figure 10.4 shows polyPEGDA particles with polygonal shapes such as triangles, squares, and hexagons, high aspect ratio posts with circular, triangular, and square cross-sections, and non-symmetrical or curved objects. All the particles showed good fidelity to the original mask features (shown in insets). The resolution of the method was roughly equal to the wavelength of light used for irradiation, however, the size of particles was down to only 3 μm.

In another variant of the projection lithography method, namely, stop-flow lithography (SFL), polymer particles with a variety of shapes were synthesized with feature sizes down to 1 μm (Dendukuri *et al.*, 2007). Further modification of the CFL method utilized the deformation of PDMS devices under external pressure, in order to synthesize three-dimensional and multifunctional particles (Bong, Pregibon, and Doyle, 2009). Particles with non-spherical shapes were "locked" in a flow, and subsequently forced out of the microfluidic reactor by using a high-pressure pulse that deformed the PDMS device and released the particles. Three-dimensional PDMS molds, containing positive relief structures protruding from the ceiling, were used to synthesize particles that would normally be "locked" in a flow. This method enabled the synthesis of composite particles in which the distinct sections had complicated overlaps.

Dynamic masks were exploited to change the shape of particles on demand and synthesize particles with three-dimensional structures (Jang *et al.*, 2007). The method relied on interference lithography, in which laser light passing through a phase (a mask producing phase differences in the light exiting the mask) generates maxima and minima in

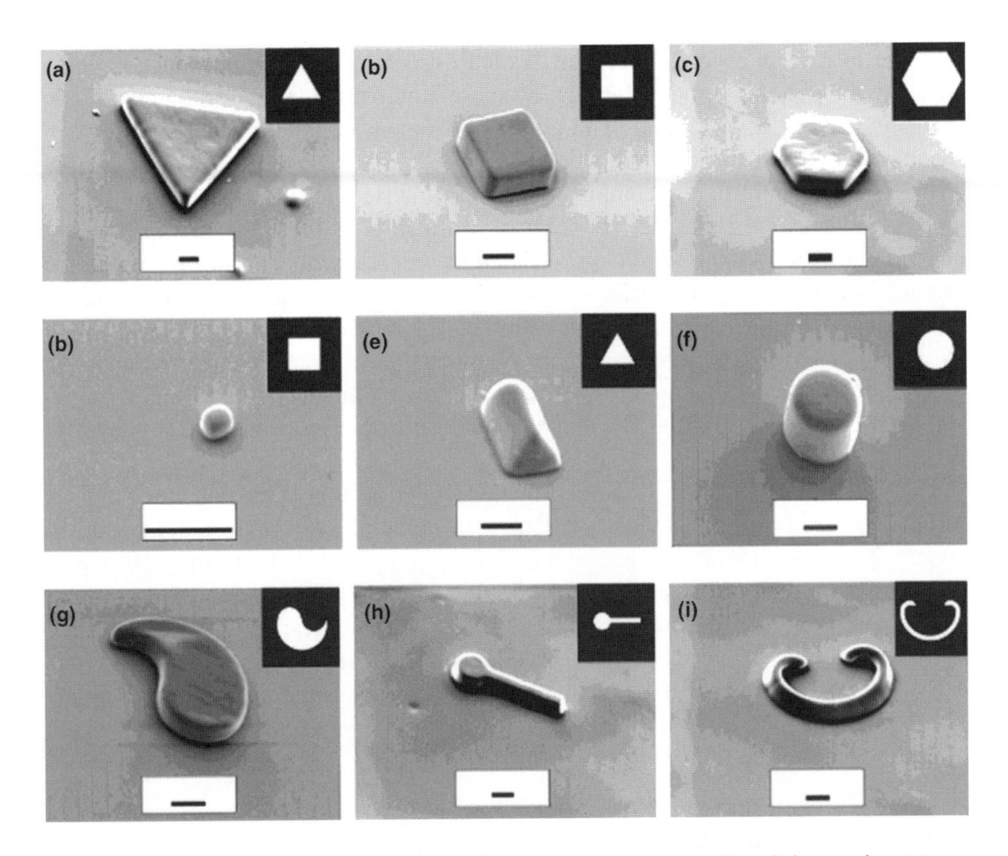

Figure 10.4 SEM images of particles synthesized by continuous flow lithography. Microparticles formed using a 20× objective were washed before being observed using SEM. The scale bars in all the figures are 10 μm. (a–c) Flat polygonal structures that were formed in a channel of height 20 μm. (d) A colloidal cuboid that was formed in a channel of height 10 μm. (e–f) High aspect ratio structures with different cross-sections that were formed in a channel of height 40 μm. (g–i) Curved particles that were all formed in a channel of height 20 μm. The inset in each image shows the transparency mask feature that was used to make the corresponding particle. Adapted, with permission, from Dendukuri et al., 2006. Reprinted with permission from Lab on a Chip, Stop flow lithography in a microfluidic device by D. Dendukuri, S. S. Gu, D. C. Pregibon et al., 7, 818–828 Copyright (2007) Royal Society of Chemistry

the light intensity in the direction of propagation of light. Under these conditions, photopolymerization yields particles with 3D structures. By combing phase mask-based interference lithography with SFL, a new process called stop flow interference lithography (SFIL) was developed (Jang *et al.*, 2007). In the "maskless lithography system" by using digital micromirror devices the pattern could be pre-programmed and changed, thereby leading to a change in the shapes of particles on demand (Chung *et al.*, 2007). Using high-speed two-dimensional spatial light modulators, polymeric microstructures with a variety of shapes were synthesized from PEGDA. The resolution in the particle synthesis was down to 1.54 μm × 1.54 μm.

In the next step, a new concept of "railed microfluidics" was introduced, which enabled the guidance and assembly of these microstructures inside microfluidic channels (Chung *et al.*, 2008). The guided movement of microstructures in microchannels was achieved by fabricating grooves ("rails") on the top surface of the channels and synthesizing complementary polymeric microstructures that fit the grooves (Chung *et al.*, 2007). A schematic drawing of the cross-sectional shape matching between a microfluidic channel and the microstructures flowing through the channel is shown in Figure 10.5a. A groove ("rail") was formed on the top surface of the channel using two-step mold fabrication and standard soft lithography. By filling the channel with an ultraviolet-curable oligomer solution and polymerizing it via optofluidic maskless lithography, a polymeric microstructure with a fin (a "microtrain") was created (Figure 10.5b). The finned microtrain object on the rail is shown in Figure 10.5c.

The matching of the fin and the rail provided guided motion of the polymeric objects in the flow field and allowed the assembly of complex one- and two-dimensional systems. Complex hierarchical assemblies were composed of more than 50 microstructures with dimensions smaller than 50 μm. Figure 10.5d,e shows representative structures assembled with the assistance of railed microfluidics: the Eiffel tower and a skeleton, respectively. Furthermore, it was shown that the finned microtrain objects can carry a microscale cargo to a designated location in the microfluidic channel (Chung *et al.*, 2007). For example, a microbead carrier was loaded with microbeads, which were transported along the rail via the transporting flow.

10.1.3 Microfluidics-Assisted Assembly of Colloid Particles

An example of "railed" assembly of three-dimensional polymer objects synthesized by microfluidic synthesis (Chung *et al.*, 2007) is given in the previous section. A different approach to the microfluidic assembly of complex three-dimensional suprastructures utilizes pre-formed (pre-existing) polymer particles. The strategy employs confinement of particles in templates. For example, microchannels can be exploited as templates that impart directionality on particle assembly, thereby generating chains of colloids. Alternatively, encapsulation of particles in droplets during microfluidic emulsification and subsequent solvent evaporation leads to particle crystallization thereby producing complex three-dimensional superstuctures.

In earlier work, chains of polystyrene (PS) particles loaded with nanoparticles of iron oxide were assembled under the action of a magnetic field applied to the suspension of microbeads (Furst *et al.*, 1998). The nanoparticles aligned along the lines of the field and the PS particles acquired a magnetic moment. Owing to the interaction of the induced dipoles via an anisotropic potential, the PS microbeads formed chains in the direction of the field, given that the interactions between them exceeded the thermal energy kT. Chains of PS microspheres formed between the two solid substrates, so that the length of the chain could be controlled by the distance between the surfaces. Following the assembly, the particles in the chains were permanently bonded by introducing a cross-linking agent into the system.

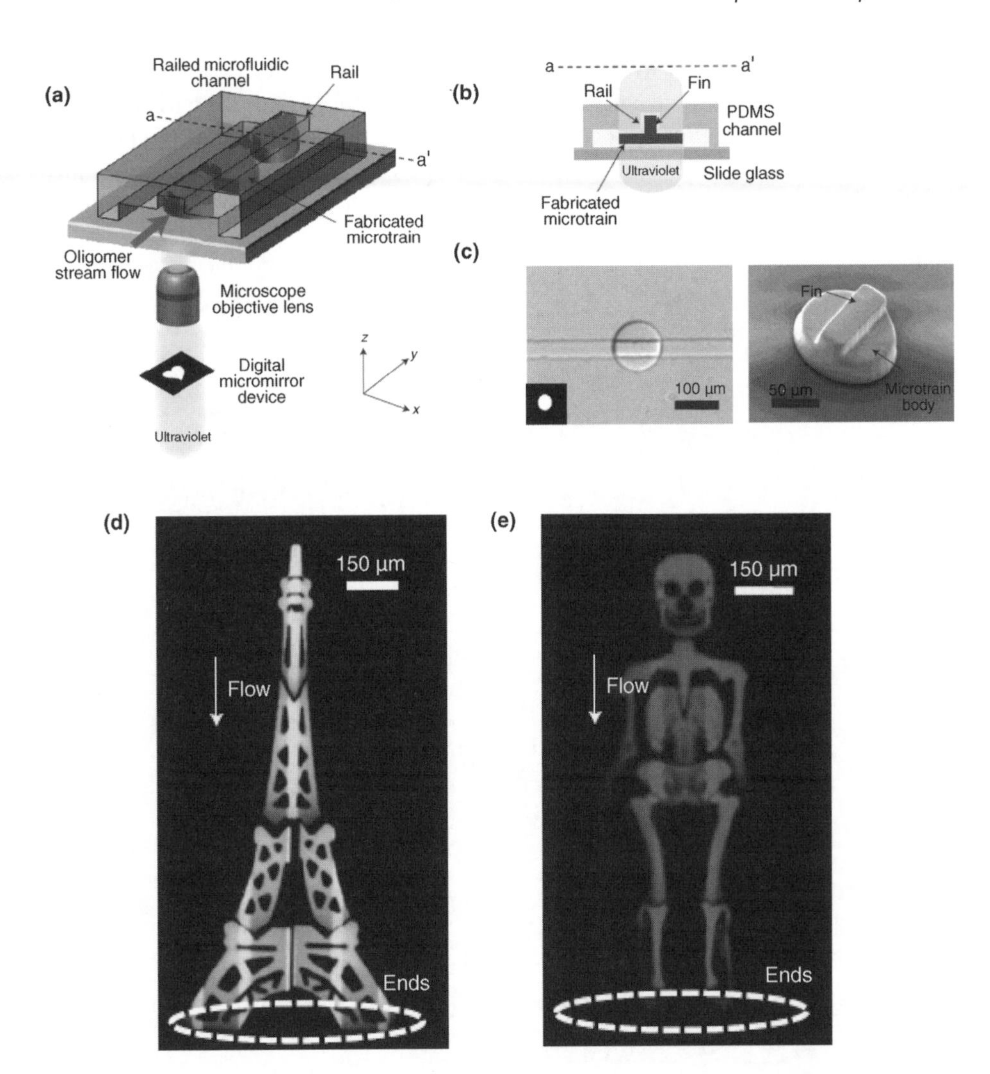

Figure 10.5 *Concept of railed microfluidics and guiding mechanism. (a) Schematic diagram of a railed microfluidic channel. (b) Cross-section of the polydimethylsiloxane railed microfluidic channel and a finned microtrain cut at a–a′ from (a). (c) Fabricated microtrain. Left: the top view of the microtrain in a differential interference contrast image. The top-view shape of the microtrain is determined by a pattern specified on a digital micromirror device (inset). Right: a corresponding scanning electron microscope image. Complex self-assembly in railed microfluidics. The complex self-assembly is completed simultaneously by applying fluidic force inside a railed microfluidic channel. Assembled structures of Eiffel tower (d) and a skeleton (e). Adapted, with permission, from Chung et al., 2007. Reprinted with permission from Applied Physics Letters, Optofluidic maskless lithography system for real-time synthesis of photopolymerized microstructures in microfluidic channels by S. Chung, W. Park, H. Park et al.,* ***91****, 4, 041106 Copyright (2007) American Institute of Physics*

Recently, microchannels were used as geometric templates for producing polymer aggregates with configurable anisotropy (Sung *et al.*, 2008). Figure 10.6 illustrates the concept of this method. Precursor colloid particles were packed into a narrow, dead-end channel using a combination of confinement and microfluidics. Following the assembly, the microbeads were bonded by thermal fusion to generate larger objects with complex

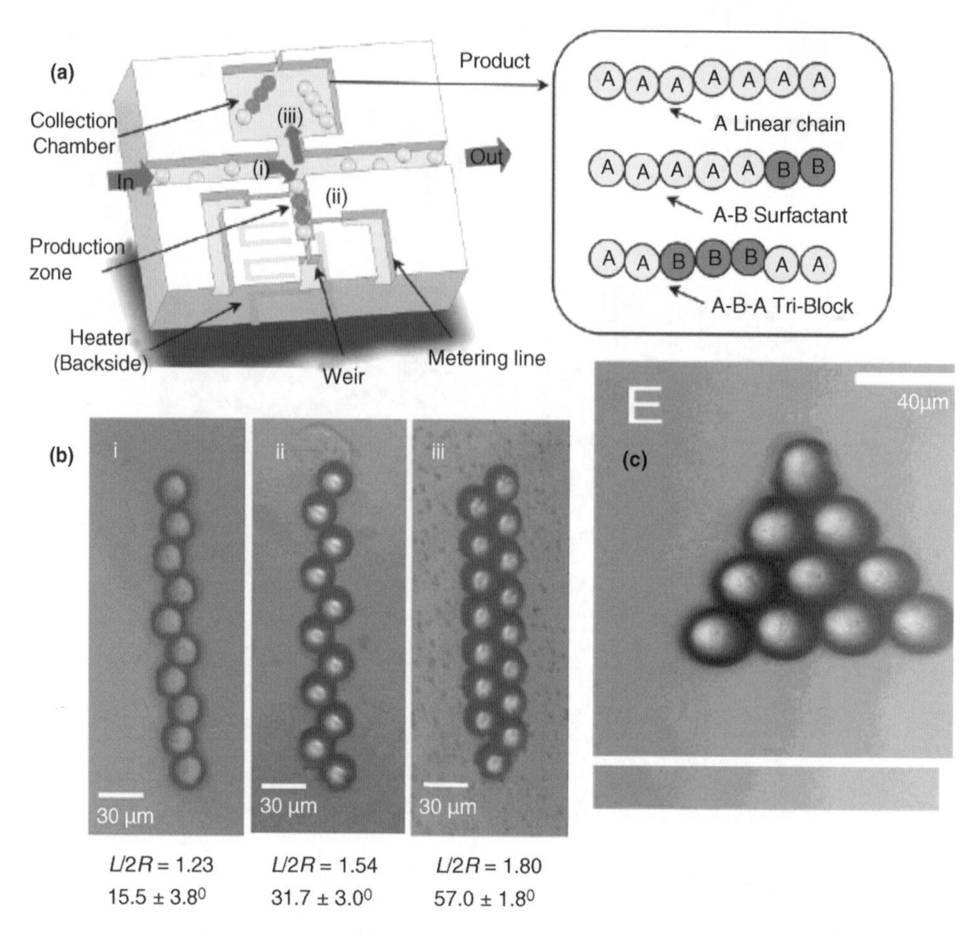

Figure 10.6 *(a) Schematic diagram of the continuous fabrication process. Polystyrene particles in DMSO flow to the production zone (i) where they undergo fusion (ii) and collection (iii). The length and width of the production zone and the actuation of input and metering lines control the sequence synthesized. Depending on actuation, each of three types of chains with different bond angles can be synthesized (b). The number of "monomer" particles in each chain is consistent with the production zone geometry (i.e., width and length). (c) Homogeneous triangular prisms produced in the tapered microchannel. Adapted, with permission, from Sung et al., 2008. Reprinted with permission from Journal of the American Chemical Society, Programmable Fluidic Production of Microparticles with Configurable Anisotropy by K.E. Sung, S.A. Vanapalli, D. Mukhija et al., **130**, 4, 1335 Copyright (2008) American Chemical Society*

shapes, which were subsequently released by reversing the flow in the production zone of the microfluidic device.

By varying the types and dimensions of microbeads introduced into the microchannel, this method allowed control of the particle sequences in the resulting linear assemblies, thereby producing homogeneous (type "A"), surfactant-like (type "A-B") or triblock (type "A–B–A") chains, as shown in Figure 10.6. In addition, the width and the length of the microchannel relative to the microbead size determined the arrangement of the particles in their chains, which had controlled repeatable bond angles α. The stacking of the individual particles of radius R in a channel of width L was determined as $\sin\alpha = L/2R - 1$, where R is the angle of a segment connecting the centroids of two adjacent particles relative to the centerline of the chain (Kumacheva *et al.*, 2003). Figure 10.6b shows that the variation in the ratio L/R led to the production of chains with α varying from $15.5 \pm 3.8°$ to $57.0 \pm 1.8°$.

The use of microchanels as geometric templates allowed the production of particles with a broad range of shapes. For example, triangular prisms with a uniform roughness were formed in microchannels with tapered geometry, as shown in Figure 10.6c.

In a different approach to producing polymer particles with non-spherical shapes, droplets generated by microfluidic emulsification were used as compartments for the crystallization of colloid mcrobeads. This approach resembled the earlier work of Velev, Lenhoff, and Kaler (2000) in which colloidal crystals formed in aqueous droplets suspended on fluorinated oil. Following the evaporation of water, assemblies with remarkable structural stability and shapes varying from spheres through ellipsoids to toroids were produced. Multibead structures had dimensions of approximately 500 μm but the number of self-assembled particles was not controlled.

In contrast, microfluidic emulsification allowed good statistical control over the number of particles compartmentalized in droplets (Yi *et al.*, 2003a; Yi et al., 2003b), due to the narrow distribution of sizes of the droplets. Water droplets compartmentalizing latex particles were dispersed in a non-polar phase using a T-junction microfluidic device. As the droplets moved through the downstream microchannel, water was slowly extracted in the continuous phase, and the droplets were transformed into consolidated colloidal assemblies. In this method, the selection of the continuous phase – a silicone oil or a fluorinated silicone oil (no surfactants used) – was determined by the strong affinity of the latex particles for the aqueous droplet phase (containing a surfactant), so that during the removal of water the microbeads did not escape in the continuous phase. The number of latex particles captured by the droplets exhibited statistical variation. For the number of compartmentalized microbeads as high as a few hundred, the fluctuations relative to the average number of particles were a small percentage. This feature paved the way to the production of colloidal assemblies with a controlled number of constituent particles.

The most interesting situation occurred when the number of particles in the droplets was small, although the distribution relative to the mean number of compartmentalized beads had broadened. Typical images of the assemblies of 2 μm-diameter polymer particles are shown in Figure 10.7, along with the assemblies formed by 15 and 4 spheres. The particles assembled into defect-free colloidal structures with shapes

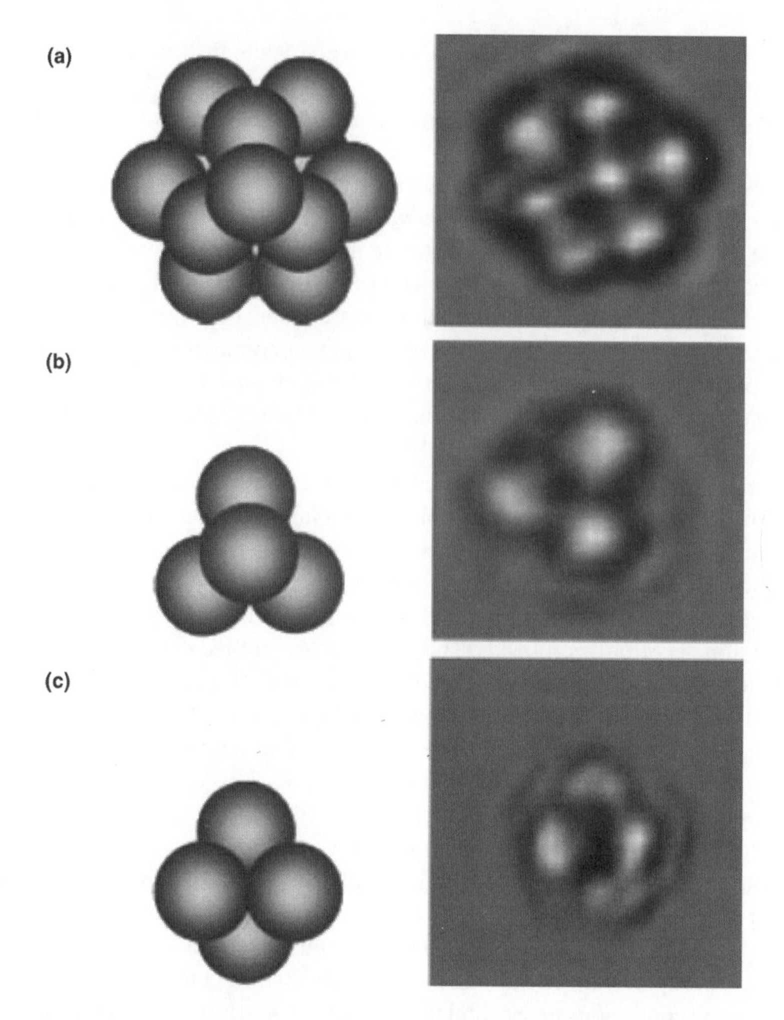

*Figure 10.7 Model assemblies (left) and optical microscopy images (right) of the colloidal assemblies, which are composed of 15 polystyrene beads (a) and 4 polystyrene beads (b, c). The diameter of the beads is 2 μm. Adapted, with permission, from Yi, 2003b. Reprinted from Advanced Materials, Generation of Uniform Colloidal Assemblies in Soft Microfluidic Devices by G. R. Yi, T. Thorsen, V. N. Manoharan et al., **15**, 1300 (2003). Copyright Wiley-VCH Verlag GmbH & Co. KGaA. Reproduced with permission.*

controlled by the ratio between the droplet and particle diameters and the number of particles encapsulated in the droplet.

10.2 Synthesis of Janus and Triphasic Particles

Janus particles (JPs) are two-faced particles named after the Roman god of doorways. These particles contain two halves that are typically formed from two different materials, or from the same material that carries on the surface two different types of functional groups and/or is loaded with two different organic or inorganic additives. Potential applications of JPs include studies of self-assembly, stabilization of emulsions, and the production of dual-functionalized optical, electronic, and sensor devices (Hong *et al.*, 2008; Perro *et al.*, 2005; Vanakaras, 2006). Janus particles with various dimensions have been derived from block copolymer micelles (Wang, Liu, and Rivas, 2003; Erhardt, *et al.*, 2003), submicrometer-size particles (Du *et al.*, 2004; Roh, Martin, and Lahann, 2005), and micrometer-size beads (Cayre, Paunov, and Velev, 2003; Love *et al.*, 2002; Takei and Shimizu, 1997; Hugonnot *et al.*, 2003).

Janus particles can be generated by a broad range of techniques, including hydrodynamic and electrohydrodynamic methods (Nisisako, Torii, and Higuchi, 2004; Roh, Martin, and Lahann, 2005); and controlled coalescence of two distinct droplets (Millman *et al.*, 2005; Fialkowski, Bitner, and Grzybowski, 2005); phase separation (Gu *et al.*, 2004; Akiva and Margel, 2005); template-directed self-assembly (Yin, Lu, and Xia, 2001; Xia *et al.*, 2003); and controlled surface nucleation (Reculusa *et al.*, 2002; Yu *et al.*, 2005).

Microfluidic synthesis is an efficient route to producing polymer Janus particles, which enables precise control over microbead morphology, that is, the relative fractions of two phases. Similarly to the preparation of particles with varying shapes, JPs were synthesized via two-phase and one-phase microfluidic synthesis.

The two-phase approach relied on the generation of droplets from a stream of co-flowing liquids and the rapid solidification of these droplets. The liquids could be of the same nature, such as a solution of the same polymer, but carry a different additive, such as a dye or a pigment. Because of the slow mixing between the two liquids, a distinct distribution of the additive in the corresponding "halves" of the resulting particles was preserved, although the sharpness of the interface between the phases was somewhat compromised. Alternatively, two co-flowing liquids could be largely immiscible, at least on the time scale of experiment. In this case, the resulting particles exhibited a sharp, well-defined interface between the two phases.

The first aproach to two-phase microfluidic synthesis of bicolored polymer Janus microbeads was demonstrated for poly(isobornyl acrylate) particles (Nisisako, Torii, and Higuchi, 2004). Figure 10.8a shows the process of breakup of two co-flowing streams of isobornyl acrylate, one containing 10 wt% of particles of carbon black pigment and the other, 10 wt% of particles of titanium dioxide pigment. The two-color stream reached the sheath-flow junction and broke up to release bicolored droplets serving as precursors for the Janus particles. The relative flow rates of the two liquids were an important factor in controlling the morphology of the droplets. When the flow rates of the liquids were close, a relatively clear boundary existed between the two halves of the droplets; however, when the flow rates were considerably different, convection occurring between the two liquids led to the mixing of the color.

The bichromic droplets containing an initiator for polymerization were dispersed in an aqueous 2 wt% solution of poly(vinyl acetate). The resulting droplets were collected in a beaker and solidified *off-chip* by thermally-initiated polymerization. Figure 10.8b shows highly monodisperse bichromal beads with dimensions in the order of 100 μm.

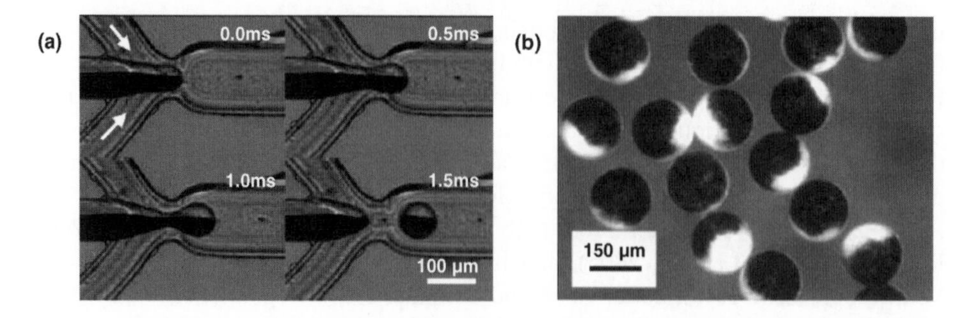

Figure 10.8 *Two-phase microfluidic synthesis of Janus particles. (a) Formation of bicolored droplets at the sheath-flow junction. To capture clear images of the color edge, white pigments were not used in this case. Frames were captured at 2000 fps by a high-speed video camera. (b) Janus particles comprising carbon black and titanium oxide in two phases. Adapted, with permission, from Nisisako, Torii, and Higuchi, 2004*

In some of the particles, the interface between the carbon black- and titanium oxide-containing phases was not sharp. In addition, in a notable fraction of the particles the hemispherical morphology was not preserved. These effects were ascribed to convection and diffusion occcurring in the droplets, prior to the polymerization of isobornyl acrylate. It was suggested that rapid solidification of the droplets can suppress these effects.

Later, the same conceptual approach was utilized for the generation of Janus microgels filled with silica particles (Shepherd *et al.*, 2006). Two parallel streams of aqueous solutions of acrylamide (containing two types of silica particles) formed a composite liquid thread, which broke up in the orifice of the microfluidic flow-focusing device and released Janus droplets. The droplets were rapidly gelled by photoinitiated polymerization of acrylamide. The concentrations of the photoinitiator and acrylamide were important in preserving the biphasic morphology of the particles: gelation had to occur within a few seconds after the formation of the droplets, in order to suppress mixing of two aqueous solutions. Because the photoinitiator showed a trend to rapidly diffuse from the droplets to the continuous phase, this effect was suppressed by introducing a photoinitiator into both the droplet and the continuous phase. Therefore, the concentration of the initiator in the droplets was sufficent to achieve rapid gelation and produce microgel particles with two well-defined phases.

Further development of the approach to producing Janus particles from two miscible liquids carrying different additives was undertaken by Roh *et al.* (2005). The biphasic morphology of Janus particles was realized by electrohydrodynamic jetting of two parallel streams of aquous polymer solutions carrying two distinct solutes, for example, dyes or dye-labeled low-molecular weight molecules. Subsequent formation of Janus droplets by breaking these two streams and rapid evaporation of water from the droplets resulted in the generation of particles in which the distinct molecules were preferentially compartmentalized in two corresponding phases. Electrohydrodynamic jetting allowed for the production of small, submicrometer-size Janus particles;

however, the microbeads had a broad distribution of sizes. A similar approach was used to generate triphasic particles with dimensions in the submicrometer size range and a broad polydispersity (Roh, Martin, and Lahann, 2006).

Formation of biphasic particles with a compromised sharpness of the interface between the phases or even gradients in composition can limit some of their applications, such as their use as solar cell systems, in which the existence of a sharp interface between the phases is crucial. Therefore, for some applications it is imperative to form droplets from monomers that are largely immiscible. Such monomers form a co-stream, which releases precursor droplets with a sharp interface between the liquids. Rapid solidification of the Janus droplets yields particles with two well-defined, distinct phases (Nie *et al.*, 2006; Nisisako and Torii, 2007).

Figure 10.9 illustrates the approach to the generation of Janus particles from immiscible monomers (Nie *et al.*, 2006). Methacryloxypropyl dimethylsiloxane was used as the first liquid (M_1); a mixture of pentaerythritol triacrylate, poly(ethylene glycol) diacrylate, and acrylic acid was used as the second liquid (M_2). Both liquids contained a photoinitiator. The volume of Janus droplets was controlled by changing the flow-rate ratio between a continuous aqueous phase and the total flow rate of M_1 and M_2. The volumetric fraction of each monomer in the droplet was precisely controlled by tuning the ratio of flow rates of M_1 and M_2. Assuming that within the droplet each phase formed a truncated sphere, a simple relationship was derived that presented the ratio of the volumes, V_{M_1}/V_{M_2}, of each liquid phase in an individual droplet with a radius R as the ratio of flow rates, Q_{M_1}/Q_{M_2} (see Figure 10.9a).

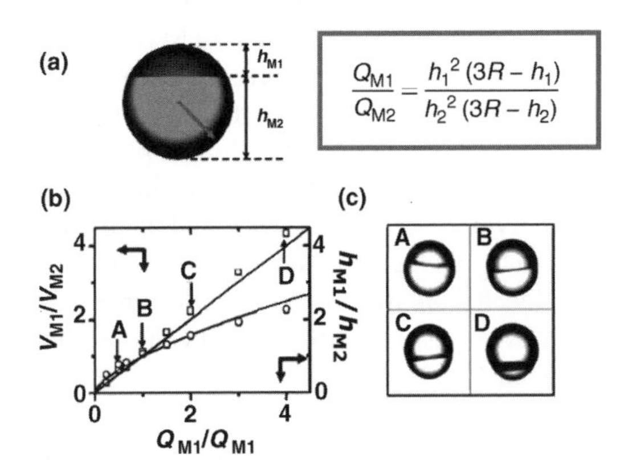

Figure 10.9 *(a) Schematic of a Janus droplet and the relationship between the ratio of monomer flow rates (Q_{M_1}/Q_{M_2}) with the ratio of volumes (V_{M_1}/V_{M_2}) and the ratio of heights (h_{M_1}/h_{M_2}) of the constituent parts in Janus droplets. (b) Variation in volume ratio V_{M_1}/V_{M_2} (□) and height ratio h_{M_1}/h_{M_2} (○) plotted versus the ratio of flow rates of M_1 and M_2. The solid lines represent theoretical calculations; the empty symbols correspond to experimental results. (c) Optical microscopy images of droplets obtained under conditions indicated in (b) as A, B, C, and D, respectively. The top part of the droplets is formed by M_1. Reproduced, with permission, from Nie et al., 2006*

Good correlation was reached between the experimental and predicted variations in the volume fractions of the liquids in the Janus droplet (and hence, in the correspoding particles) with increasing ratio of monomer flow rates (Figure 10.9b). The representative optical microscopy images of Janus droplets with volume ratios of monomer phases of 2/1, 1/1, 1/2, and 1/4 are shown in Figure 10.9c. It should be noted that the interfacial energies of the selected liquids precluded the formation of core–shell droplets.

The use of largely immiscible liquids was further extended to the generation of triphasic particles (Nie *et al.*, 2006). In this work, formation of particles was governed by the combination of hydrodynamic and thermodynamic factors. To generate precursor droplets, the monomer mixture M_2 was introduced in the central channel, M_1 was supplied to two intermediate channels, and an aqueous continuous phase was injected in the side channels of the microfluidic droplet generator. Triphasic droplets were produced in the jetting regime by breaking up a stream of three co-flowing liquids. Morphology of precursor ternary droplets (and the corresponding Janus particles) was governed by a combination of hydrodynamic and thermodynamic factors. The latter were determined by the values of interfacial energies between the liquids (characterized by the spreading coefficient, S_i, of each phase). The value of S_i was defined as $S_i = \sigma_{jk} - (\sigma_{ij} + \sigma_{ik})$ (Torza and Mason, 1969), where the subscripts i, j, k represent M_1, M_2, and the aqueous continuous phase (w), respectively, without repetition; and σ was the value of interfacial tension between the two subscripted phases. For $S_w > 0$ ($S_{M_1} < 0$, $S_{M_2} < 0$), two "halves" of the Janus droplet did not adhere to each other and produced individual droplets of M_1 and M_2. Ternary droplets formed only when M_1 (the monomer with a lower value of S_i) was introduced in the intermediate channels: if M_1 was supplied to the central channel, the side streams of M_2 merged at the tip of the liquid thread and engulfed M_1, thereby generating a core–shell droplet. The relative affinity of the monomers to the surface of microfluidic droplet generator played a crucial role in the formation of ternary droplets: monomer M_1 (the droplet phase with a lower affinity to the surface of the device), was injected in the intermediate channels.

Free-radical *on-chip* polymerization of ternary and Janus droplets with different volume fractions of M_1 and M_2 yielded polymer multiphase particles (Nie *et al.*, 2006). Janus particles were produced from the droplets with volume ratios of monomers M_1/M_2 of 1/1, 1/3, and 4/1. In ternary particles, the typical volume ratio of polymers derived from M_1 and M_2 varied from 1/2 to 2/1, respectively. All resulting particles had a sharp interface between the phases.

Owing to their amphiphilic nature, Janus particles assembled at the water–oil interface in such a way that their hydrophobic part (polymerized methacryloxypropyl dimethylsiloxane) was immersed in the oil phase (Nie *et al.*, 2006). In the homogeneous polar liquid medium – a water–methanol mixture – the Janus particles formed clusters with an aggregation number determined by the ratio of volume fractions of the hydrophilic and hydrophobic segments in the microspheres. The extent of the aggregation increased with increasing fraction of the hydrophobic polymer in the particle. This result was in agreement with the simulations which showed that the probability of forming adhesive contacts by colloidal particles depends on the fraction of particle surface covered with bridging groups (Moncho-Jorda *et al.*, 2003). Thus the microfluidic synthesis offers a route to designing colloid particles and the corresponding self-assembled structures by producing building blocks with a controlled fraction of the aggregating constituent.

Furthermore, the surface of the polymer Janus particles was selectively post-functionalized after the microfluidic synthesis (Nie *et al.*, 2006). A hydrophilic part of the microbeads was synthesized from pentaerythritol triacrylate mixed with 16 wt% of glycidyl methacrylate. The surface epoxy groups present in the hydrophilic compartment of the microbeads were used for protein immobilization via reactions with nucleophilic groups of proteins. Following microfluidic formation of the precursor Janus droplets and *on-chip* free-radical polymerization of the constituent monomers, bovine serum albumin covalently labeled with fluorescein isothiocynate was attached to the surface of the hydrophilic compartment of the particles. The reaction of protein coupling was carried out for 3 h at room temperature at pH = 6. Precise control of the morphology of the Janus particles allowed control over the surface area carrying the protein.

Polymer particles with multiple compartments were also produced by one-phase microfluidic microscope projection photolithography synthesis (Dendukuri *et al.*, 2005). Rectangular polymer Janus particles were generated by polymerizing across the interface of the co-flowing monomer and rhodamine-labelled monomer streams (Figure 10.10). Particles with variable volume fractions of two compartments with different compositions were synthesized by exploiting the limited diffusion-controlled mixing in laminar flow and by controlling the location of the interface. Furthermore, it was speculated that the utilization of co-flow of multiple, concurrent, laminar streams through a microfluidic channel and polymerization conducted across the streams, could facilitate the synthesis of polymer particles with multiple phases.

10.3 Other Particles with "Non-Conventional" Morphologies

The continuous flow projection lithography method was further modified to synthesize polymer particles for multiplexed high-throughput screening that can be applied in genetic analysis, combinatorial chemistry, and clinical diagnostics (Pregibon, Toner, and Doyle, 2007). In a typical experiment, two monomer streams (one loaded with a fluorescent dye and the other with an acrylate-modified probe) were introduced side-by-side in a microfluidic channel, as shown in Figure 10.11a. Continuous-flow lithography (Dendukuri *et al.*, 2006) was used to polymerize particles by exposing them to 30 ms UV-irradiation across the streams. The resulting particles had a fluorescent, graphically encoded compartment and a compartment loaded with probe molecules. The particles had two-dimensional shape, as shown in Figure 10.11b. The morphology and the composition of the particles were determined by design of the mask and the composition of the co-flowing streams, respectively.

The particles were derived from polyethylene glycol, a biocompatible and transparent polymer (the latter feature allowed transmission of the fluorescent signal through

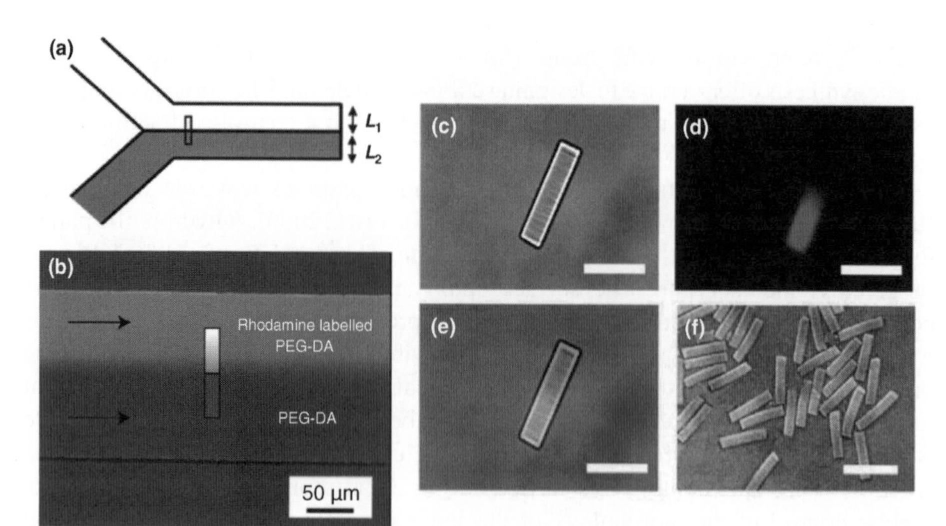

Figure 10.10 (a) A cartoon showing the synthesis of Janus particles. The widths of the streams, L_1 and L_2 can be altered by changing the flow rates of the streams. (b) Two streams containing polyethyleneglycol-diacrylate (PEG-DA) (grey) and PEG-DA with rhodamine labeled cross-linker (white) are co-flowed through a channel. A cartoon representing the formation of a bar-shaped particle 130 μm in length and 20 μm width is overlaid on the picture. (c) Differential interference contrast (DIC) image of a Janus particle. (d) Fluorescence microscopy image of the particle in (c). The rhodamine labeled portion is seen. (e) An overlaid image of the entire particle showing both the fluorescently labeled and the non-labeled sections. The scale bar in figures (c–e) is 50 μm. (f) Multiple Janus particles with the fluorescent portion shown. The scale bar is 100 μm. Reprinted, with permission, from Dendukuri et al., 2006. See Plate 12

both particle compartments). The dot-coding pattern on the surface of the particles could bear over a million codes (Figure 10.11c). The codes were "read" along five lanes oriented parallel to the long axis of the particles, with alignment indicators that were used to identify the position and direction of the code (Figure 10.11c). Figure 10.11e–g shows exemplary particles with a single-probe region, multiple-probe regions, and probe-region gradients synthesized in channels with varying designs by selectively labeling monomer streams with a fluorophore.

The described approach enabled one-step synthesis of particles for high-throughput screening in biological analysis and combinatorial chemistry by using multiplexing, which allowed for the simultaneous assay of several analytes (Pregibon, Toner, and Doyle, 2006). The authors demonstrated the use of encoded particle libraries for the multiplexed, single-fluorescence highly sensitive detection of DNA oligomers with encoded particle libraries in a flow-through microfluidic channel.

Another interesting application of microfluidic synthesis was demonstrated for the production of suprastructures ("photonic balls"), that is, the particles comprising colloid crystals of smaller microbeads (Kim *et al.*, 2008a; Sun *et al.*, 2008). In this work, the

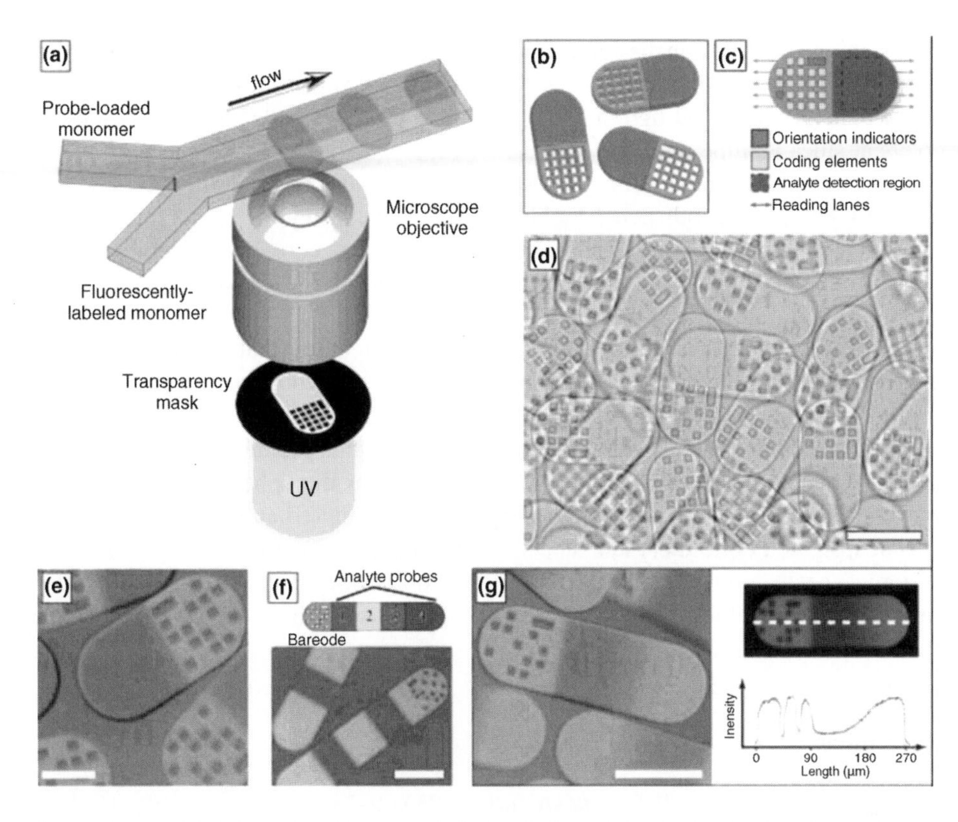

Figure 10.11 *(a) Schematic diagram of dot-coded particle synthesis showing polymerization across two adjacent laminar streams to make single-probe, half-fluorescent particles [shown in (b)]. (c) Diagrammatic representation of particle features for encoding and analyte detection. The encoding scheme shown allows the generation of 220 (1 048 576) unique codes. (d) Differential interference contrast (DIC) image of particles generated by using the scheme shown in (a). (e–g) Overlap of fluorescence and DIC images of single-probe (e), multiprobe (f, bottom), and probe-gradient (g, left) encoded particles. Shown also is a schematic representation of multiprobe particles (f, top) and a plot of fluorescent intensity along the center line of a gradient particle (g, right). Scale bars indicate 100 mm in (d), (f), and (g) and 50 mm in (e). Adapted, with permission, from Pregibon, Toner, and Doyle, 2006. See Plate 13*

emulsification of a dispersion of silica particles in the monomer ethoxylated trimethylolpropane triacrylate (ETPTA) resin was followed by the rapid photopolymerization of ETPTA. In the dispersion, at moderately high concentrations of silica microbeads, they spontaneously organized into a three-dimensional crystal, due to the strong repulsive forces relative to a weaker van der Waals attraction. The highly viscous nature of the ETPTA resin precluded disturbances caused in the colloid crystal by external flows, Brownian diffusion, and microemulsification. Owing to the rapid polymerization of the ETPTA monomer, the fluid colloid crystals that formed in the droplets were trapped in the solid state and preserved in the polymer particles. Because of the existing refractive index contrast between the polymerized ETPTA and the silica particles, the resulting composite material diffracted light and the resulting "photonic balls" exhibited iridescent colors at silica volume fractions of 10%.

The method of preparation of the "photonic balls" was further modified by generating highly monodisperse double emulsions encapsulating crystalline colloidal arrays (Kim *et al.*, 2008b). The shell phase of the core–shell droplets was a photocurable resin that was photopolymerized *in situ* in the downstream channel of the microfluidic device.

The generation of these particles possessed all the features that originated from the microfluidic generation of double emulsions, namely, hydrodynamically mediated control of the shell thickness and the number of cores per droplet. The encapsulated colloid crystals exhibited photonic band gaps for normal incident light, which were independent of the position of the beam of light on the spherical surface of the particles. This property was explained by the face-centered cubic (fcc) structure of the encapsulated colloid crystal. The solidified shells did not permit the penetration of ionic species into the particle interior and the encapsulated fluid crystalline structures were stable in a continuous aqueous phase of high ionic strength for at least one month.

References

Akiva, U. and Margel, S. (2005) *Colloids Surf., A*, **253**, 9–13.
Alargova, R.G., Bhatt, K.H., Paunov, V.N., and Velev, O.D. (2004) *Adv. Mater.*, **16**, 1653–1657.
Alargova, R.G., Paunov, V.N., and Velev, O.D. (2006) *Langmuir*, **22**, 765–774.
Bon, S.A.F., Mookhoek, S.D., Colver, P.J., Fischer, H.R., and van der Zwaag, S. (2007) *Eur. Polym. J.*, **43**, 4839–4842.
Bong, K.W., Pregibon, D.C., and Doyle, P.S. (2009) *Lab Chip*, **9**, 863–866.
Bowden, N., Choi, I.S., Grzybowski, B.A., and Whitesides, G.M. (1999) *J. Am. Chem. Soc.*, **121**, 5373–5391.
Breen, T.L., Tien, J., Oliver, S.R.J., Hadzic, T., and Whitesides, G.M. (1999) *Science*, **284**, 948–951.
Cayre, O., Paunov, V.N., and Velev, O.D. (2003) *J. Mater. Chem.*, **13**, 2445–2450.
Cho, I. and Lee, K.-W. (1985) *J. Appl. Polym. Sci.*, **30**, 1903.
Chung, S.E., Park, W., Park, H., Yu, K., Park, N., and Kwon, S. (2007) *Appl. Phys. Lett.*, **91**, 1693–1695.
Chung, S.E., Park, W., Shin, S., Lee, S.A., and Kwon, S. (2008) *Nat. Mater.*, **7**, 581–587.
Decker, C. and Jenkins, A.D. (1985) *Macromolecules*, **18**, 1241.
Dendukuri, D., Tsoi, K., Hatton, T.A., and Doyle, P.S. (2005) *Langmuir*, **21**, 2113.
Dendukuri, D., Pregibon, D.C., Collins, J., Hatton, T.A., and Doyle, P.S. (2006) *Nat. Mater.*, **5**, 365.
Dendukuri, D., Gu, S.S., Pregibon, D.C., Hatton, T.A., and Doyle, P.S. (2007) *Lab Chip*, **7**, 818–828.
Du, Y.Z., Tomohiro, T., Zhang, G., Nakamura, K., and Kodaka, M. (2004) *Chem. Commun.* **5**, 616–617.
Erhardt, R., Zhang, M.F., Böker, A., Zettl, H., Abetz, C., Frederik, P., Krausch, G., Abetz, V., and Müller, A.H.E. (2003) *J. Am. Chem. Soc.* **125**, 3260–3267.

Fialkowski, M., Bitner, A., and Grzybowski, B.A. (2005) *Nat. Mater.*, **4**, 93–97.
Furst, E.M., Suzuki, C., Fermigier, M., and Gast, A.P. (1998) *Langmuir*, **14**, 7334–7336.
Gu, H., Zheng, R., Zhang, X., and Xu, B. (2004) *J. Am. Chem. Soc.*, **126**, 5664–5665.
Hong, L., Cacciuto, A., Luijten, E., and Granick, S. (2008) *Langmuir*, **24**, 621–625.
Hosein, I.D., Ghebrebrhan, M., Joannopoulos, J.D., and Liddell, C.M. (2009) *Langmuir*, **26**, 2151–2159.
Hugonnot, E., Carles, A., Delville, M.H., Panizza, P., and Delville, J.P. (2003) *Langmuir*, **19**, 226–229.
Jang, J., Dendukuri, D., Hatton, T.A., Thomas, E.L., and Doyle, P.S. (2007) *Angew. Chem. Int. Ed.*, **46**, 9027–9031.
Kim, S.-H., Jeon, S.-J., Yi, G.-R., Heo, C.-J., Choi, J.H., and Yang, S.-M. (2008a) *Adv. Mater.*, **20**, 1649–1655.
Kim, S.-H., Jeon, S.-J., and Yang, S.-M. (2008b) *J. Am. Chem. Soc.*, **130**, 6040–6046.
Kumacheva, E., Garstecki, P., Wu, H., and Whitesides, G.M. (2003) *Phys. Rev. Lett.*, **91**, 128301–128304.
Love, J.C., Gates, B.D., Wolfe, D.B., Paul, K.E., and Whitesides, G.M. (2002) *Nano Lett.*, **2**, 891–894.
Lu, Y., Yin, Y., and Xia, Y. (2001) *Adv. Mater.*, **13**, 415–420.
Millman, J.R., Bhatt, K.H., Prevo, B.G., and Velev, O.D. (2005) *Nat. Mater.*, **4**, 98–102.
Moncho-Jorda, A., Odriozola, G., Tirado-Miranda, M., Schmitt, A., and Hidalgo-A'Ivarez, R. (2003) *Phys. Rev. E*, **68**, 011404.
Nie, Z., Xu, S., Seo, M., Lewis, P.C., and Kumacheva, E. (2005) *J. Am. Chem. Soc.*, **127**, 8058–8063.
Nie, Z., Li, W., Seo, M., Xu, S., and Kumacheva, E. (2006) *J. Am. Chem. Soc.*, **128**, 9408–9412.
Nisisako, T., Torii, T., and Higuchi, T. (2004) *Chem. Eng. J.*, **101**, 23–29.
Nisisako, T. and Torii, T. (2007) *Adv. Mater.*, **19**, 1489–1493.
Okubo, M., Fujibayashi, T., and Terada, A. (2005) *Colloid Polym. Sci.*, **283**, 793.
Perro, A., Reculusa, S., Ravaine, S., Bourgeat-Lami, E., and Duguet, E. (2005) *J. Mater. Chem.*, **15**, 3745–3760.
Pregibon, D.C., Toner, M., and Doyle, P.S. (2007) *Science*, **315**, 1393–1396.
Reculusa, S., Poncet-Legrand, C., Ravaine, S., Mingotaud, C., Duguet, E., and Bourgeat-Lami, E. (2002) *Chem. Mater.*, **14**, 2354–2359.
Roh, K.H., Martin, D.C., and Lahann, J. (2005) *Nat. Mater.*, **4**, 759–763.
Roh, K.H., Martin, D.C., and Lahann, J. (2006) *J. Am. Chem. Soc.*, **128**, 6796–6797.
Seo, M., Xu, S., Nie, Z., Lewis, P.C., Graham, R., Mok, M., and Kumacheva, E. (2005) *Langmuir*, **21**, 4773–4775.
Shepherd, R.F., Conrad, J.C., Rhodes, S.K., Link, D.R., Marquez, M., Weitz, D.A., and Lewis, J.A. (2006) *Langmuir*, **22**, 8618–8622.
Skjeltorp, A.T., Ugelstad, J., and Ellingsen, T. (1986) *J. Colloid Interface Sci.*, **113**, 577.
Sugiura. S., Oda, T., Aoyagi, Y., Satake, M., Ohkohchi, N., and Nakajima, M. (2008) *Lab Chip*, **8**, 1255.
Sun, C., Zhao, X.-W., Zhao, Y.-J., Zhu, R., and Gu, Z.-Z. (2008) *Small*, **4**, 1–5.
Sung, K.E., Vanapalli, S.A., Mukhija, D., McKay, H.A., Millunchick, J.M., Burns, M.A., and Solomon, M.J. (2008) *J. Am. Chem. Soc.*, **130**, 1335.
Takei, H. and Shimizu, N. (1997) *Langmuir*, **13**, 1865–1868.
Torza, S. and Mason, S.G. (1969) *Science*, **163**, 813–814.
Vanakaras, A.G. (2006) *Langmuir*, **22**, 88–93.
Velev, O.D., Lenhoff, A.M., and Kaler, E.W. (2000) *Science*, **287**, 2240–2243.
Wang, J., Liu, G., and Rivas, G. (2003) *Anal. Chem.*, **75**, 4667–4671.
Xia, Y., Yin, Y., Lu, Y., and McLellan, J. (2003) *Adv. Funct. Mater.*, **13**, 907–918.
Xu, S., Nie, Z., Seo, M., Lewis, P.C., Kumacheva, E., Garstecki, P., Weibel, D., Gitlin, I., Whitesides, G.M., and Stone, H.A. (2005) *Angew. Chemie Int. Ed.*, **44**, 724–728.

Yi, G.R., Jeon, S.J., Thorsen, T., Manoharan, V.N., Quake, S.R., Pine, D.J., and Yang, S.M. (2003a) *Synth. Metals*, **139**, 803.

Yi, G.R., Thorsen, T., Manoharan, V.N., Hwang, M.J., Jeon, S.J., Pine, D.J., Quake, S.R., and Yang, S.M. (2003b) *Adv. Mater.*, **15**, 1300.

Yin, Y., Lu, Y., and Xia, Y. (2001) *J. Am. Chem. Soc.*, **123**, 771–772.

Yu, H., Chen, M., Rice, P.M., Wang, S.X., White, R.L., and Sun, S. (2005) *Nano. Lett.*, **5**, 379–382.

Summary and Outlook

We have reviewed the methods of formation of droplets and the microfluidic synthesis of polymeric particles. Clearly, the academic demonstrations have opened doors to applications of the method. This prompts the questions on the next important steps that have to be completed, in order to put into operation a microfluidic technological platform for the production of polymer particles?

One of the differences between the science and engineering is dominated by the requirements to the materials and processing conditions. An academic demonstration often directs the way and proves concepts, but it leaves an ocean of challenges when it comes to formulating and synthesizing specific polymer particles at a defined efficiency and cost of operation. These challenges can be addressed by bringing together investors, engineers and scientists. Below we list a number of directions that we believe are important in the development of a microfluidic technology for the production of polymer particles.

(i) *Microfluidic synthesis of high-value polymer particles.* Whereas the properties and several "proof-of-concept" applications of these microbeads have been demonstrated, to the best of our knowledge, no application has been realized beyond the lab bench. Partly, this is explained by the low productivity of microfluidic synthesis, yet we believe that niches should be found in applications of high value polymer particles where the throughput of production is a second consideration. These applications may be the first to be commercially realized with the use of microfluidics.

(ii) *Sophisticated syntheses.* When material properties are of outermost importance, microfluidic synthesis of polymer particles may become the method of choice. Multi-step reactions, well-defined reaction conditions and complicated synthetic schemes can be executed in a single- or multiphase flow and can produce microbeads with beneficial chemical and structural properties. In addition to the new synthetic chemistries that can be explored and implemented in microreactors, the ease of control of kinetics of reactions in microfluidic reactors enable multi-step reactions. These may become attractive even at lower throughput-rates, if compared to classical synthetic routes.

Microfluidic Reactors for Polymer Particles. Eugenia Kumacheva and Piotr Garstecki.

(iii) *Fabrication of microreactors.* Further progress in microfluidic synthesis of polymer particles will depend on the ability to make this process cost-effective, robust and reliable. Fabrication of high-throughput microreactors will become critically important in addressing these requirements. For example, microfluidic reactors would have to be fabricated in materials that can accommodate multi-hour synthesis without chemical or physical degradation. Along with chemical stability of such materials, their surface properties, as well as the fidelity, ease and cost of fabrication will play a major role. Other important criteria for the selection of materials for microfabrication will depend on the particular synthetic method. For example, photoinitiated microfluidic synthesis requires transparency of the reactor material in the wavelength range used for photoinitiation. Good thermostability of the material will be required for high-temperature synthesis of polymer microbeads.

(iv) *Scaled up productivity.* Although continuous microfluidic synthesis of polymer particles paved the way for the production of high value micorbeads, industrial applications of this technology platform will depend on the ability to increase the productivity of this process. The most straightforward strategy is to increase the number of microfluidic reactors working in parallel. This approach is called "scaling up by numbering out." To avoid high cost and practical complications associated with the utilization of many tubes, connectors, and pumps, liquid reagents will be supplied to multiple parallel reactors using 2-3 pumps. The scaled up approach would require suppressed feedback between adjacent individual reactors and good fidelity in fabricating microreactors with identical geometries. Without addressing these two challenges, polydispersity of polymer particles generated in multiple parallel microfluidic reactors will be increased. It is imperative to use microfluidic reactors with a modular design, that is, many individual modules containing a particular number of microfluidic reactors, in order to control the productivity of particle synthesis and the ability to easily remove clogged reactors.

(v) *Optimization of formulations.* Microfluidic synthesis allows high throughput screening of formulations used for the synthesis of polymer particles. The composition and concentration of liquid reagents supplied in a microfluidic reactor can be tuned by the varying flow rates of the carrier liquids supplied to the reactor. More work should be done in this direction, in order to exploit the exploratory side of microfluidics. Development of tools for in-line chemical and physical characterization of reagents and products will benefit the optimization of reaction conditions. In particular, in-line spectroscopies (photoluminescence and absorbance), light scattering and gel permission chromatography will be important in microfluidic synthesis of polymer particles.

We are confident that new fascinating opportunities in microfluidic technologies for synthesis of new materials will be discovered in the near future, and we hope that the list presented above will fall short of the richness of creativity of scientists and engineers working hand in hand in order to bring the new technology to the industrial realm.

Index

Bold page numbers indicate tables. *Italic* numbers indicate figures.

Microfluidic Reactors for Polymer Particles. Eugenia Kumacheva and Piotr Garstecki.